Mitochondrial Disorders Caused by Nuclear Genes

Lee-Jun C. Wong
Editor

Mitochondrial Disorders Caused by Nuclear Genes

Editor
Lee-Jun C. Wong
Department of Molecular and Human Genetics
Baylor College of Medicine
Houston, TX
USA

ISBN 978-1-4899-9241-3   ISBN 978-1-4614-3722-2 (eBook)
DOI 10.1007/978-1-4614-3722-2
Springer New York Dordrecht Heidelberg London

© Springer Science+Business Media, LLC 2013
Softcover reprint of the hardcover 1st edition 2013
All rights reserved. This work may not be translated or copied in whole or in part without the written permission of the publisher (Springer Science+Business Media, LLC, 233 Spring Street, New York, NY 10013, USA), except for brief excerpts in connection with reviews or scholarly analysis. Use in connection with any form of information storage and retrieval, electronic adaptation, computer software, or by similar or dissimilar methodology now known or hereafter developed is forbidden.
The use in this publication of trade names, trademarks, service marks, and similar terms, even if they are not identified as such, is not to be taken as an expression of opinion as to whether or not they are subject to proprietary rights.

Printed on acid-free paper

Springer is part of Springer Science+Business Media (www.springer.com)

Contents

Part I Overview

1 **The Clinical Spectrum of Nuclear DNA-Related Mitochondrial Disorders** 3
   Salvatore DiMauro and Valentina Emmanuele

2 **Biochemical and Molecular Methods for the Study of Mitochondrial Disorders** 27
   Lee-Jun C. Wong

Part II Genes Involved in Mitochondrial DNA Biogenesis and Maintenance of Mitochondrial DNA Integrity

3 **Mitochondrial Disorders Associated with the Mitochondrial DNA Polymerase γ: A Focus on Intersubunit Interactions** 49
   Matthew J. Young and William C. Copeland

4 **Alpers–Huttenlocher Syndrome, Polymerase Gamma 1, and Mitochondrial Disease** 73
   Russell P. Saneto and Bruce H. Cohen

5 **Deoxyguanosine Kinase** 91
   David Paul Dimmock

6 ***MPV17*-Associated Hepatocerebral Mitochondrial DNA Depletion Syndrome** 103
   Ayman W. El-Hattab

7 **Mitochondrial DNA Depletion due to Mutations in the *TK2* Gene** 113
   Fernando Scaglia

8 **Mitochondrial DNA Multiple Deletion Syndromes, Autosomal Dominant and Recessive (POLG, POLG2, TWINKLE and ANT1)** ... 123
   Margherita Milone

9 Defects in Mitochondrial Dynamics and Mitochondrial DNA Instability 141
Patrick Yu-Wai-Man, Guy Lenaers and Patrick F. Chinnery

10 Depletion of mtDNA with MMA: *SUCLA2* and *SUCLG1* 163
Nelson Hawkins Jr and Brett H. Graham

11 *RRM2B*-Related Mitochondrial Disease 171
Gráinne S. Gorman, Robert D. S. Pitceathly, Douglass M. Turnbull and Robert W. Taylor

Part III Complex Subunits and Assembly Genes

12 Complex Subunits and Assembly Genes: Complex I 185
Ann Saada (Reisch)

13 Mitochondrial Respiratory Chain Complex II 203
Jaya Ganesh, Lee-Jun C. Wong and Elizabeth B. Gorman

14 Mitochondrial Complex III Deficiency of Nuclear Origin: Molecular Basis, Pathophysiological Mechanisms, and Mouse Models 219
Alberto Blázquez, Lorena Marín-Buera, María Morán, Alberto García-Bartolomé, Joaquín Arenas, Miguel A. Martín and Cristina Ugalde

15 Mitochondrial Cytochrome *c* Oxidase Assembly in Health and Human Diseases 239
Flavia Fontanesi and Antoni Barrientos

Part IV Mitochondrial Protein Translation Related Diseases

16 Mitochondrial Aminoacyl-tRNA Synthetases 263
Henna Tyynismaa

17 Mitochondrial Protein Translation-Related Disease: Mitochondrial Ribosomal Proteins and Translation Factors 277
Brett H. Graham

18 Disorders of Mitochondrial RNA Modification 287
William J. Craigen

Part V Others

19 Pyruvate Dehydrogenase Complex Deficiencies 301
Suzanne D. DeBrosse and Douglas S. Kerr

20 Nuclear Genes Causing Mitochondrial Cardiomyopathy 319
Stephanie M. Ware and Jeffrey A. Towbin

21 Mitochondrial Diseases Caused by Mutations in Inner Membrane Chaperone Proteins 337
Lisbeth Tranebjærg

Index 367

Contributors

Joaquín Arenas Instituto de Investigación, Hospital Universitario 12 de Octubre, Avda. de Córdoba s/n, 28041 Madrid, Spain

Centro de Investigación Biomédica en Red de Enfermedades Raras (CIBERER), 46010 Valencia, Spain

Antoni Barrientos Department of Neurology, Biochemistry & Molecular Biology, Universtiy of Miami Miller School of Medicine, 1600 NW 10th Ave., RMSB # 2067, Miami, FL 33136, USA
e-mail: abarrientos@med.miami.edu

Alberto Blázquez Instituto de Investigación, Hospital Universitario 12 de Octubre, Avda. de Córdoba s/n, 28041 Madrid, Spain

Centro de Investigación Biomédica en Red de Enfermedades Raras (CIBERER), 46010 Valencia, Spain

Patrick F. Chinnery Wellcome Trust Centre for Mitochondrial Research, Institute of Genetic Medicine, International Centre for Life, Newcastle University, Newcastle Upon Tyne, NE1 3BZ, UK

Department of Neurology, Royal Victoria Infirmary, Newcastle Upon Tyne, NE1 4LP, UK

Bruce H. Cohen NeuroDevelopmental Science Center, Neurology Division, Children's Hospital Medical Center of Akron, Northeast Ohio Medical University, Akron, OH 44087, USA

William J. Craigen Department of Molecular and Human Genetics, Baylor College of Medicine, S842, One Baylor Plaza, Houston, TX 77030, USA
e-mail: wcraigen@bcm.edu

Suzanne D. DeBrosse Pediatric Endocrinology and Metabolism, Department of Pediatrics, Case Western Reserve University, Rainbow Babies and Children's Hospital, 11100 Euclid Avenue, Cleveland, OH 44106–6004, USA

Salvatore DiMauro College of Physicians & Surgeons, 630 West 168th Street, New York, NY 10032, USA
e-mail: sd12@columbia.edu

Neurology Department, H. Houston Merritt Clinical Research Center, Columbia University Medical Center, New York, NY, USA

David Paul Dimmock Department of Pediatrics, Children's Hospital of Wisconsin, Medical College of Wisconsin, 9000 W.Wisconsin Ave., MS716, Milwaukee, WI 53226, USA
e-mail: ddimmock@mcw.edu

Ayman W. El-Hattab Medical Genetics Section, Department of Pediatrics, The Children's Hospital at King Fahad Medical City and King Saud bin Abdulaziz University for Health Science, P. O. Box 59046, Riyadh 11525, Kingdom of Saudi Arabia
e-mail: elhattabaw@yahoo.com

Valentina Emmanuele Department of Neurology, Columbia University Medical Center, New York, NY, USA

Flavia Fontanesi Department of Neurology, University of Miami Miller School of Medicine, 1600 NW 10th Ave., RMSB # 2067, Miami, FL 33136, USA
e-mail: ffontanesi@med.miami.edu

Jaya Ganesh Clinical Pediatrics, Section of Metabolic Diseases, The Children's Hospital of Philadelphia, Philadelphia, PA, USA

Alberto Garcia-Bartolomé Instituto de Investigación, Hospital Universitario 12 de Octubre, Avda. de Córdoba s/n, 28041 Madrid, Spain

Centro de Investigación Biomédica en Red de Enfermedades Raras (CIBERER), 46010 Valencia, Spain

Elizabeth B. Gorman Medical Genetics Laboratories, Baylor College of Medicine, One Baylor Plaza, NAB 2015, Houston, TX 77030, USA
e-mail: egorman@bcm.edu

Gráinne S. Gorman Wellcome Trust Centre for Mitochondrial Research, Institute for Ageing and Health, The Medical School, Newcastle University, Newcastle upon Tyne, NE2 4HH, UK

Brett H. Graham Department of Molecular and Human Genetics, Baylor College of Medicine, One Baylor Plaza, MS:BCM225, Houston, TX 77030, USA
e-mail: bgraham@bcm.edu

Medical Genetics, Texas Children's Hospital, Houston, TX, USA

Nelson Hawkins Jr Department of Molecular and Human Genetics, Baylor College of Medicine, One Baylor Plaza, MS:BCM225, Houston, TX 77030, USA

Douglas S. Kerr Pediatric Endocrinology and Metabolism, Department of Pediatrics, Case Western Reserve University, Rainbow Babies and Children's Hospital, 11100 Euclid Avenue, Cleveland, OH 44106–6004, USA
e-mail: douglas.kerr@case.edu

Guy Lenaers Neuropathies Optiques Héréditaires, Institut des Neurosciences de Montpellier, Université Montpellier I et II, INSERM U1051, Montpellier, France

Patrick Yu-Wai-Man Wellcome Trust Centre for Mitochondrial Research, Institute of Genetic Medicine, International Centre for Life, Newcastle University, Newcastle Upon Tyne, NE1 3BZ, UK
e-mail: Patrick.Yu-Wai-Man@ncl.ac.uk

Department of Ophthalmology, Neuro-Ophthalmology Division, Royal Victoria Infirmary, Newcastle Upon Tyne, NE1 4LP, UK

Margherita Milone Department of Neurology, Neuromuscular Division, Mayo Clinic, 200 First St., SW, Rochester, MN 55905, USA
e-mail: Milone.Margherita@mayo.edu

Lorena Marín-Buera Instituto de Investigación, Hospital Universitario 12 de Octubre, Avda. de Córdoba s/n, 28041 Madrid, Spain

Centro de Investigación Biomédica en Red de Enfermedades Raras (CIBERER), 46010 Valencia, Spain

Miguel A. Martín Instituto de Investigación, Hospital Universitario 12 de Octubre, Avda. de Córdoba s/n, 28041 Madrid, Spain

Centro de Investigación Biomédica en Red de Enfermedades Raras (CIBERER), 46010 Valencia, Spain

María Morán Instituto de Investigación, Hospital Universitario 12 de Octubre, Avda. de Córdoba s/n, 28041 Madrid, Spain

Centro de Investigación Biomédica en Red de Enfermedades Raras (CIBERER), 46010 Valencia, Spain

Robert D. S. Pitceathly MRC Centre for Neuromuscular Diseases, UCL Institute of Neurology and National Hospital for Neurology and Neurosurgery, Queen Square, London WC1N 3BG, UK

Ann Saada (Reisch) Department of Genetic and Metabolic Diseases and Monique and Jacques Roboh

Department of Genetic Research, Hadassah Hebrew University Medical Center, P. O. Box 12000, 91120 Jerusalem, Israel
e-mail: Annsr@hadassah.org.il

Russell P. Saneto Neurology Department, Division of Pediatric Neurology, Seattle Children's Hospital, 4800 Sand Point Way NE, Seattle, WA 98105, USA
e-mail: russ.saneto@seattlechildrens.org

Fernando Scaglia Department of Molecular and Human Genetics, Baylor College of Medicine and Texas Children's Hospital, Clinical Care Center, Suite 1560, 6701 Fannin Street, Mail Code CC1560, Houston, TX 77030, USA
e-mail: FSCAGLIA@bcm.edu

Robert W. Taylor Wellcome Trust Centre for Mitochondrial Research, Institute for Ageing and Health, The Medical School, Newcastle University, Newcastle upon Tyne, NE2 4HH, UK
e-mail: robert.taylor@ncl.ac.uk

Jeffrey A. Towbin Cincinnati Children's Hospital Medical Center and the University of Cincinnati College of Medicine, 240 Albert Sabin Way, MLC 7020 USA, Cincinnati, OH 45229, USA

Lisbeth Tranebjærg Department of Audiology, Bispebjerg Hospital, 2400 Copenhagen, NV, Denmark
e-mail: Tranebjaerg@sund.ku.dk

Wilhelm Johannsen Centre for Functional Genome Research, Department of Cellular and Molecular Medicine (ICMM), The Panum Institute, University of Copenhagen, 2200 Copenhagen NV, Denmark

Douglass M. Turnbull Wellcome Trust Centre for Mitochondrial Research, Institute for Ageing and Health, The Medical School, Newcastle University, Newcastle upon Tyne, NE2 4HH, UK

Henna Tyynismaa Research Program of Molecular Neurology, Biomedicum Helsinki, Haartmaninkatu 8, 00014 University of Helsinki, Finland
e-mail: henna.tyynismaa@helsinki.fi

Cristina Ugalde Instituto de Investigación, Hospital Universitario 12 de Octubre, Avda. de Córdoba s/n, 28041 Madrid, Spain
e-mail: cugalde@h12o.es

Centro de Investigación Biomédica en Red de Enfermedades Raras (CIBERER), 46010 Valencia, Spain

Stephanie M. Ware Cincinnati Children's Hospital Medical Center and the University of Cincinnati College of Medicine, 240 Albert Sabin Way, MLC 7020 USA, Cincinnati, OH 45229, USA
e-mail: stephanie.ware@cchmc.org

Lee-Jun C. Wong Department of Molecular and Human Genetics, Baylor College of Medicine, One Baylor Plaza, NAB 2015, Houston, TX 77030, USA
e-mail: ljwong@bcm.edu

Matthew J. Young Laboratory of Molecular Genetics, National Institute of Environmental Health Sciences, National Institutes of Health, DHHS, Research Triangle Park, NC 27709, USA

William C. Copeland Laboratory of Molecular Genetics, National Institute of Environmental Health Sciences, National Institutes of Health, DHHS, Research Triangle Park, NC 27709, USA
e-mail: copelan1@niehs.nih.gov

Part I
Overview

Chapter 1
The Clinical Spectrum of Nuclear DNA-Related Mitochondrial Disorders

Salvatore DiMauro and Valentina Emmanuele

Introduction

Mitochondrial diseases are conventionally defined as clinical disorders due to defects in the mitochondrial respiratory chain (RC), the terminal pathway of oxidative phosphorylation (OXPHOS), where most cellular ATP is generated. This is also the only metabolic pathway that depends on two genomes: mitochondrial DNA (mtDNA) encodes 13 of the ∼85 RC proteins and nuclear DNA (nDNA) encodes all remaining proteins (Fig. 1.1). The RC is composed of seven discrete functional entities, five multimeric complexes (I–V) and two mobile electron carriers, coenzyme Q_{10} (CoQ_{10}) and cytochrome c (cyt c). Among the 13 subunits encoded by mtDNA, seven are components of complex I, one (cytochrome b, cyt b) is part of complex III, three are subunits of complex IV (cytochrome c oxidase, COX), and two are subunits of complex V (ATP synthetase). Complex II (succinate dehydrogenase, SDH) is entirely encoded by nDNA.

Due to its dual genetic control, RC defects can be due to mutations in mtDNA or in nDNA. The first mitochondrial diseases to be understood at the molecular level were both due to mtDNA mutations [1, 2], probably because the small mtDNA (16,569 bp) is easily sequenced and because the overt maternal inheritance in one family with Leber hereditary optic neuropathy (LHON) offered a strong clue to the faulty DNA.

These seminal papers opened a floodgate of research that associated multiple clinical syndromes as well as nonsyndromic symptoms and signs to mtDNA large-scale rearrangements or point mutations. In short order, mtDNA became a veritable Pandora's box of human disorders, mostly encephalomyopathies, leading one of the authors to title a review published in 2000: "Mutations in mtDNA: Are we scraping

---

S. DiMauro (✉)
College of Physicians & Surgeons, 630 West 168th Street, New York, NY 10032, USA
e-mail: sd12@columbia.edu

Neurology Department, H. Houston Merritt Clinical Research Center,
Columbia University Medical Center, New York, NY, USA

V. Emmanuele
Department of Neurology, Columbia University Medical Center, New York, NY, USA

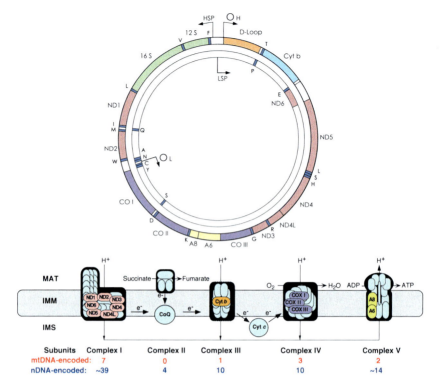

Fig. 1.1 Superimposed schematic representations of mitochondrial DNA (mtDNA) and the mitochondrial RC. The mtDNA genes are matched in color with their encoded proteins. The subunits shown in *aqua* are encoded by nuclear DNA (nDNA). (From [164]. Reprinted with permission from Wiley-Blackwell)

the bottom of the barrel?" [3]. Of course, we were far from scraping the bottom of the mtDNA barrel, but for many years there was a huge gap between our knowledge of mtDNA- and nDNA-related mitochondrial disorders.

Although two novel paradigms of Mendelian mitochondrial disorders were identified in 1989 (multiple mtDNA deletions) [4] and 1991 (mtDNA depletion) [5], the first mutation in a nuclear gene encoding a subunit of the RC was not described until 1995 [6]. The patients were two girls with Leigh syndrome (LS) and the mutant protein was the flavoprotein subunit of the small complex II (four subunits, all encoded by nDNA). This was followed by an avalanche of mutations identified in highly conserved genes of the behemoth complex I (45 subunits, 38 encoded by nDNA), mostly in patients with LS [7, 8]. "Direct hits" (i.e., mutations in genes encoding RC subunits) are apparently rare occurrences for complex III [9] and complex IV [10] whereas "indirect hits" (i.e., mutations in genes that do not encode RC subunits but rather factors needed for the assembly of functionally competent RC complexes) have been reported for all five complexes and are especially important in causing defects of complexes III, IV, and V (for review, see [11]).

General Clinical Considerations

In general, mtDNA-related diseases are clinically more heterogeneous than Mendelian mitochondrial disorders. This is easily explained by the rather "loose" rules of mitochondrial genetics, especially heteroplasmy, the threshold effect, and mitotic segregation [12]. Thus, different mutation loads in tissues and individuals account for different and variably severe clinical expressions even in members of the same family, as exemplified by the NARP/MILS syndrome [13, 14]. Similarly, longitudinal changes in the mutation load explain how the clinical phenotype can change with the passing of time, as in children who unfortunately transition from Pearson syndrome (PS) to Kearns–Sayre syndrome (KSS) [15, 16].

Conversely, the "tighter" rules of Mendelian genetics explain why nDNA-related disorders tend to be more stereotypical, to appear earlier in life, and to be generally more severe and often lethal, as illustrated by complex I deficiency [7, 8].

However, an important distinction has to be made here between the two major groups of nDNA mutations causing mitochondrial diseases, those that affect the RC directly ("direct hits") or indirectly ("indirect hits") and those that affect mtDNA maintenance (resulting in mtDNA multiple deletions) or mtDNA replication (resulting in mtDNA depletion). The latter group is commonly classified under the rubric "defects of intergenomic communication" and, although inheritance is unequivocally Mendelian, these disorders share much of the clinical heterogeneity of primary mtDNA-related diseases, presumably because the polyploid mtDNA is involved in both conditions.

Thus, for example, the degree of mtDNA depletion may vary not only in different tissues but also in the same tissue in different patients, explaining the extreme clinical heterogeneity of the mtDNA depletion syndromes [17]. Similarly, even before specific mutated nuclear genes were identified, it was noted that patients with autosomal-dominant progressive external ophthalmoplegia (AD-PEO) had higher proportions of mtDNA deletions in muscle than those with autosomal-recessive PEO (AR-PEO) [18].

Discovery of the genes responsible for multiple mtDNA deletions and for mtDNA depletion has revealed that they involve the same pathogenic pathway controlling the mitochondrial nucleotide pool [17, 19]. Thus, the two conditions may coexist, as in skeletal muscle from patients with mitochondrial neurogastrointestinal encephalomyopathy (MNGIE) [20].

Also, mutations in the same gene, such as *TK2*, cause predominantly mtDNA depletion with infantile or childhood weakness but occasionally multiple mtDNA deletions with adult-onset AR-PEO [21]. Conversely, mutations in the *PEO1* gene encoding the Twinkle helicase usually cause AD-PEO, but can also cause autosomal-recessive hepatocerebral syndrome with mtDNA depletion in liver [22, 23].

The ultimate example of a single gene causing either mtDNA depletion or multiple mtDNA deletions resulting in extremely diverse syndromes is *POLG* [24–27]. Depending largely on which of the three domains of the *POLG* (polymerase, exonuclease, linker region) harbors the mutation(s), the clinical phenotype spans from

Table 1.1 Classification of nuclear DNA-related mitochondrial disorders

Mutations in RC subunits ("direct hits")
Complex I
Complex II
Complex III
Complex IV
Complex V

Mutations in RC ancillary proteins ("indirect hits")
Complex I
Complex II
Complex III
Complex IV
Complex V

Defects of intergenomic communication
Multiple mtDNA deletions
mtDNA Depletion
Defects of mtDNA translation
− Structural mitochondrial ribosomal proteins
− Mitochondrial ribosomal assembly factors
− Mitochondrial translation factors
− Mitochondrial tRNA modifying factors
− Mitochondrial aminoacyl tRNA synthetases
− Mitochondrial translational activators or mRNA stabilizers

Defects of the lipid milieu
Barth syndrome
Congenital megaconial myopathy

RC Mitochondrial respiratory chain complexes

a severe hepatocerebral disorder (Alpers syndrome) [28, 29] to AD- or AR-PEO [30], to Parkinsonism [31, 32], and to several other clinical phenotypes, including sensory ataxic neuropathy, dysarthria and ophthalmoparesis (SANDO) [33], and mitochondrial-recessive ataxia syndrome (MIRAS) [34, 35].

The Clinical Spectrum

To help us sort out the vast clinical spectrum of Mendelian mitochondrial diseases, we will follow the classification proposed in Table 1.1 and consider sequentially mutations in RC subunits, mutations in ancillary proteins, defects of intergenomic signaling, and defects of the lipid milieu.

Mutations in RC Subunits ("Direct Hits")

To repeat, these disorders are due to mutations in genes that encode subunits of RC complexes.

Complex I As mentioned above, complex I (NADH-ubiquinone oxidoreductase) is the Goliath of the RC, containing 45 subunits, of which 38 are encoded by nDNA. It is also an energetic giant, generating about 40 % of the proton-motive force eventually harnessed by ATP synthetase (complex V). Given its size, it is not too surprising that isolated complex I deficiency is the most frequently encountered RC defect [36].

Pathogenic mutations have been identified in all of the 14 evolutionarily conserved subunits that comprise the catalytic core. The resulting clinical pictures are remarkably homogeneous and to a large extent can be assimilated to LS, which reflects the ravages of energy shortage on the developing nervous system. LS is defined neuropathologically (or neuroradiologically) by bilateral symmetrical lesions all along the nervous system, but especially in the basal ganglia, thalamus, brainstem, and cerebellar roof nuclei. Microscopically, there is neuronal loss, proportionate loss of myelin, reactive astrocytosis, and proliferation of cerebral microvessels.

Clinically, these children have psychomotor retardation or regression, respiratory abnormalities, hypotonia, failure to thrive, seizures, dystonia, and blindness. These are exactly the symptoms most commonly encountered by Distelmaier et al. [7] in a study of 15 children with complex I deficiency due to mutations in nDNA-encoded genes, and their findings were reinforced by a review of 26 additional patients reported in the literature.

Besides LS, four distinct presentations had been attributed to Mendelian complex I deficiency: (1) fatal infantile lactic acidosis (FILA), (2) neonatal cardiomyopathy with lactic acidosis, (3) leukodystrophy with macrocephaly, and (4) hepatopathy with renal tubulopathy [37, 38]. However, no specific genotype/phenotype emerged for the different variants and it is more likely that they represent variations on a common LS theme than separate clinical entities [7]. After all, complex I deficiency is generalized and differences in residual activity may account for the apparent differential tissue involvement.

One important clinical feature common to these disorders is that onset is invariably in infancy or childhood but almost never in fetal life [39]. Usually, these children are fine at birth and early in life, although onset of symptoms is followed by a rapid and relentless progression leading to death usually before 1 year of age [7, 8]. Lactic acid is consistently and markedly elevated both in blood and in the cerebrospinal fluid (CSF). In agreement with the predominant clinical expression of complex I mutations as LS, brain MRI studies of ten patients with nDNA mutations showed bilateral brainstem lesions in all, putaminal lesions in seven, and leukoencephalopathy in five [40]. Finally, from a genetic point of view, it is noteworthy that mutations in *NDUFA1* cause X-linked LS [41], a most unusual hereditary pattern for RC defects.

Complex II Complex II (succinate-ubiquinone oxidoreductase) is the smallest multimeric component of the RC, comprising only four subunits. Complex II deficiency is also probably the least common RC defect [42]. Although only a handful of patients have been described, clinical presentations vary somewhat.

Four patients had typical LS [6, 43, 44], whereas one child at 5 months developed hypotonia, respiratory distress, hepatosplenomegaly, and cardiomegaly. Sudden apnea during a lumbar puncture required artificial ventilation, cardiac dysrhythmia ensued and the child died [45]. Two children with psychomotor regression and spastic quadriparesis showed diffuse leukodystrophy by MRI and a third child presented in the first year of life with stunted growth and lactic acidosis without neurological symptoms. Interestingly, all three children stabilized or improved after riboflavin administration [46]. Finally, two sisters heterozygous for a mutation in the gene encoding the flavoprotein subunit and showing a partial defect of succinate dehydrogenase (SDH) had late-onset optic atrophy, ataxia, and myopathy [47].

Coenzyme Q_{10} Deficiency Enzyme defects that impair the biosynthesis of the small electron carrier CoQ_{10} can be considered "direct hits" because they drastically reduce the concentration of CoQ_{10}, thus blocking the flux of electrons in the RC between complexes I and II and complex III (see Fig. 1.1).

CoQ_{10} is a lipophilic molecule composed of a redox active benzoquinone ring conjugated to an isoprenoid chain, which in humans contains ten isoprenyl units. The quinone group is synthesized in mitochondria from para-hydroxy-benzoate, a catabolite of tyrosine, whereas the polyisoprene tail is synthesized in the cytoplasm starting from acetyl-CoA through the mevalonate pathway [48, 49]. CoQ_{10} biosynthesis is carried out by a set of at least seven enzymes encoded by nuclear genes named *COQ*; an eighth gene, *ADCK3/CABC1*, the human ortholog of the yeast *COQ8* gene, encodes a putative regulatory protein. Pathogenic mutations have been identified in six genes and cause heterogeneous clinical conditions. Five main clinical syndromes have been associated with CoQ_{10} deficiency in skeletal muscle.

The first syndrome is an encephalomyopathy described in 1989 by Ogasahara et al. [50] in two sisters with mitochondrial myopathy and recurrent myoglobinuria, seizures, mental retardation, and ataxia. The triad of mitochondrial myopathy, myoglobinuria, and encephalopathy was then reported in a few other patients [51–54] but responsible genes remain elusive.

The second syndrome is childhood-onset cerebellar ataxia and atrophy, with variable additional symptoms, including neuropathy, seizures, mental retardation, and muscle weakness [55–58]. The molecular basis of this syndrome was identified in some but not all patients and the mutated gene was *ADCK3/CABC1* [59, 60].

The third syndrome is an infantile encephalomyopathy, first described by Rotig et al. [61] in three siblings with nystagmus, optic atrophy, sensorineural hearing loss, ataxia, dystonia, weakness, and nephropathy. We found a homozygous missense mutation in the *COQ2* gene in an infant boy who presented with nystagmus at 2 months and, at 12 months, had hypotonia, psychomotor delay and severe nephrotic syndrome requiring renal transplantation. His neurological condition deteriorated, he developed psychomotor regression, tremor, weakness, and status epilepticus. Brain MRI showed cerebral and cerebellar atrophy and stroke-like lesions. His younger sister had severe nephrosis at 12 months but never developed neurological symptoms. Oral CoQ_{10} administration improved the boy's condition dramatically and presumably protected his sister from developing neurological problems [62].

A similar but rapidly fatal condition was associated with *COQ2* mutations in two siblings, a girl with neonatal neurological distress, nephrosis, hepatopathy, pancytopenia, diabetes, seizures, and lactic acidosis, leading to death at 12 days [63]. Her older brother had anemia, liver failure, and nephropathy and survived only 1 day.

COQ1 is composed of two subunits, PDSS1 and PDSS2. Mutations in *PDSS2* were responsible for typical Leigh syndrome and nephrotic syndrome in an infant boy who died at the age of 8 months of intractable status epilepticus despite CoQ$_{10}$ supplementation [64]. Mutations in *PDSS1* were associated with early deafness, encephaloneuropathy, obesity, valvulopathy, livedo reticularis, and mental retardation in two siblings, who were 14 and 22 years old at the time of publication [63]. Mutations in *COQ2* were described in two much more severely affected siblings, who died at 1 and 12 days with neonatal neurologic distress, liver failure, nephrotic syndrome, diabetes, seizures, and pancytopenia [63].

The fourth syndrome is dominated by severe nephrotic syndrome. Among two patients with mutations in *COQ2*, one had isolated nephropathy but the other had also epileptic encephalopathy and died of renal failure at 6 months [65]. Early-onset steroid-resistant nephritic syndrome, due to mutations in the *COQ6* gene, was associated with sensorineural deafness in 13 patients from seven families [66].

The fifth syndrome, characterized by isolated myopathy, is controversial because muscle CoQ$_{10}$ deficiency was found in some patients with multiple acyl-CoA dehydrogenase deficiency (MADD) due to mutations in the gene (*ETFDH*) encoding the electron-transferring flavoprotein dehydrogenase [67] but not in others [68, 69].

Complex III Two distinct "direct hits" for complex III (ubiquinol-cytochrome *c* reductase) have been described one in a single patient, the other in a large consanguineous family. A deletion in the *UQCRB* gene encoding the ubiquinone-binding protein (QP-c subunit or subunit VII) was identified in an 8-month-old girl with moderate hepatomegaly, hypoglycemia, metabolic acidosis, and lactic acidosis [9]. During a fasting test, she developed hypoglycemia after 19 hours, which became symptomatic after 21 hours. Lactic acid was also markedly increased (4.16 mmol/L) and ketogenesis was impaired, suggesting a functional defect of fatty acid oxidation. At the age of 4 years, the child was normal and the hepatomegaly had regressed. Complex III activity was decreased in liver, lymphocytes, and fibroblasts.

Twenty members of an Israeli-Bedouin family presented with a uniform encephalomyopathy that was severe but compatible with long survival. These patients were normal at birth and for the first few months, then showed psychomotor retardation and developed extrapyramidal signs, including dystonia, athetoid movements, ataxia, and dementia [70]. They shared a homozygous mutation in *UQCRQ*, ubiquinol-cytochrome *c* reductase, subunit VII [70].

Complex IV It is likely that most "direct hits" in nuclear encoded subunits of complex IV (cytochrome *c* oxidase) are incompatible with life because years of research have revealed mutations in only one of the ten nDNA-encoded subunits, COX6B1 [10]. Although a sister of the patients had died immediately after birth and the mother had had a second trimester miscarriage, of the two affected brothers one had died at 10 years of age and the other was 8 years old at the time of publication. Both children had had failure to thrive but normal psychomotor development until late childhood, when they developed muscle weakness, cognitive deterioration, visual problems, and lactic acidosis. MRI showed severe cavitating leukodystrophy in both [10].

Complex V As for complex IV, "direct hits" of complex V appear to be very rare, having been reported only in one patient, who harbored a homozygous mutation in the *ATP5E* gene and—paradoxically—had a more benign course than patients with "indirect hits" (see below) [71].

This girl (patient 3 in [72]), was small at birth and had poor suck, respiratory distress, lactic acidosis, and 3-methylglutaconic aciduria. Although she had recurrent metabolic crises, her clinical course stabilized at 5–6 years and she completed school and was gainfully employed. At the age of 17 years, she had mild ataxia, horizontal nystagmus, exercise intolerance, mixed axonal and demyelinating polyneuropathy, and mild left ventricular hypertrophy. Brain MRI had shown hyperintense lesions in the caudate and lentiform nuclei at the age of 14, but had normalized by the age of 17.

Mutations in RC Ancillary Proteins ("Indirect Hits")

Pathogenic mutations in proteins needed for the assembly, stabilization, and functional regulation have been identified for all five complexes of the RC.

Complex I After several mutations were found in the 14 core subunits crucial for catalytic function (7 mtDNA- and 7 nDNA-encoded), attention was directed to the remaining 31 "accessory" subunits, at least some of which are presumably involved in assembly [8, 73]. Pathogenic mutations have been reported in six assembly subunits [74–79] and in five chaperone factors [80–86] and many more mutated genes in both groups will be identified through the increased use of exome sequencing [87, 88].

The clinical manifestations tend to be more heterogeneous than those associated with "direct hits" in complex I, although there is, understandably, a good deal of overlap. All the patients had encephalopathy clinically resembling LS but often with leukodystrophy rather than gray matter involvement [75, 76, 78]. Cardiomyopathy was more common [74, 79, 80, 85, 89] and sometimes the dominating feature [77]. Although onset was in infancy or early childhood and early death was common, the course was often prolonged and in some cases fluctuating [78]. Death occurred between 5 and 13 years of age in three patients [76, 80, 86] and a few patients were alive at ages 4, 5, and 20 years at the time of publication [76, 83, 85].

Three siblings in a doubly consanguineous Dutch family and an unrelated patient, who harbored mutations in the *ACAD9* gene had isolated myopathy with exercise intolerance and lactic acidosis responsive to riboflavin [84]. The slower progression in some of these cases may have been due to residual complex I activity [76].

Complex II Even the tiny complex II requires assembly factors and mutations have been described in one of them (SDHAF1), which contains a LYR motif and is presumably involved in Fe–S metabolism [90]. After developing normally for the first 6–11 months, five children from two consanguineous families developed acute psychomotor regression, impaired growth, spastic quadriplegia, and moderate cognitive decline. One child died at 18 months but the others were alive and 6 to 12-year-old at the time of publication. MRI and positron MRS of the brain showed leukoencephalopathy, sparing U fibers and increased lactate and succinate peaks [90].

Complex III The first assembly defect in complex III was identified in 2002 in Finnish infants with an extremely severe syndrome named GRACILE, which summarizes the main symptoms and signs: growth retardation, aminoaciduria, cholestasis, iron overload, and early death (before 5 months of age). The mutated protein, BCS1L is a member of the AAA family of ATPases needed for insertion of the Rieske FeS subunit into the complex [91]. Interestingly, these children did not have neurological impairment nor overt complex III deficiency [92].

In contrast, British patients with GRACILE syndrome had both neurological involvement (hypotonia, seizures) and decreased enzyme activity. In even sharper contrast, five Turkish children from four unrelated but consanguineous families developed severe hepatopathy (with complex III deficiency in liver) soon after birth, proximal renal tubulopathy, and encephalopathy compatible with LS in two cases [93]. Three of these children died before 2 years of age, one was severely retarded at 9 years, and one, who had had an episode of myoglobinuria at the age of 1 month, was lost to follow-up at 5 months [93].

Two unrelated patients with mutations in *BCS1L* suffered from severe encephalopathy without visceral involvement: both showed respiratory distress at birth and developed psychomotor delay or regression, spastic quadriparesis, and seizures in one case [91]. Both patients had cerebral atrophy by MRI and, curiously, both had brittle hair. One patient with a novel homozygous *BCS1L* mutation had a mild presentation with infantile onset of hypotonia, psychomotor retardation, coarse facial features, and hirsutism [94].

The most benign expression of mutations in *BCS1L* is Björnstad syndrome, characterized by congenital sensorineural hearing loss and pili torti, resulting in brittle hair, usually recognized in childhood [95]. It is notable that brittle hair had also been noted in two children with severe encephalopathy [91].

A second assembly factor for complex III is tetratricopeptide repeat 19 (TTC19), a protein of the inner mitochondrial membrane, where it interacts with complex III [96]. Four Italian patients with homozygous mutations in *TTC19* had severe encephalopathy. Two siblings presented in childhood with neurological symptoms that progressed for years and included mental retardation, ataxia, nystagmus, dysphagia, dysphonia, tremor, dystonia, and hearing loss. The sister was alive but confined to a wheelchair at the age of 30 and her younger brother was totally incapacitated at the age of 20. A third patient also had childhood-onset mental retardation, dystonia, and ataxia. At the age of 19, she was bedridden, on a ventilator, and in a fluctuating comatose state. The fourth patient did not seek medical attention until 42 years of age, when he developed limb weakness and fasciculations, apraxia of gait and hand movements, dysarthria, dystonia, and paraparesis. He died at the age of 45. The first and fourth patients had lactate peaks by 1H-MRS of the brain and brain MRI in the first patient showed bilateral necrotic lesions of the caudate nuclei, cerebellar atrophy, and leukodystrophy. The third patient's brain MRI showed LS-like lesions affecting medulla, inferior olives, pons, cerebellum, cerebral peduncles, periaqueductal grey matter, and bilateral thalami. The fourth patient had diffuse cortical atrophy and necrotic lesions in the right caudate and both putamina [96].

Complex IV It has long been known that cytochrome *c* oxidase (COX, complex IV) deficiency is a common cause of LS [97, 98]. Yet, years went by before a mutated gene was associated with LS: the gene (*SURF1*) is essential for correct COX assembly and this was the first example of an "indirect hit" to affect the RC [99, 100]. Mutations in *SURF1* were found in numerous patients with typical LS and COX deficiency [101–106] and by 2001, 30 different mutations had been identified [107].

It has been suggested that brain MRI can be a clue to the molecular diagnosis when it combines symmetrical lesions in the subthalamic nuclei, the medulla, the inferior cerebellar peduncles, and the substantia nigra [108], although one case was characterized by leukodystrophy [109], and another had an unusually mild course with normal MRI at 3 years and lesions limited to the brainstem and the cerebellum at 8 years [110]. Three of four cases with typical LS also had proximal renal tubular acidosis and the fourth child had ragged-red fibers in his muscle biopsy, a most unusual finding in LS [111].

Mutations in the COX assembly gene *SCO2*, which encodes a metallochaperone involved in mitochondrial copper delivery, cause a much more severe clinical phenotype that combines neonatal hypertrophic cardiomyopathy with encephalopathy and is fatal in the first weeks or months of life [104, 106, 112–115] and may be associated with early fetal lethality [116]. The fatal infantile presentation is typical of compound heterozygous patients, who always harbor the "common" E140K mutation [117]. Homozygosity for the same mutation causes delayed onset and longer survival [114, 118].

The differential features between the clinical phenotypes associated with *SURF1* and *SCO2* mutations have been summarized by Sue et al. [106]: all patients with SURF1 mutations had typical LS whereas patients with SCO2 mutations had early onset and rapidly fatal cardiomyopathy and encephalopathy without the typical neuropathological features of LS.

Copper supplementation in the culture medium rescued COX activity in cultured fibroblasts and myoblasts from patients with *SCO2* mutations [119, 120], which led to the treatment of a girl homozygous for the E140K mutation with subcutaneous and oral administration of Cu-His [121]. There was objective improvement of the cardiomyopathy but not of the encephalopathy and the girl died at 42 months of pneumonia.

Importantly, *SCO2* mutations (both in compound heterozygosity and in homozygosity) can simulate spinal muscular atrophy (SMA) and the neurogenic, often SMA-like, histological pattern of the muscle biopsy further complicates the differential diagnosis [122–124]. One postmortem study of the spinal cord showed severe neuronal loss and astrocytosis in the anterior horns [123]. Two distinctive features orient toward the correct diagnosis: clinically, the severe cardiomyopathy; and, morphologically, the lack of COX stain in the muscle biopsy. However, in patients with the clinical picture of SMA but without mutations in the *SMN* gene, it is important to consider *SCO2* mutations.

A second and even more deceitful mitochondrial cause of the SMA phenocopy is mtDNA depletion due to *TK2* mutations (briefly discussed above under "General Clinical Considerations") [125].

Mutations in *SCO1*, a second metallochaperone needed for copper insertion into the COX holocomplex, were first published in 2000 and reported to affect multiple members of a large family [126]. The child described in that paper was hypotonic and lethargic at birth, had severe metabolic acidosis, and needed respiratory assistance. Liver function tests were severely altered at day 4, followed by hepatomegaly. He developed axial hypotonia, episodes of apnea and bradycardia and died at 2 months. A liver biopsy obtained immediately postmortem showed swollen hepatocytes with microvesicular steatosis, and a muscle biopsy showed lipid storage. A sibling presented with metabolic acidosis and "severe neurological distress" and died at 5 days.

A comparative study of *SCO2* and *SCO1* transcription and protein expression in different human and mouse tissues revealed that both genes are expressed ubiquitously but the expression of *SCO1* is especially robust in blood vessels and in liver, thus explaining the apparent "hepatoencephalopathy" manifested by patients with *SCO1* deficiency [127]. The report of a single family suggests that *SCO1* mutations are either incompatible with life or very rare, a conclusion supported by several genetic screening of patients with COX deficiency [103, 104, 128, 129].

Mutations in *COX10*, which encodes a factor involved in the first step of the mitochondrial heme-A biosynthesis, were described in five patients from three families and with different clinical presentations [130, 131]. Three siblings from a consanguineous family died at 2, 3, and 5 years of age: the child described in more detail was normal until 18 months, when he developed ataxia, weakness, ptosis, pyramidal signs, and status epilepticus. He had lactic acidosis and increased urinary amino acids, suggesting proximal renal tubulopathy. His younger sister was reported to have similar "progressive neurological deterioration" [130]. One of two unrelated children had lactic acidosis, hypotonia, growth retardation, sensorineural hearing loss, transfusion-dependent macrocytic anemia, biventricular hypertrophic cardiomyopathy, and died at 5 months [131]. The other child had a sudden arrest of her psychomotor development at 1.5 months, developed transfusion-dependent anemia, mild splenomegaly, and hypotonia. Brain MRI showed symmetrical lesions in the basal ganglia typical of LS. She died of central respiratory failure at 4 months [131].

COX15, like *COX10*, is involved in heme-A biosynthesis and mutations in *COX15* caused hypotonia, seizures, and lactic acidosis in an infant girl, who had a normal echocardiography at 6 days of age but developed acute biventricular hypertrophy by 22 days and died 2 days later. Postmortem examination revealed massive accumulation of morphologically abnormal mitochondria in the heart [132, 133].

The heart was not affected in a second patient who presented with typical LS: he was normal at birth and for the first seven months, then developed hypotonia, nystagmus, leg spasticity, progressive microcephaly, and retinopathy. His condition deteriorated dramatically when he developed bloody diarrhea at 2 years 10 months, and he died of pneumonia a year later. Lactic acid was increased both in blood and in the CSF and brain MRI showed multiple symmetrical lesions in the basal ganglia, dorsal midbrain, cerebral peduncles, and periaqueductal grey matter [134]. A third patient also had a diagnosis of LS, but her course was unusually protracted and, at the time of publication, she was still alive at the age of 16 [135].

Considering that at least 30 chaperones are needed for the assembly of complex IV in yeast, it is not surprising that the list of "indirect hits" responsible for COX deficiency in humans keeps getting longer. On the other hand, it is difficult to make general statements on clinical phenotypes because mutations in some genes have been reported in single or very few patients, often siblings from consanguineous marriages. This is the case for a homozygous mutation in *C2orf64*, which encodes a factor involved in the early steps of COX assembly [136]. Two siblings had onset of cardiomyopathy *in utero* with fetal distress and isolated biventricular hypertrophic cardiomyopathy causing death 8 and 10 days after birth. Postmortem examination of the heart revealed mitochondrial proliferation and lipid storage.

Two siblings from another consanguineous family had a slower but devastating disorder dominated by encephalopathy, with severely delayed psychomotor development, optic atrophy, intractable seizures leading to status epilepticus in the affected boy, and left hemiplegia in the affected girl. At 14 years of age, the girl was unable to stand or walk and had a 20-word vocabulary. The boy was lost to follow-up at the age of 4 years, when he was bedridden and incapable of communication or purposeful movement [134]. The mutant gene (*KIAA0971*), identified through homozygosity mapping, encodes a protein known as fas-activated serine-threonine kinase domain 2 (FASTKD2) that plays a role in mitochondrial apoptosis rather than in COX assembly [137]. Although COX activity was markedly decreased in muscle, COX and SDH histochemical stains were normal.

Through integrated genomics, based on bioinformatics-generated intersection of DNA, mRNA, and protein data sets, Mootha et al. [138] identified the gene (*LRPPRC*) responsible for the French Canadian type of LS (LSFC) with COX deficiency [139]. As the LRPPRC (leucine-rich pentatricopeptide repeat cassette) protein has a role in the translation or stability of the mRNA of mtDNA-encoded COX subunits [140], LDFC might be more appropriately classified among the defects of mtDNA translation (see below).

Another defect of mitochondrial translation affecting COX, and specifically the COX I subunit, is caused by mutations in the gene *TACO1* (translational activator of COX I) [141]. Five children of a consanguineous Turkish family were affected with slowly progressive LS, more severe in girls than in boys [142]. Onset was in childhood or adolescence and progression was slow: all patients were alive at the time of publication and three were in their 20s. The clinical picture was dominated by small stature, mental retardation, dystonia, dysarthria, spasticity with pyramidal signs, and optic atrophy. Lactic acid was inconsistently increased, and MRI of the brain showed bilateral basal ganglia lesions in all patients.

Integrative genomics also facilitated identification of the *ETHE1* gene in children with ethylmalonic encephalomyopathy (EE), an early-onset disorder with microangiopathy, chronic diarrhea, and greatly increased levels of ethylmalonic acid and short-chain acylcarnitines in body fluids [143]. Based on their studies of patients and *Ethe1*-null mice, Valeria Tiranti and Massimo Zeviani introduced a new paradigm of "indirect hit," which can be called "toxic indirect hit." They documented that ETHE1 is a matrix thioesterase and its dysfunction leads to excessive accumulation of sulfide, a powerful COX inhibitor [144]. The clinical presentation has been confirmed and the number of mutations extended in a study of 14 patients with EE [145].

Complex V Mutations in two ancillary proteins cause severe infantile disorders. A homozygous missense mutation in *ATP12* (now known as *ATPAF2*) was first reported in 2004 in an infant girl who died at 14 months [146]. She had dysmorphic features, arthrogryposis, hepatomegaly, hypoplastic kidneys, fluctuating lactic acidosis, and, terminally, seizures. Brain MRI showed cortical-subcortical atrophy, dysgenesis of the corpus callosum, and hypoplastic white matter.

A homozygous mutation (c.317-2A > G) in the assembly gene TMEM70 was identified in 25 infants, 24 of whom were of Roma ethnic origin, with severe multisystem symptoms, lactic acidosis, and 3-methylglutaconic aciduria [72]. The clinical picture was described in detail in a subsequent paper [147] and was characterized by neonatal hypotonia, apneic spells, hypertrophic cardiomyopathy, severe lactic acidosis, and hyperammonemia. Ten patients died within 2 months. Those who survived showed hypotonia and psychomotor delay. The cardiomyopathy was nonprogressive and sometimes reversible. About half of the boys had hypospadia or cryptorchidism. The authors suggest that mutations in TMEM70 should be suspected in critically ill neonates with hypotonia, hypertrophic cardiomyopathy and the laboratory triad of lactic acidosis, hyperammonemia, and 3-methylglutaric aciduria [147]. The report of more patients has widened the clinical spectrum, including congenital hypertonia with multiple contractures, early-onset cataracts and gastrointestinal problems, and Reye syndrome-like episodes [148–150].

Defects of Intergenomic Communication

At least in general clinical terms, disorders associated with multiple mtDNA deletions and mtDNA depletion have been discussed above under "General Clinical Considerations." Therefore, here we will only review briefly a third group of diseases, those due to defects of mitochondrial translation.

Defects of Mitochondrial Translation Clinically, these disorders came to the attention of pediatricians because they cause severe infantile syndromes, often similar to those associated with mtDNA depletion even though mtDNA is normal both qualitatively and quantitatively. Although these disorders are usually characterized by multiple RC enzyme deficiencies, mutations in some translation factors, such as LRPPRC and TACO1, impair specifically COX activity, as described above [138, 141].

A thorough update on this rapidly expanding group of diseases can be found in a recent Workshop Report [151], in which the authors classify the proteins affected into six types: (1) structural mitochondrial ribosomal proteins; (2) mitochondrial ribosomal assembly proteins; (3) mitochondrial translation factors; (4) mitochondrial tRNA modifying factors; (5) mitochondrial aminoacyl tRNA synthetases; and (6) mitochondrial translational activators or mRNA stabilizers (including LRPPRC and TACO1 discussed under complex IV indirect hits). Age at onset was at or soon after birth and age of death varied but was usually in infancy or childhood. The clinical features listed in Table 2 of the Workshop Report [151] show that encephalomyopathy was the most common presentation, although cardiomyopathy and hepatopathy were occasionally seen, and lactic acidosis was almost invariably present.

As evidence of the vitality of this area of research, several new examples of mtDNA translation disorders have already been described after the subgroups listed above were published only 6 months ago.

Through MitoExome sequencing, pathogenic compound heterozygous mutations have been identified in the *MTFMT* gene, which encodes the modifying factor mitochondrial methionyl-tRNA formyltransferase (subgroup IV above), an enzyme needed to generate N-formylmethionine-tRNA^{Met}, which in turn is needed for translation initiation [152]. The two affected children had LS and multiple RC enzyme deficiencies.

Mutations in *DARS2* (belonging in group V above) had been associated with leukoencephalopathy and brain stem and spinal cord involvement (LBSL), a devastating encephalopathy [153], but homozygous mutations in the same gene can also cause a relatively benign condition characterized by exercise-induced paroxysmal gait ataxia and areflexia responsive to acetazolamide [154].

A homozygous mutation in the gene (*GFM1*) encoding an elongation factor (subgroup III above) caused a severe encephalopathy manifesting soon after birth in a girl with microcephaly, developmental delay, axial hypotonia, roving eye movements, intractable seizures, and hepatomegaly. Brain MRI showed atrophy of the frontal cortex and of the corpus callosum and delayed myelination. She died at 2 years from respiratory insufficiency [155].

A phenocopy of the Cornelia de Lange dysmorphic phenotype (synophrys [joined eyebrows], low implanted and posteriorly rotated ears, retrognathia, redundant skin over the neck, hypospadia, nonpitting edema of limbs, back, and eyelids) was associated with a homozygous mutation in the MRPS22 protein (subgroup I above) in an infant boy with congenital hypertrophic cardiomyopathy and third ventricle dilation. Along the years, he developed dysphagia, seizures, and Wolff–Parkinson–White syndrome. At 4 years of age, the child was unable to sit independently and did not communicate [156].

An interesting clinically reversible hepatopathy characterizes infants with mutations in the *TRMU* gene, encoding mt tRNA 2-thiouridylase (subgroup III above) [157]. This peculiar phenotype was confirmed in a 2-year-old girl who was compound heterozygous for two mutations in *TRMU* and was neurologically normal after surviving liver failure between 2 and 4 months of age and despite persistent liver cirrhosis [158]. The authors stress the importance of early diagnosis, the potential usefulness of liver transplantation, and the curious finding of secondary muscle CoQ$_{10}$ deficiency. Mutations in the same gene have been described in two patients with infantile reversible COX-deficient myopathy [159].

Defects of the Lipid Milieu

The phospholipids components of the inner mitochondrial membrane (IMM) do not simply act as a scaffold for the RC, but participate in its structural and functional integrity. This is especially true for cardiolipin, the most abundant phospholipid component of the IMM. Alterations in cardiolipin biosynthesis and remodeling have been associated with Barth syndrome, due to mutations in the *TAZ1* gene that encodes the monolysocardiolipin transacylase [160]. Barth syndrome is an X-linked disorder characterized by mitochondrial cardiac and skeletal myopathy, cyclic neutropenia, and growth retardation [161]. Two commonly associated laboratory abnormalities are urinary excretion of 3-methylglutaconic acid and hypocholesterolemia.

The importance of the mitochondrial lipid milieu was recently confirmed by the description of 15 patients with a congenital myopathy characterized clinically by early-onset muscle weakness and mental retardation with protracted course (most patients were alive at the time of publication and only four had died between the ages of 2 years 6 months and 28 years). The hallmark of the disease was the presence in muscle of greatly enlarged mitochondria displaced to the periphery of the fibers [162].

Ten of the patients were Turkish, 4 Japanese, and 1 British, and they all harbored various mutations in the gene (*CHKB*) encoding choline kinase beta, the enzyme that catalyzes the first step in the de novo biosynthesis of phosphatidyl choline (PtdCho) and phosphatidylethanolamine (PtdEtn) via the Kennedy pathway.

A new homozygous *CHKB* mutation was found in an American patient, who also had weakness and psychomotor delay and was alive at 2 years [163]. The American patient had COX deficiency in muscle and impaired RC enzyme activities were documented in muscle from mice harboring a loss-of-function mutation in the the ortholog gene, *CHKB*, and which were affected with retrocaudal muscular dystrophy (rmd) [164].

Conclusion

Although the clinical spectrum of mitochondrial diseases due to nDNA mutations is very wide, a number of general conclusions are warranted:

1. *Onset* is almost invariably in infancy or early childhood, often following several months of normal development. Prenatal manifestations are uncommon, as shown by the rare occurrence of arthrogryposis and dysmorphisms. In a few instances fetal cardiopathy is revealed by fetal bradycardia or by echocardiography. Adult onset is confined to some disorders of intergenomic communications, most notably multiple mtDNA deletions with progressive external ophthalmoplegia.
2. The *course* is usually rapidly downhill leading to death in infancy or childhood. The few disorders with prolonged survival are associated with severe disability. Benign conditions are exceptional.
3. *Tissue involvement*. These are usually multisystem disorders dominated by encephalopathy. The most common encephalopathy is LS, which reflects the extensive damage that occurs when the developing nervous system is starved for energy. Leukoencephalopathy, sometimes with cystic cavitation, is less common, but not infrequent. Three other organs are often involved, usually together with the brain but occasionally in the absence of encephalopathy. Hypertrophic cardiomyopathy can be the presenting and predominant problem. It is often rapidly

fatal but it can also be chronic as in Barth syndrome. Liver involvement is typically associated with mtDNA depletion or with defects in mtDNA translation ("hepatocerebral syndromes ") but can rarely be seen in direct or indirect hits to RC complexes. Nephropathy typically presents as proximal tubulopathy with Fanconi syndrome, but, when it presents as nephrosis with albuminuria, it should raise the diagnostic suspicion of CoQ_{10} deficiency.
4. *Laboratory examinations.* Severe lactic acidosis is extremely common but neither invariably present nor necessarily severe. Increased lactate in the CSF is frequent and detectable by 1H-MRS.
5. *Muscle biopsy.* Except for defects of intergenomic communication, ragged-red fibers (RRF) are not found in these conditions. COX-negative fibers are diffusely present (not distributed in a mosaic pattern as in mtDNA-related diseases) in both direct and indirect hits to complex IV. They are also seen in defects of mtDNA translation.

As shown by the bibliography, a rapidly increasing number of Mendelian conditions affecting the RC has been recognized in recent years, in part due to our greater awareness of their clinical presentations (summarized above), in part due to the utilization of homozygosity mapping in consanguineous families and of genome-wide or MitoExome sequencing in puzzling mitochondrial patients.

Besides their heuristic value, these discoveries have immediate practical utility because they allow us to provide prenatal diagnosis to young families in whom a child has usually succumbed to one of these devastating pediatric diseases.

Acknowledgments This work has been supported by NICHD grant P01-H23062 and by the Marriott Mitochondrial Disorder Clinical Research Fund (MMDCRF).

References

1. Holt IJ, Harding AE, Morgan Hughes JA (1988) Deletions of muscle mitochondrial DNA in patients with mitochondrial myopathies. Nature 331:717–719
2. Wallace DC, Singh G, Lott MT et al (1988) Mitochondrial DNA mutation associated with Leber's hereditary optic neuropathy. Science 242:1427–1430
3. DiMauro S, Andreu AL (2000) Mutations in mtDNA: are we scraping the bottom of the barrel? Brain Pathol 10:431–441
4. Zeviani M, Servidei S, Gellera C et al (1989) An autosomal dominant disorder with multiple deletions of mitochondrial DNA starting at the D-loop region. Nature 339:309–311
5. Moraes CT, Shanske S, Tritschler HJ et al (1991) mtDNA depletion with variable tissue expression: a novel genetic abnormality in mitochondrial diseases. Am J Hum Genet 48:492–501
6. Bourgeron T, Rustin P, Chretien D et al (1995) Mutation of a nuclear succinate dehydrogenase gene results in mitochondrial respiratory chain deficiency. Nat Genet 11:144–149
7. Distelmaier F, Koopman WJH, van den Heuvel LP et al (2009) Mitochondrial complex I deficiency: from organelle dysfunction to clinical disease. Brain 132:833–842
8. Valsecchi F, Koopman WJH, Manjeri GR et al (2010) Complex I disorders: causes, mechanisms, and development of treatment strategies at the cellular level. Devel Disabil Res Rev 16:175–182

9. Haut S, Brivet M, Touati G et al (2003) A deletion in the human QP-C gene causes a complex III deficiency resulting in hypoglycemia and lactic acidosis. Hum Genet 113:118–122
10. Massa V, Fernandez-Vizarra E, Alshahwan S et al (2008) Severe infantile encephalomyopathy caused by a mutation in COX6B1, a nucleus-encoded subunit of cytochrome c oxidase. Am J Hum Genet 82:1281–1289
11. DiMauro S (2011) A history of mitochondrial diseases. J Inher Metab Dis 34:261–276
12. DiMauro S, Davidzon G (2005) Mitochondrial DNA and disease. Ann Med 37:222–232
13. Holt IJ, Harding AE, Petty RK, Morgan Hughes JA (1990) A new mitochondrial disease associated with mitochondrial DNA heteroplasmy. Am J Hum Genet 46:428–433
14. Tatuch Y, Christodoulou J, Feigenbaum A et al (1992) Heteroplasmic mtDNA mutation (T>G) at 8993 can cause Leigh disease when the percentage of abnormal mtDNA is high. Am J Hum Genet 50:852–858
15. Larsson NG, Holme E, Kristiansson B et al (1990) Progressive increase of the mutated mitochondrial DNA fraction in Kearns-Sayre syndrome. Pediatr Res 28:131–136
16. McShane MA, Hammans SR, Sweeney M et al (1991) Pearson syndrome and mitochondrial encephalomyopathy in a patient with a deletion of mtDNA. Am J Hum Genet 48:39–42
17. Suomalainen A, Isohanni P (2010) Mitochondrial DNA depletion syndromes—many genes, common mechanisms. Neuromuscul Disord 20:429–437
18. Carrozzo R, Hirano M, Fromenty B et al (1998) Multiple mitochondrial DNA deletions features in autosomal dominant and recessive diseases suggest distinct pathogeneses. Neurology 50:99–106
19. Rotig A, Poulton J (2009) Genetic causes of mitochondrial DNA depletion in humans. Biochim Biophys Acta 1792:1103–1108
20. Hirano M, Lagier-Tourenne C, Valentino ML et al (2005) Thymidine phosphorylase mutations cause instability of mitochondrial DNA. Gene 354:152–156
21. Tyynismaa H, Sun R, Ahola-Erkkila S et al (2012) Thymidine kinase 2 mutations in autosomal recessive progressive external ophthalmoplegia with multiple mitochondrial DNA deletions. Hum Mol Genet 21:66–75
22. Hakonen AH, Isohanni P, Paetau A et al (2007) Recessive Twinkle mutations in early onset encephalopathy with mtDNA depletion. Brain 130:3032–3040
23. Sarzi E, Goffart S, Serre V et al (2007) Twinkle helicase (PEO1) gene mutation causes mitochondrial DNA depletion. Ann Neurol 62:579–587
24. Wong L-JC, Naviaux RK, Brunetti-Pierri N et al (2008) Molecular and clinical genetics of mitochondrial diseases due to POLG mutations. Hum Mutat 29:E150–E172
25. Chan SSL, Copeland WC (2009) DNA polymerase gamma and mitochondrial disease: understanding the consequence of POLG mutations. Biochim Biophys Acta 1787:312–319
26. Horvath R, Hudson G, Ferrari G et al (2006) Phenotypic spectrum associated with mutations of the mitochondrial polymerase gamma gene. Brain 129:1674–1684
27. DiMauro S, Davidzon G, Hirano M (2006) A polymorphic polymerase. Brain 126:1637–1639
28. Naviaux RK, Nguyen KV (2004) POLG mutations associated with Alpers' syndrome and mitochondrial DNA depletion. Ann Neurol 55:706–712
29. Davidzon G, Mancuso M, Ferraris S et al (2005) POLG mutations and Alpers syndrome. Ann Neurol 57:921–923
30. Van Goethem G, Dermaut B, Lofgren A et al (2001) Mutation of POLG is associated with progressive external ophthalmoplegia characterized by mtDNA deletions. Nat Genet 28:211–212
31. Davidzon G, Greene P, Mancuso M et al (2006) Early-onset familial; parkinsonism due to POLG mutations. Ann Neurol 59:859–862
32. Eerola J, Luoma PT, Peuralinna T et al (2010) POLG1 polyglutamine tract variants associated with Parkinson's disease. Neurosci Lett 477:1–5
33. Van Goethem G, Martin JJ, Dermaut B et al (2003) Recessive POLG mutations presenting with sensory and ataxic neuropathy in compound heterozygote patients with progressive external ophthalmoplegia. Neuromuscul Disord 13:133–142

34. Hakonen AH, Heiskanen S, Juvonen V et al (2005) Mitochondrial DNA polymerase W748S mutation: a common cause of autosomal recessive ataxia with ancient European origin. Am J Hum Genet 77:430–441
35. Milone M, Brunetti-Pierri N, Tang L-Y et al (2008) Sensory ataxic neuropathy with ophthalmoparesis caused by POLG mutations. Neuromuscul Disord 18:626–632
36. Smeitink J, Sengers R, Trijbels F, van den Heuvel L (2001) Human NADH: ubiquinone oxidoreductase. J Bioenerg Biomembr 33:259–266
37. Janssen RJ, Nijtmans LG, van den Heuvel LP, Smeitink JAM (2006) Mitochondrial complex I: structure, function and pathology. J Inher Metab Dis 29:499–515
38. Bugiani M, Invernizzi F, Alberio S et al (2004) Clinical and molecular findings in children with complex I deficiency. Biochim Biophys Acta 1659:136–147
39. DiMauro S, Garone C (2011) Metabolic disorders of fetal life: glycogenoses and mitochondrial defects of the mitochondrial respiratory chain. Semin Fetal Neonatal Med 16:181–189
40. Lebre AS, Rio M, Faivre d' Arcier L et al (2011) A common pattern of brain MRI imaging in mitochondrial diseases with complex I deficiency. J Med Genet 48:16–23
41. Fernandez-Moreira D, Ugalde C, Smeets R et al (2007) X-linked NDUFA1 gene mutations associated with mitochondrial encephalomyopathy. Ann Neurol 61:73–83
42. Rustin P, Rotig A (2002) Inborn errors of complex II—unusual human mitochondrial diseases. Biochim Biophys Acta 1553:117–122
43. Parfait B, Chretien D, Rotig A et al (2000) Compound heterozygous mutation in the flavoprotein gene of the respiratory chain complex II in a patient with Leigh syndrome. Hum Genet 106:236–243
44. Horvath R, Abicht A, Holinski-Feder E et al (2006) Leigh syndrome caused by mutations in the flavoprotein (Fp) subunit of succinate dehydrogenase (SDHA). J Neurol Neurosurg Psychiatry 77:74–76
45. Van Coster R, Seneca S, Smet J et al (2003) Homozygous Gly555Glu mutation in the nuclear-encoded 7-kDa flavoprotein gene causes instability of respiratory chain complex II. Am J Med Genet 120:13–18
46. Bugiani M, Lamantea E, Invernizzi F et al (2006) Effects of riboflavin in children with complex II deficiency. Brain Dev 28:576–581
47. Birch-Machin MA, Taylor RW, Cochran B et al (2000) Late-onset optic atrophy, ataxia, and myopathy associated with a mutation of a complex II gene. Ann Neurol 48:330–335
48. Quinzii CM, Hirano M (2010) Coenzyme Q and mitochondrial disease. Dev Disabil Res Rev 16:183–188
49. Trevisson E, DiMauro S, Navas P, Salviati L (2011) Coenzyme Q deficiency in muscle. Curr Opin Neurol 24:449–456
50. Ogasahara S, Engel AG, Frens D, Mack D (1989) Muscle coenzyme Q deficiency in familial mitochondrial encephalomyopathy. Proc Natl Acad Sci U S A 86:2379–2382
51. Sobreira C, Hirano M, Shanske S et al (1997) Mitochondrial encephalomyopathy with coenzyme Q10 deficiency. Neurology 48:1238–1243
52. Boitier E, Degoul F, Desguerre I et al (1998) A case of mitochondrial encephalomyopathy associated with a muscle coenzyme Q10 deficiency. J Neurol Sci 156:41–46
53. Di Giovanni S, Mirabella M, Spinazzola A et al (2001) Coenzyme Q10 reverses pathological phenotype and reduces apoptosis in familial CoQ_{10} deficiency. Neurology 57:515–518
54. Auré K, Benoist JF, Ogier de Baulny H et al (2004) Progression despite replacement of a myopathic form of coenzyme Q10 defect. Neurology 63:727–729
55. Musumeci O, Naini A, Slonim AE et al (2001) Familial cerebellar ataxia with muscle coenzyme Q10 deficiency. Neurology 56:849–855
56. Lamperti C, Naini A, Hirano M et al (2003) Cerebellar ataxia and coenzyme Q10 deficiency. Neurology 60:1206–1208
57. Gironi M, Lamperti C, Nemni R et al (2004) Late-onset cerebellar ataxia with hypogonadism and muscle coenzyme Q10 deficiency. Neurology 62:818–820

58. Artuch R, Brea-Calvo G, Briones P et al (2006) Cerebellar ataxia with coenzyme Q10 deficiency: diagnosis and follow-up after coenzyme Q10 supplementation. J Neurol Sci 246:153–158
59. Lagier-Tourenne C, Tazir M, Lopez LC et al (2008) ADCK3, an ancestral kinase, is mutated in a form of recessive ataxia associated with coenzyme Q10 deficiency. Am J Hum Genet 82:661–672
60. Mollet J, Delahodde A, Serre V et al (2008) CABC1 gene mutations cause ubiquinone deficiency with cerebellar ataxia and seizures. Am J Hum Genet 82:623–630
61. Rotig A, Appelkvist EL, Geromel V et al (2000) Quinone-responsive multiple respiratory-chain dysfunction due to widespread coenzyme Q10 deficiency. Lancet 356:391–395
62. Montini G, Malaventura C, Salviati L (2008) Early coenzyme Q10 supplementation in primary coenzyme Q10 deficiency. N Engl J Med 358:2849–2850
63. Mollet J, Giurgea I, Schlemmer D et al (2007) Prenyldiphosphate synthase (PDSS1) and OH-benzoate prenyltransferase (COQ2) mutations in ubiquinone deficiency and oxidative phosphorylation disorders. J Clin Invest 117:765–772
64. Lopez LC, Schuelke M, Quinzii C et al (2006) Leigh syndrome with nephropathy and CoQ_{10} deficiency due to decaprenyl diphosphate synthase subunit 2 (PDSS2) mutations. Am J Hum Genet 79:1125–1129
65. Diomedi-Camassei F, Di Giandomenico S, Santorelli F et al (2007) COQ2 nephropathy: a newly described inherited mitochondriopathy with primary renal involvement. J Am Soc Nephrol 18:2773–2780
66. Heeringa SF, Chernin G, Chaki M et al (2011) COQ6 mutations in human patients produce nephrotic syndrome with sensorineural deafness. J Clin Invest 121:2013–2024
67. Gempel K, Topaloglu H, Talim B et al (2007). The myopathic form of coenzyme Q10 deficiency is caused by mutations in the electron-transferring-flavoprotein dehydrogenase (ETFDH) gene. Brain 130:2037–2044
68. Liang W-C, Ohkuma A, Hayashi YK et al (2009) ETFDH mutations, CoQ_{10} levels, and respiratory chain activities in patients with riboflavin-responsive multiple acyl-CoA dehydrogenase deficiency. Neuromuscul Disord 19:212–216
69. Ohkuma A, Noguchi S, Sugie H et al (2009) Clinical and genetic analysis of lipid storage myopathies. Muscle Nerve 39:333–342
70. Barel O, Shorer Z, Flusser H et al (2008) Mitochondrial complex III deficiency associated with a homozygous mutation in UQCRQ. Am J Hum Genet 82:1211–1216
71. Mayr JA, Havlickova V, Zimmermann F et al (2010) Mitochondrial ATP synthase deficiency due to a mutation in the ATP5E gene for the F1 epsilon subunit. Hum Mol Genet 19:3430–3439
72. Cizkova A, Stranecky V, Mayr JA et al (2008) TMEM70 mutations cause isolated ATP synthase deficiency and neonatal mitochondrial encephalomyopathy. Nat Genet 40:1288–1290
73. Lazarou M, Thorburn DR, Ryan MT, McKenzie M (2009) Assembly of mitochondrial complex I and defects in disease. Biochim Biophys Acta 1793:78–88
74. Hoefs SJ, Dieteren CEJ, Distelmaier F et al (2008) NDUFA2 complex I mutation leads to Leigh disease. Am J Hum Genet 82:1306–1315
75. Saada A, Vogel RO, Hoefs SJ et al (2009) Mutations in NDUFAF3 (C3ORF60), encoding an NDUFAF4 (C6ORF66)-interacting complex I assembly protein, cause fatal neonatal mitochondrial disease. Am J Hum Genet 84:718–727
76. Hoefs SJG, Skjeldal OH, Rodenburg RJ et al (2010) Novel mutations in the NDUFS1 gene cause low residual activities in human complex I deficiencies. Mol Genet Metab 100:251–256
77. Fassone E, Taanman J-W, Hargreaves IP et al (2011) Mutations in the mitochondrial complex I assembly factor NDUFAF1 cause fatal infantilehypertrophic cardiomyopathy. J Med Genet 48:691–697
78. Ferreira M, Torraco A, Rizza T et al (2011) Progressive cavitating leukoencephalopathy associated with respiratory chain complex I deficiency and a novel mutation in NDUFS1. Neurogenetics 12:9–17
79. Berger I, Hershkovitz E, Shaag A et al (2008) Mitochondrial complex I deficiency caused by a deleterious NDUFA11 mutation. Ann Neurol 63:405–408

80. Ogilvie I, Kennaway NG, Shoubridge EA (2005) A molecular chaperone for mitochondrial complex I assembly is mutated in a progressive encephalopathy. J Clin Invest 115:2784–2792
81. Sugiana C, Pagliarini DJ, McKenzie M et al (2008) Mutation of C20orf7 disrupts complex I assembly and causes lethal neonatal mitochondrial disease. Am J Hum Genet 83:468–478
82. Saada A, Edvardson S, Rapoport M et al (2008) C6ORF66 is an assembly factor of mitochondrial complex I. Am J Hum Genet 82:32–38
83. Dunning CJ, McKenzie M, Sugiana C et al (2007) Human CIA30 is involved in the early assembly of mitochondrial complex I and mutations in its gene cause disease. EMBO J 26:3227–3237
84. Gerards M, van den Bosch BJC, Danhauser K et al (2011) Riboflavin-responsive oxidative phosphorylation complex I deficiency caused by defective ACAD9: new function for an old gene. Brain 134:210–219
85. Haack TB, Danhauser K, Haberberger B et al (2010) Exome sequencing identifies ACAD9 mutations as a cause of complex I deficiency. Nat Genet 42:1131–1134
86. Saada A, Edvardson S, Shaag A et al (2012) Combined OXPHOS complex I and IV defect, due to mutated complex I assembly factor C20ORF7. J Inherit Metab Dis 35:125–131
87. Calvo SE, Tucker EJ, Compton A et al (2010) High-throughput, pooled sequencing identifies mutations in NUBPL and FOXRED1 in human complex I deficiency. Nat Genet 42:851–858
88. Pagliarini DJ, Calvo SE, Chang B et al (2008) A mitochondrial protein compendium elucidates complex I disease biology. Cell 134:112–123
89. Ghezzi D, Goffrini P, Uziel G et al. (2009) SDHAF1, encoding a LYR complex II specific assembly factor, is mutated in SDH-defective infantile leukoencephalopathy. Nat Genet 41:654–656
90. Fernandez-Vizarra E, Bugiani M, Goffrini P et al (2007) Impaired complex III assembly associated with BCS1L gene mutations in isolated mitochondrial encephalomyopathy. Hum Mol Genet 16:1241–1252
91. Visapaa I, Fellman V, Vesa J et al (2002) GRACILE syndrome, a lethal metabolic disorder with iron overload, is caused by a point mutation in BCS1L. Am J Hum Genet 71:863–876
92. de Lonlay P, Valnot I, Barrientos A et al (2001) A mutant mitochondrial respiratory chain assembly protein causes complex III deficiency in patients with tubulopathy, encephalopathy and liver failure. Nat Genet 29:57–60
93. Blazquez A, Gil-Borlado MC, Moran M et al (2009) Infantile mitochondrial encephalomyopathy with unusual phenotype caused by a novel BCS1L mutation in an isolated complex III-deficient patient. Neuromuscul Disord 19:143–146
94. Hinson JT, Fantin VR, Schonberger J et al (2007) Missense mutations in the BCS1L gene as a cause of the Bjornstad syndrome. N Engl J Med 356:809–819
95. Ghezzi D, Arzuffi P, Zordan M et al (2011) Mutations in TTC19 cause mitochondrial complex III deficiency and neurological impairment in human and flies. Nat Genet 43:259–263
96. DiMauro S, Servidei S, Zeviani M et al (1987) Cytochrome c oxidase deficiency in Leigh syndrome. Ann Neurol 22:498–506
97. Van Coster R, Lombes A, DeVivo DC et al (1991) Cytochrome c oxidase-associated Leigh syndrome: phenotypic features and pathogenetic speculations. J Neurol Sci 104:97–111
98. Zhu Z, Yao J, Johns T et al (1998) SURF1, encoding a factor involved in the biogenesis of cytochrome c oxidase, is mutated in Leigh syndrome. Nat Genet 20:337–343
99. Tiranti V, Hoertnagel K, Carrozzo R et al (1998) Mutations of SURF-1 in Leigh disease associated with cytochrome c oxidase deficiency. Am J Hum Genet 63:1609–1621
100. Tiranti V, Jaksch M, Hofmann S et al (1999) Loss-of-function mutations of SURF-1 are specifically associated with Leigh syndrome with cytochrome c oxidase deficiency. Ann Neurol 46:161–166
101. Moslemi AR, Tulinius M, Darin N et al (2003) SURF1 gene mutations in three cases with Leigh syndrome and cytochrome c oxidase deficiency. Neurology 61:991–993
102. Coenen MJH, Smeitink JAM, Pots JM et al (2006) Sequence analysis of the structural nuclear encoded subunits and assembly genes of cytochrome c oxidase in a cohort of 10 isolated complex IV-deficient patients revealed five mutations. J Child Neurol 21:508–511

103. Bohm M, Pronicka E, Karczmarewicz E et al (2006) Retrospective, multicentric study of 180 children with cytochrome c oxidase deficiency. Pediatr Res 59:21–26
104. Darin N, Moslemi AR, Lebon S et al (2003) Genotypes and clinical phenotypes in children with cytochrome c oxidase deficiency. Neuropediatrics 34:311–317
105. Sue CM, Karadimas C, Checcarelli N et al (2000) Differential features of patients with mutations in two COX assembly genes, SURF-1 and SCO2. Ann Neurol 47:589–595
106. Pequignot MO, Dey R, Zeviani M et al (2001) Mutations in the SURF1 gene associated with Leigh syndrome and cytochrome c oxidase deficiency. Hum Mutat 17:374–381
107. Rossi A, Biancheri R, Bruno C et al (2003) Leigh syndrome with COX deficiency and SURF1 gene mutations: MR imaging findings. Am J Neuradiol 24:1188–1191
108. Rahman S, Brown RM, Chong WK et al (2001) A SURF1 gene mutation presenting as isolated leukodystrophy. Ann Neurol 49:797–800
109. Salviati L, Freehauf C, Sacconi S et al (2004) Novel SURF1 mutation in a child with subacute encephalopathy and without the radiological features of Leigh syndrome. Am J Med Genet 128A:195–198
110. Tay SKH, Sacconi S, Akman HO et al (2005) Unusual clinical presentation in four cases of Leigh disease, cytochrome c oxidase deficiency, and SURF1 gene mutations. J Child Neurol 20:670–674
111. Papadopoulou LC, Sue CM, Davidson MM et al (1999) Fatal infantile cardioencephalomyopathy with COX deficiency and mutations in SCO2, a COX assembly gene. Nat Genet 23:333–337
112. Jaksch M, Ogilvie I, Yao J et al (2000) Mutations in SCO2 are associated with a distinct form of hypertrophic cardiomyopathy and cytochrome c oxidase deficiency. Hum Mol Genet 9:795–801
113. Vesela K, Hulkova H, Hansikova H et al (2008) Structural analysis of tissues affected by cytochrome c oxidase deficiency due to mutations in the SCO2 gene. APMIS 116:41–49
114. Knuf M, Faber J, Huth RG et al (2007) Identification of a novel compound heterozygote SCO2 mutation cytochrome c oxidase deficient fatal infantile cardioencephalomyopathy. Acta Paediatr 96:128–134
115. Tay SKH, Shanske S, Kaplan P, DiMauro S (2004) Association of mutations in SCO2, a cytochrome c oxidase assembly gene, with early fetal lethality. Arch Neurol 61:950–952
116. Leary SC, Mattman A, Wai T et al (2006) A hemizygous SCO2 mutation in an early onset rapidly progressive, fatal cardiomyopathy. Mol Genet Metab 89:129–133
117. Jaksch M, Horvath R, Horn N et al (2001) Homozygosity (E140K) in SCO2 causes delayed infantile onset of cardiomyopathy and neuropathy. Neurology 57:1440–1446
118. Jaksch M, Paret C, Stucka R et al (2001) Cytochrome c oxidase deficiency due to mutations in SCO2, encoding a mitochondrial copper-binding protein, is rescued by copper in human myoblasts. Hum Mol Genet 10:3025–3035
119. Salviati L, Hernandez-Rosa E, Walker WF et al (2002) Copper supplementation restores cytochrome c oxidase activity in cultured cells from patients with SCO2 mutations. Biochem J 363:321–327
120. Freisinger P, Horvath R, Macmillan C et al (2004) Reversion of hypertrophic cardiomyopathy in a patient with deficiency of the mitochondrial copper binding protein Sco2: is there a potential effect of copper? J Inherit Metab Dis 27:67–79
121. Tarnopolsky MA, Bourgeois JM, Fu MH et al (2004) Novel SCO2 mutation (G1521A) presenting as a spinal muscular atrophy type I phenotype. Am J Med Genet A 125A:310–314
122. Salviati L, Sacconi S, Rasalan MM et al (2002) Cytochrome c oxidase deficiency due to a novel SCO2 mutation mimics Werdnig-Hoffmann disease. Arch Neurol 59:862–865
123. Pronicki M, Kowalski P, Piekutowska-Abramczuk D et al (2010) A homozygous mutation in the SCO2 gene causes a spinal muscular atrophy like presentation with stridor and respiratory insufficiency. Eur J Paediatr Neurol 14:253–260
124. Oskoui M, Davidzon G, Pascual J et al (2006) Clinical spectrum of mitochondrial DNA depletion due to mutations in the thymidine kinase 2 gene. Arch Neurol 63:1122–1126

125. Valnot I, Osmond S, Gigarel N et al (2000) Mutations of the SCO1 gene in mitochondrial cytochrome c oxidase deficiency with neonatal-onset hepatic failure and encephalopathy. Am J Hum Genet 67:1104–1109
126. Brosel S, Yang H, Tanji K et al (2010) Unexpected vascular enrichment of SCO1 over SCO2 in mammalian tissues. Am J Pathol 177:2541–2548
127. Horvath R, Lochmuller H, Stucka R et al (2000) Characterization of human SCO1 and COX17 genes in mitochondrial cytochrome-c-oxidase deficiency. Biochem Biophys Res Commun 276:530–533
128. Sacconi S, Salviati L, Sue CM et al (2003) Mutation screening in patients with isolated cytochrome c oxidase deficiency. Pediatr Res 53:224–230
129. Valnot I, von Kleist-Retzow JC, Barrientos A et al (2000) A mutation in the human heme-A:farnesyltransferase gene (COX 10) causes cytochrome c oxidase deficiency. Hum Mol Genet 9:1245–1249
130. Antonicka H, Leary SC, Guercin GH et al (2003) Mutations in COX10 result in a defect in mitochondrial heme A biosynthesis and acount for multiple, early-onset clinical phenotypes associated with isolated COX deficiency. Hum Mol Genet 12:2693–2702
131. Kennaway NG, Carrero-Valenzuela RD, Ewart G et al (1990) Isoforms of mammalian cytochrome c oxidase: correlation with human cytochrome c oxidase deficiency. Pediatr Res 28:529–535
132. Antonicka H, Mattman A, Carlson CG et al (2003) Mutations in COX15 produce a defect in the mitochondrial heme biosynthetic pathway, causing early-onset fatal hypertrophic cardiomyopathy. Am J Hum Genet 72:101–114
133. Oquendo CE, Antonicka H, Shoubridge EA et al (2004) Functional and genetic studies demonstrate that mutation in the COX15 gene can cause Leigh syndrome. J Med Genet 41:540–544
134. Bugiani M, Tiranti V, Farina L et al (2005) Novel mutations in COX15 in a long surviving Leigh syndrome patient with cytochrome c oxidase deficiency. J Med Genet 42:e28
135. Huigsloot M, Nijtmans LGJ, Szklarczyk R et al (2011) A mutation in C2orf64 causes impaired cytochrome c oxidase assembly and mitochondrial cardiomyopathy. Am J Hum Genet 88:488–493
136. Ghezzi D, Saada A, D'Adamo P et al (2008) FASTKD2 nonsense mutation in an infantile mitochondrial encephalomyopathy associated with cytochrome c oxidase deficiency. Am J Hum Genet 83:415–423
137. Mootha VK, Lepage P, Miller K et al (2003) Identification of a gene causing human cytochrome c oxidase deficiency by integrative genomics. Proc Nat Acad Sci U S A 100:605–610
138. Morin C, Mitchell G, Larochelle J et al (1993) Clinical, metabolic, and genetic aspects of cytochrome c oxidase deficiency in Saguenay-Lac-Saint-Jean. Am J Hum Genet 53:488–496
139. Xu F, Morin C, Mitchell G et al (2004) The tole of LRPPRC (leucine-rich pentatricopeptide repeat cassette) gene in cytochrome oxidase assembly: mutation causes lowered levels of COX (cytochrome c oxidase) I and COX III mRNA. Biochem J 382:331–336
140. Weraarpachai W, Antonicka H, Sasarman F et al (2009) Mutation in TACO1, encoding a translational activator of COX I, results in cytochrome c oxidase deficiency and late-onset Leigh syndrome. Nat Genet 41:833–837
141. Seeger J, Schrank B, Pyle A et al (2010) Clinical and neuropathological findings in patients with TACO1 mutations. Neuromuscul Disord 20:720–724
142. Tiranti V, D'Adamo P, Briem E et al (2004) Ethylmalonic encephalopathy is caused by mutations in ETHE1, a gene encoding a mitochondrial matrix protein. Am J Hum Genet 74:239–252
143. Tiranti V, Viscomi C, Hildebrandt T et al (2009) Loss of ETHE1, a mitochondrial dioxygenase, causes fatal sulfide toxicity in ethylmalonic encephalopathy. Nat Med 15:200–205
144. Mineri R, Rimoldi M, Burlina AB et al (2008) Identification of new mutations in the ETHE1 gene in a cohort of 14 patients presenting with ethylmalonic encephalopathy. J Med Genet 45:473–478

145. De Meirleir L, Seneca S, Lissens W et al (2004) Respiratory chain complex V deficiency due to a mutation in the assembly gene ATP12. J Med Genet 41:120–124
146. Honzik T, Tesarova M, Mayr JA et al (2010) Mitochondrial encephalocardio-myopathy with early neonatal onset due to TMEM70 mutation. Arch Dis Child 95:296–301
147. Cameron JM, Levandovskiy V, MacKay N et al (2011) Complex V TMEM70 deficiency results in mitochondrial nucleoid disorganization. Mitochondrion 11:191–199
148. Shchelochkov OA, Li FY, Wang J et al (2010) Milder clinical course of type IV 3-methylglutaconic aciduria due to a novel mutation in TMEM70. Mol Genet Metab 101:282–285
149. Spiegel R, Khayat M, Shalev SA et al (2011) TMEM70 mutations are a common cause of nuclear encoded ATP synthase assembly defect: further delineation of a new syndrome. J Med Genet 48:177–182
150. Chrzanowska-Lightowlers ZMA, Horvath R, Lightowlers RN (2011) 175th ENMC International workshop: mitochondrial protein synthesis in health and disease, 25–27th June 2010, Naarden, The Netherlands. Neuromusc Dis 21:142–147
151. Tucker EJ, Hershman SG, Kohrer C et al (2011) Mutations in MTFMT underlie a human disorder of formylation causing impaired mitochondrial translation. Cell Metab 14:428–434
152. Scheper GC, van der Klok T, van Andel RJ et al (2007) Mitochondrial aspartyl-tRNA synthetase deficiency causes leukoencephalopathy with brain stem and spinal cord involvement and lactate elevation. Nat Genet 39:534–539
153. Synofzik M, Schicks J, Lindig T et al (2011) Acetazolamide-responsive exercise-induced episodic ataxia associated with a novel homozygous DARS2 mutation. J Med Genet 48:713–715
154. Smits P, Antonicka H, van Hasselt PM et al (2011) Mutation in subdomain G1 is associated with combined OXPHOS deficiency in fibroblasts but not in muscle. Eur J Hum Genet 19:275–279
155. Smits P, Saada A, Wortmann SB et al (2011) Mutation in mitopchondrial ribosomal protein MRPS22 leads to Cornelia de Lange phenotype, brain abnormalities and hypertrophic cardiomyopathy. Eur J Hum Genet 19:394–399
156. Zeharia A, Shaag A, Pappo O et al (2009) Acute infantile liver failure due to mutations in the TRMU gene. Am J Hum Genet 85:401–407
157. Schara U, von Kleist-Retzow J-C, Lainka E et al (2011) Acute liver failure with subsequent cirrhosis as the primary manifestation of TRMU mutations. J Inherit Metab Dis 34:197–201
158. Uusimaa J, Jungbluth H, Fratter C et al (2011) Reversible infantile respiratory chain deficiency is a unique, genetically heterogeneous mitochondrial disease. J Med Genet 48:660–668
159. Claypool SM, Koehler CM (2012) The complexity of cardiolipin in health and disease. Trends Biochem Sci 37:32–41
160. Schlame M, Ren M (2006) Barth syndrome, a human disorder of cardiolipin metabolism. FEBS Lett 580:5450–5455
161. Mitsuhashi S, Ohkuma A, Talim B et al (2011) A congenital muscular dystrophy with mitochondrial structural abnormalities caused by defective de novo phosphatidylcholine biosynthesis. Am J Hum Genet 88:845–851
162. Gutierrez-Rios P, Kalra AA, Wilson JD et al (2012) Congenital megaconial myopathy due to a novel defect in the choline kinase beta (CHKB) gene. Arch Neurol 69(5):657–661
163. Mitsuhashi S, Hatakeyama H, Karahashi M et al (2011) Muscle choline kinase beta defect causes mitochondrial dysfunction and increased mitophagy. Hum Mol Genet 20:3841–3851
164. DiMauro S, Hirano M, Schon EA (2006) Approaches to the treatment of mitochondrial diseases. Muscle Nerve 34:265–283

Chapter 2
Biochemical and Molecular Methods for the Study of Mitochondrial Disorders

Lee-Jun C. Wong

Introduction

Mitochondrial disorders are a group of genetically heterogeneous complex diseases [1–6]. Although mitochondrial structure and function involve two genomes, the biogenesis of mitochondrion and more than 99 % of its protein contents are encoded by the nuclear genome [7]. As a result, the majority of the mitochondrial disorders are caused by molecular defects in the nuclear genome [5, 6, 8–10]. Early studies of mitochondrial disorders focused on mutations in the 16.6 kb tiny mitochondrial DNA (mtDNA) because of its genomic simplicity and maternal inheritance [3, 11]. In the past decade, autosomal recessive mitochondrial disorders have been increasingly recognized [4, 5, 8, 10, 12–22]. Recent studies through integrated genomics and system biology identified approximately 1,500 nuclear-encoded proteins targeted to mitochondria [7], although mutations have only been identified in about 150 nuclear genes. The biogenesis of mtDNA requires nuclear-encoded genes, including DNA polymerase gamma (*POLG*) and DNA helicase (*TWINKLE*) for replication as well as purine/pyrimidine nucleoside kinases, *DGUOK* and *TK2*, for salvage synthesis and maintenance of deoxynucleotide (dNTP) pools. Defects in any of these genes cause mtDNA depletion [23, 24] or mtDNA multiple deletions [25, 26]. In addition, mtDNA transcription and mitochondrial protein biosynthesis require nuclear DNA (nDNA)-encoded factors [7, 27–29], including RNA polymerase, various transcription factors, small and large subunits (> 80) of mitochondrial ribosomal proteins, translation initiation and elongation factors, and all mitochondrial aminoacyl tRNA synthetases [22, 30–35]. The majority of mitochondrial respiratory chain complex subunits and all complex assembly factors are also nuclear encoded [28, 36–44]. Furthermore, the roles of mitochondria in autophagy, apoptosis, and reactive oxygen species (ROS) production have recently proven to be indispensible in the pathogenic mechanisms of neurodegenerative diseases (Parkinson, Alzheimer, Huntington, etc.)

L.-J. C. Wong (✉)
Department of Molecular and Human Genetics, Baylor College of Medicine,
One Baylor Plaza, NAB 2015, Houston, TX 77030, USA
e-mail: ljwong@bcm.edu

and cancer [45–47]. It is, therefore, conceivable that defects in any of the ~ 1,500 genes targeted to mitochondria are potentially detrimental to mitochondrial structure and function [3, 7, 28, 36–44, 47].

Due to the heterogeneity and complexity of mitochondrial disorders, it is crucial to have simple biochemical and molecular screening methods that will help pinpoint a specific group of candidate genes leading to the final diagnosis. This chapter will briefly review the currently available biochemical and molecular methods that can help with the diagnosis of mitochondrial disorders.

Diagnostic Criteria and Diagnostic Algorithms

Criteria for the Diagnosis of Mitochondrial Respiratory Chain Disorders

Based on the review of 118 patients referred for the diagnosis of mitochondrial respiratory chain disorders (MRCD) and comprehensive reevaluation of previously proposed diagnostic criteria [48], consensus general criteria have been developed [49]. The modified diagnostic criteria, separated into major and minor, for MRCD are listed in Table 2.1. If a patient's clinical, histological, enzymological, functional, molecular, and metabolic evaluations meet two major criteria or one major and two minor criteria, he or she is classified as having a definite diagnosis of MRCD (Table 2.2) [49]. If the patient meets one major and one minor, or three minor criteria, then he or she is classified as having a probable diagnosis (Table 2.2). These criteria are useful in assessing patients who have a diagnosis of MRCD. Using these criteria to evaluate 1,500 patients suspected of having mitochondrial syndromic hearing loss, 45 patients having both a definite diagnosis of MRCD and syndromic hearing loss were evaluated for their molecular defects [50]. Among this subset of patients, 18 harbored undisputed mtDNA mutations, and 11 had mtDNA multiple deletions and/or mtDNA copy number abnormalities [50].

Diagnostic Algorithms

Since mitochondrial disorders are complex and a large number of genes are involved, disease diagnosis is often challenging, requiring the incorporation of clinical, biochemical, and histochemical evaluations, followed by final confirmation of the diagnosis by the identification of deleterious mutations in causative genes. Wong et al. [5, 6, 8–10] proposed an algorithm that could be used to help diagnose mitochondrial disease. Briefly, a patient is first brought to the clinician's attention of having a mitochondrial disorder because of family history and/or clinical presentation. Clinical evaluation includes recognizable syndromes caused by common mtDNA mutations or nuclear gene defects [5, 6]. Minimally invasive assessment may include imaging,

Table 2.1 Modified diagnostic criteria based on Walker et al. [48], Bernier et al. [49], and Scaglia et al. [50]

	Major criteria	Minor criteria
Clinical	Clinically complete RC encephalomyopathy, *recognizable mitochondrial syndromes*[a], or a mitochondrial cytopathy fulfilling three conditions[b]	Symptoms compatible with an RC defect[c]
Histology	Greater than 2 % ragged red fibers (RRF) in skeletal muscle	Smaller numbers or RRF, SSAM, or widespread electron microscopy abnormalities of mitochondria
Enzymology	Cytochrome c oxidase negative fibers or residual activity of an RC complex < 20 % in a tissue; < 30 % in a cell line, or < 30 % in two or more tissues	Antibody-based demonstration of an RC defect or residual activity of an RC complex 20–30 % in a tissue, 30–40 % in a cell line, or 30–40 % in two or more tissues
Functional	Fibroblast ATP synthesis rates > 3 SD below mean	Fibroblast ATP synthesis 2–3 SD below mean, or fibroblasts unable to grow in galactose media
Molecular	Nuclear or mtDNA mutation of undisputed pathogenicity[d]	Nuclear or mtDNA mutation of probable pathogenicity[d]
Metabolic		One or more metabolic indicators of impaired metabolic function

The details of this modified diagnostic criteria are described by Bernier et al. [49]
ATP adenosine triphosphate, *SSAM* subsarcolemmal accumulation of mitochondria
[a]Leigh disease, Alpers disease, Lethal infantile mitochondrial disease, *mtDNA deletion syndromes*, including Pearson syndrome and Kearns–Sayre syndrome, MELAS, MERRF, NARP, MNGIE, LHON, mtDNA depletion syndrome, *mtDNA multiple deletion syndromes* including progressive external ophthalmoplegia (PEO) and myopathy. Italicized parts are modifications from Bernier et al. [49]
[b](1) Multisystemic symptoms typical for an RC disorder; (2) a progressive clinical course with episodes of exacerbation due to infection or drug toxicity, or *a family history*; and (3) other possible metabolic or nonmetabolic disorders have been excluded by appropriate testing
[c]Added pediatric features: stillbirth associated with a paucity of intrauterine movement, neonatal death or collapse, movement disorder, severe failure to thrive, neonatal hypotonia, and neonatal hypertonia as minor clinical criteria
[d]See Table 2.2 for the definition of definite and probably mutations

Table 2.2 Classification of diagnosis

Diagnosis	Definition by number of criteria met
Definite	Two major criteria or one major plus two minor
Probable	One major plus one minor or three minor
Possible	One major or two minor, one of which must be clinical

and blood and urine chemistry (Biochemical Assessment Section). Molecular sequence analysis of genes responsible for recognizable syndromes can be performed on blood DNA. With the development of next-generation sequencing, it is possible to analyze the entire mitochondrial genome comprehensively or the analysis of specific panels of nuclear genes, all mitochondrial-targeted nuclear genes, the whole exome, or the whole genome directly using the DNA extracted from blood samples [51–53].

Invasive investigation, on the other hand, may include biochemical, histochemical, and molecular studies of liver, skin, or muscle biopsy. The type of test to be performed depends on the type of tissue available from the patient. The results from invasive tissue studies may support the diagnosis and/or point to a group of genes responsible for the disease and direct the sequence analysis of a specific group of genes.

Biochemical Assessment

Lactic Acid and Pyruvate

Elevation of lactic acid in blood or CSF, although nonspecific, is a common phenomenon in patients with MRCD, particularly patients with severe autosomal recessive type [16, 18, 21, 32, 35, 54–58]. Severe infantile lactic acidosis has been reported in patients with mutations in numerous genes including *MPV17* [21], *SUCLA2* [54, 59], *SUCLG1* [18], *BCS1L* [30], *DARS2* [22], *PUS1* [30], *TK2* [60, 61]. The blood lactate/pyruvate ratios indirectly reflect the NADH/NAD+ cytoplasmic redox state [62], therefore, an accurate measurement of pyruvate is also necessary. Elevated plasma alanine is an indication of the accumulation of pyruvate. Defects in pyruvate metabolism such as pyruvate dehydrogenase complex deficiency, pyruvate carboxylase deficiency, or biotinidase deficiency will cause pyruvate elevation in blood and/or CSF.

Plasma/CSF Amino Acids

Amino acid analysis is usually performed using ion-exchange chromatography followed by postcolumn derivetization in an automated amino acid analyzer. Recently, tandem mass spectrometry (MS/MS) is also being used. Elevated alanine is defined as having an alanine:lysine ratio > 3:1 and alanine:phenylalanine+tyrosine ratio > 4:1, or an absolute elevation in alanine above 450 μ(mu)M [63]. Elevation of tyrosine and/or phenylalanine without the presence of succinylacetone may be an indication of the hepatocerebral form of mtDNA depletion syndrome [8]. Generalized aminoaciduria with renal tubular acidosis and glycosuria similar to Fanconi syndrome has also been observed in patients with mtDNA deletions [64, 65].

Urine Organic Acid

Urine organic acids reflect the catabolites of amino acids, carbohydrates, and fatty acids. Analysis is usually performed by gas chromatography followed by mass spectrometry. Lactic aciduria is not a good discriminator for mitochondrial disorders.

Elevations in TCA cycle intermediates, ethylmalonic acid, and 3-methyl glutaconic acid are common but rarely diagnostic of a specific mitochondrial disease [66–68]. Nevertheless, elevations of 3-methyl glutaconic acid and 3-methyl glutaric acid have been reported in multiple patients with mutations in *TMEM70*, a chaperone protein needed for proper assembly of complex V, ATP synthase [69, 70]. Dicarboxylic aciduria is another common finding in mitochondrial disorders due to impairment of fatty acid β-oxidation. In particular, moderate elevations in methylmalonic acid (MMA) have been observed in patients with mutations in the *SUCLA2* and *SUCLG1* genes, which cause mtDNA depletion [16, 18, 54, 71].

Carnitine and Acylcarnitine

Carnitine, β-hydroxy-γ-trimethylammonium butyrate, is a positively charged molecule required for the transfer of long-chain fatty acids from the cytoplasm to the mitochondrial matrix for β-oxidation. A deficiency in any step of β-oxidation results in abnormal levels of total and free carnitine in plasma, and an abnormal acyl-carnitine profile suggesting a specific defect of fatty acid oxidation. Carnitine deficiency and fatty acid oxidation defects may be secondary to MRCD. Quantitative analysis of acyl-carnitine is usually performed by tandem mass spectrometry (MS/MS) or HPLC followed by electrospray ionization (ESI)-MS/MS.

Histochemistry and Immunohistochemistry

Mitochondrial proliferation in skeletal myofiber is suggestive of a respiratory chain disorder, which is revealed as ragged red fibers (RRF) on modified Gomori trichrome staining as red granular deposits of mitochondrial in the subsarcolemmal space [72]. The RRFs are shown as ragged blue fibers (RBFs) on histochemical staining of succinate dehydrogenase (SDH) [72]. RRFs or RBFs can also appear in myopathic forms of mtDNA depletion syndromes [25, 26, 73]. SDH histochemistry is also useful for the diagnosis of complex II deficiency. Another useful histochemical evaluation is the staining for cytochrome c oxidase (COX). Muscle fibers with normal COX activity appear brown, while negative fibers stain poorly. However, sequential staining with SDH causes the COX deficient fibers to stain dark blue. Immunohistochemical staining uses antibodies against specific protein subunits of the respiratory chain complexes to reveal the defective genes. For example, a muscle biopsy from a patient with Kearns–Sayre syndrome (KSS) may show marked reduction of COXII (mtDNA encoded) but normal COXIV (nDNA encoded) immunohistochemical staining. Anti-DNA antibodies can be used to identify muscle fibers with abnormal mtDNA content by showing normal nDNA staining but reduced mtDNA staining suggesting mtDNA depletion [72].

Electron Microscopy (EM)

EM studies can reveal abnormal ultrastructural changes of mitochondria including mitochondrial number, shape, size, absence of cristae, and paracrystalline inclusions. However, these ultrastructural changes are not specific for mitochondrial disorders.

Assay of Electron Transport Chain Activities (ETC)

Mitochondrial respiratory chain activities are studied by spectrophotometric assay of the ETC enzyme complexes: complex I (NADH-ubiquione oxidoreductase), complex II (succinate-ubiquinone oxidoreductase), complex III (decylubiquinone-cytochrome c reductase), complex IV (cytochrome c oxidase), complex I+III (NADH-cytochrome c reductase), or complex II+III (succinate-cytochrome c reductase). The measurement of ETC complex activities is based on the absorbance change of the substrate, either NADH or cytochrome c, depending on the complex being assayed [74, 75]. Mutations in specific complex protein subunits or complex assembly factors cause an isolated complex deficiency, while mutations in nuclear genes responsible for mtDNA biosynthesis and maintenance of mtDNA integrity result in multicomplex deficiencies [8, 21]. Mutations in enzymes involved in coenzyme Q_{10} (CoQ_{10}) biosynthesis result in CoQ_{10} deficiencies and exhibit reduced activities of combined complexes I+III and II+III [17, 76, 77]. The spectrophotometric assay of complex V (F1-ATPase) in tissue homogenate measures in the reverse direction by linking the hydrolysis of ATP to the oxidation of NADH using pyruvate kinase (PK) and lactate dehydrogenase (LDH) [74, 75]. The rate of NADH oxidation is monitored by measuring the absorbance decrease at 340 nm. Assays of complex V may also be linked to complex I [74, 75].

Oxygen Consumption Rate (OCR)

Live cells and isolated mitochondria from fresh muscle tissue offer the advantage of assaying integrated mitochondrial function such as oxygen consumption and oxidative ATP synthesis rates [78]. By using complex I substrates (glutamate, malate, and pyruvate) or complex II substrate (succinate), ATP synthesis rate in the presence (state III, ADP-stimulating) or absence (state IV, ADP-limiting) of ADP is measured. The ratio of state III rate to state IV rate is a good indicator of the integrity of the inner membrane of the isolated mitochondria. The ratio of the amount of ADP usage and oxygen consumption directly reflects the efficiency of oxidative phosphorylation and can indicate abnormalities of ATP synthase or coupling [79, 80].

Analysis of Respiratory Chain Protein Complexes using Blue Native Gel

Blue native polyacrylamide gel electrophoresis (BN-PAGE) is a powerful diagnostic tool for the detection of assembly defects in the enzyme complexes of oxidative phosphorylation [81–84]. This method separates all five complexes on the gel, which is followed by immunoblotting with commercially available antibodies [85]. If there is an assembly defect, the complex with the defect will be absent. Using established in-gel activity staining [86, 87], residual enzymatic activity of the oxidative phosphorylation complexes can also be measured. Mitochondria contain their own protein translation machinery. Defects in mitochondrial protein translation result in decreased mitochondrial protein synthesis. The molecular defects may be in the mitochondrial genome, such as mutations in mitochondrial tRNA, rRNA, or mRNA. Mutations in nuclear genes encoding mitochondrial aminoacyl tRNA synthetases; posttranscription RNA-modification enzymes; translation initiation, elongation, or termination factors, or mitochondrial ribosomal protein subunits; can all potentially cause impaired mitochondrial protein translation. Therefore, specific radioactive labeling of the mitochondrial translation protein products in cultured cells followed by the analysis of the proteins on blue native gel is useful in the study of disease-causing mutations in both the mitochondrial and the nuclear genomes [82, 84, 88].

Coenzyme Q_{10} (CoQ_{10})

Coenzyme Q_{10} serves as a mobile electron carrier to shuttle electrons from complex I or II to complex III. Mutations in genes involved in CoQ_{10} biosynthesis cause primary CoQ_{10} deficiency. Conversely, secondary CoQ_{10} deficiency in muscle may be due to defects in mtDNA or nuclear genes such as electron-transferring flavoprotein dehydrogenase (*ETFDH*). Intracellular levels of coenzyme Q and its redox status are important markers of oxidative stress associated with MRCD [89–92]. The measurement of coenzyme Q concentrations in biological specimens including plasma, white blood cells, or tissues is performed by high-performance liquid chromatography (HPLC) methods with electrochemical (EC) detection [93] or liquid chromatography-tandem mass spectrometry [94, 95].

Molecular Evaluation

Various methods are used for the detection of molecular defects. Depending on the purposes, these methods can be used alone or in combination. These methods will be discussed in two categories: the detection of mutations in the mitochondrial genome and in the nuclear genes.

Analysis of Alterations in the Mitochondrial Genome

Detection of Recurrent Common Mutations

The recurrent common mtDNA point mutations are usually screened by the PCR/RFLP or PCR/ASO (allele-specific oligonucleotide) method [96–99]. The former is performed by testing each known mutation individually, while the latter is carried out with multiplex PCR of the regions containing the common point mutations followed by dot-blot ASO analysis [96–99]. Other mutation detection methods including TaqMan allele discrimination assays or Sanger sequencing of the specific region of the mitochondrial genome can also be used [100–102].

Detection of Rare or Unknown Point Mutations in the Mitochondrial Genome

Historically, there are several mutation detection methods for screening unknown mutations including single-strand conformation polymorphism (SSCP) [103], temporal temperature gradient gel electrophoresis (TTGE) [104, 105], temperature gradient gel electrophoresis (TGGE) [106], denaturant gradient gel electrophoresis (DGGE) [107, 108] and denaturing high-performance liquid chromatography (cHPLC) [109]. Since Sanger sequencing is required to confirm the exact mutations resulting from the detected changes, these indirect mutation detection methods have been replaced by direct Sanger sequencing of the target regions using specific primers [100–102, 110]. Sanger sequencing analysis of the whole mitochondrial genome is usually performed by PCR amplification using 24–36 pairs of primers amplifying overlapping fragments covering the entire mitochondrial genome followed by sequencing using specific primers [110]. Although this method is straightforward, it does not, however, detect large deletions or quantify degree of mutation heteroplasmy. In addition, due to the extensive polymorphic features of the mtDNA, the high frequency of single nucleotide polymorphisms (SNPs) in the multiple primer sites, and the interference of nuclear mtDNA homology sequences, the true mtDNA fragments may not be efficiently amplified or not amplified at all leading to calling bases not truly representative of the mtDNA sequence. Yet, these obstacles can be resolved by long-range PCR of the entire mitochondrial genome followed by massively parallel sequencing, which has now become routinely performed [111]. The advantages of this newly developed approach is the simultaneous detection of mtDNA point mutations and large deletions while also acquiring quantitative heteroplasmy information for every single nucleotide of the entire 16.569 kb genome, deletion breakpoints, and deletion heteroplasmy (Comprehensive one-step analysis of the mitochondrial genome section) [111].

Quantification of the Point Mutation Heteroplasmy

Knowing the degree of mutation heteroplasmy and its tissue distribution is crucial for correlating clinical phenotype with test results of patients with mtDNA disorders [5, 6, 11, 96, 112]. Traditionally, this is performed by the addition of γ-^{32}P-ATP to the PCR mixture at the last cycle followed by restriction fragment length polymorphism (RFLP) analysis and quantification of DNA bands by Phosphor-Imager [96]. However, due to the disadvantages of using radioactive material and time-consuming RFLP procedures, this method is gradually being replaced by the development of a nonradioactive, more accurate and rapid real-time amplification refractory mutation systems quantitative PCR (ARMS-qPCR) method, which provides simultaneous detection and quantification of heteroplasmic mtDNA point mutations [113–116].

Detection of Large mtDNA Deletions

Large single deletions of mtDNA are usually detected by Southern blot analysis [11, 96, 99, 117–119]. However, procedures for quantification of deletion heteroplasmy and determination of the breakpoints are tedious and may involve the usage of radioactive material similar to the procedures described in Sect. 4.1.3 except that the restriction fragments are hybridized with nick-translated α-^{32}P-dCTP or non-radioactively labeled probes such as digoxygenine [11, 96, 117–119]. The recent development of custom oligonucleotide array comparative genome hybridization (aCGH), MitoMet© allows reliable detection of mtDNA deletions with elucidation of deletion breakpoints and the percentage of deletion heteroplasmy [118, 120–122].

Measurement of mtDNA Copy Number

Alteration in mtDNA copy number is an indication of mitochondrial disorder. Increased mtDNA content suggests a compensatory mechanism due to deficient mitochondrial function [123], while reduced mtDNA content implies defects in mtDNA biosynthesis, usually secondary to mutations in nuclear genes [8–10, 16, 18, 21, 59, 112, 124–126]. Measurement of mtDNA copy number is performed by real-time quantitative PCR using an mtDNA probe and a unique nuclear gene reference [112, 116, 127]. The copy number ratio of mtDNA/nDNA is a measure of mtDNA content [112, 127]. Since mtDNA content varies among different tissues and in some tissues it also changes with age, the mtDNA copy number of an individual is compared with the mean value of tissue and age matched controls [8, 21, 112]. The mtDNA content in muscle tissue from patients with encephalomyopathic mtDNA depletion syndrome is < 50 % of that of tissue and age-matched control mean, while the mtDNA content in the liver specimen from patients with hepatocerebral or infantile hepatic form of mtDNA depletion syndrome is < 20 % of that of the matched control [8, 21, 112]. Measurement of mtDNA content in an affected tissue may assist with narrowing the cause of the disease to a group of nuclear genes for sequence analysis [5, 6]. Therefore, assessment of mtDNA copy number in affected tissues such as muscle and liver is a screening method for the identification of mtDNA depletion syndromes or mtDNA compensatory overamplification [25, 73, 123].

Comprehensive One-Step Analysis of the Mitochondrial Genome

As described in the above analysis of alterations in the mitochondrial genome sections, it is clear that comprehensive analysis of the mitochondrial genome requires the application of multiple techniques to detect point mutations, deletions, mutation heteroplasmy, deletion breakpoints, and mtDNA depletion [5, 96, 123]. This step-wise approach complicates the molecular diagnosis of the disease. Recent advancement of next-generation sequencing enables the analysis of the entire mitochondrial genome comprehensively by massively parallel sequencing [52, 111]. A recent report by Cui et al. [111] demonstrates that massively parallel sequencing strategy can provide complete qualitative and quantitative information on mtDNA point mutations, single and multiple deletions, degree of mutation heteroplasmy, deletion breakpoints, and heteroplasmy in one step [111, 128, 129].

Analysis of Nuclear Genes Causing Mitochondrial Disorders

Sanger Sequencing

Sanger sequencing has been used as the gold standard to detect mutations in candidate genes by using specific primers to amplify the coding exons and ~ 50 nucleotides of the flanking intronic regions of the gene of interest followed by sequencing of the PCR fragments. Selection of candidate genes for sequencing is a challenge because mitochondrial disorders are genetically and clinically heterogeneous as briefly mentioned in "Diagnostic Criteria and Diagnostic Algorithms" [5]. An estimated 1,500 nuclear-encoded proteins are associated with mitochondrial biogenesis, structure, and function (Table 2.3) [7]; however, mutations in only about 200 have been reported to cause disease [5]. The selection of candidate genes for diagnostic sequencing analysis is generally based on the patient's clinical presentation and family history along with the molecular, imaging, or biochemical characteristics [5, 6]. For example, if the mtDNA content in the affected tissue, such as muscle or liver, is showing depletion by the qPCR method [112], then a group of genes responsible for the hepatocerebral form of mtDNA depletion syndrome may be sequenced [5, 6, 8–10, 21, 112]. If there is an indication of myopathy and/or progressive external ophthalmoplegia, and multiple mtDNA deletions, then the genes responsible for these phenotypes may be sequenced [10, 12, 25, 26, 130].

Oligonucleotide Array Comparative Genomic Hybridization (aCGH)

Although sequencing analysis detects point mutations and small insertion/deletions, it does not detect large intragenic or whole gene deletions. A custom-designed oligonucleotide array-based comparative genomic hybridization (aCGH) method has been developed to provide both tiled coverage of the entire 16.6 kb mitochondrial genome and high-density coverage of nuclear genes involved in mitochondrial biogenesis, structure, and function [120]. Several studies have demonstrated that this design is able to detect large deletions in both nuclear and mitochondrial genomes [9, 120, 121, 131–134]. Most importantly, the breakpoints and degree of heteroplasmy of large mtDNA deletions can be estimated [120, 121, 135]. In addition,

Table 2.3 Categories of nuclear-encoded proteins targeting to mitochondria

Nuclear genes targeted to mitochondria							
Category	Number	Category	Number	Category	Number	Category	Number
Apoptosis	32	Metabolism[a]	356[a]	Proteases	24	aa-tRNA Synthase	18
Chaperone	11	Mito dynamics	12	Nucleases	2	Signal trans duction	34
Cytochromes	12	Replication	4	Translation[b]	34	Protein folding and modification	21
Cytoskeletal	11	Transcription	15	MRPLs	47	Protein import: TIMMs and TOMMs	28
DNA repair	11	Rregulation	7	MRPSs	29	RC complex and assembly	107
Fe–S cluster	4	Porins	3	Transporters	80	RNA modification and processing	7
Immune	3	ROS	16	Others	31	Unknown	113

[a] Metabolic pathways include amino acids, carbohydrates, creatine, iron, ketones, Kreb cycle, lipids, nucleotide, phospholipid, porphyrin, steroid, tetrahydrofolate, tRNA, CoQs, pyruvate, etc.
[b] *Translation* initiation, elongation, termination proteins, and factors.
[c] *MRPLs* mitochondrial ribosomal proteins large subunits, *MRPSs* mitochondrial ribosomal proteins small subunits.
[d] *Transporters* ATP-binding cassette, iron, nucleotides, solutes, and uncouplers.

for patients with mtDNA depletion syndromes, this array can detect copy number changes in both nuclear genes and mtDNA (reduced mtDNA copy number). A good example is that of a baby boy who developed liver failure during infancy. A liver biopsy revealed severe mtDNA depletion (only 6 % of tissue matched control). His older brother also died from liver failure during infancy. Sequence analysis found a heterozygous mutation, p.E227K, in the *DGUOK* gene. Since the hepatocerebral form of mtDNA depletion syndrome is an autosomal recessive disorder, the family history and the mtDNA content in liver suggested that a second mutant allele must be present. The aCGH analysis revealed a heterozygous deletion of exon 4 in the proband and the carrier mother [133], and the mtDNA profile showed an approximate 97 % reduction in copy number, consistent with mtDNA depletion [133]. Several other examples showing large deletions in genes, including POLG, DGK, TK2, TP, and MPV17, responsible for mtDNA depletion, have also been reported [9, 120, 121, 131–134].

Next-Generation Massively Parallel Sequencing

Step-wise sequence analysis of several individual candidate genes (Table 2.3) by Sanger sequencing is tedious, time consuming, and costly. However, the newly developed massively parallel sequencing (MPS) allows simultaneous sequencing of multiple genes at high coverage and low cost. The small 16.6 kb mtDNA genome can be amplified with overlapping primers followed by MPS [52]. This method can reliably detect mtDNA heteroplasmy at a 5 % level, but it, unfortunately, does not detect large deletions. Conversely, the method described by Cui et al. detects both point mutations and large deletions with heteroplasmy and exact deletion breakpoints in a one-step comprehensive approach (comprehensive one-step analysis of the mitochondrial genome section) [111]. As for nuclear genes, a proof-of-concept study was conducted by Vasta et al. [53] by using microarrays to capture the mtDNA genome along with 362 nuclear genes related to mitochondrial disorders, followed by MPS. However, this study did not report the identification of causative mutations in unknown patients with mitochondrial disorders. More recently, Calvo et al [51] performed "MitoExome" MPS analysis of mtDNA and coding exon sequences of ~ 1,000 nuclear genes encoding mitochondrial-targeted proteins. These investigators were able to identify reported mutations in ten patients and novel mutations that require confirmation of pathogenicity in 13 others [51]. The MPS approach is promising, but in order to apply it to a clinical diagnostic laboratory, many obstacles remain, including false negative, false positive, variant interpretation, and functional confirmation of pathogenicity. Also, even by capture of ~ 1,000 genes with MPS, about 50–75 % of patients with mitochondrial disorders remain undiagnosed [51]. The utilization of whole exome or whole human genome sequencing analysis may be necessary in order to find the molecular cause of the disease in these situations [136–138]. Table 2.3 lists the categories of ~ 1,500 genes with mitochondrial involvement. Due to the extreme clinical and genetic heterogeneity of the mitochondrial disorders, exome and/or whole genome sequence analyses may be the ultimate solution.

Interpretation of Sequence Variants

With the improvement in next-generation sequencing, a large number of novel or rare sequence variants have been rapidly discovered. To understand the clinical and/or functional significance of these variants, they should be properly classified based on family history, inheritance, clinical correlation, functional studies, and literature reports [139, 140]. The classification and guidelines published by ABMG in 2008 for the interpretation of novel variants using ACMG's Standards and Guidelines should be followed [141]. Detailed procedures have also been published in *Methods in Molecular Biology* [139, 140]. Several publically available databases and softwares that are accessible on the internet include Human Genome Mutation Database (HGMD), Single Nucleotide Polymorphism Database dbSNP, PubMed, Online Mendelian Inheritance in Man (OMIM), Sorting Intolerant From Tolerant (SIFT), Polymorphism Phenotypng (PolyPhen), ESE, NetGene, and BDGP (Table 2.4) [139, 140]. Although all these databases and in silico analytical algorithms help in the interpretation of novel and rare variants, the most challenging task is to provide experimental functional evidences to support the pathogenicity of the sequence variants. Hopefully, advances in functional genomics will greatly facilitate the interpretation of sequence variants.

Table 2.4 Internet accessible databases and analytical software

Name of databases	Abbreviation	Web address
Databases		
Human Genome Mutation Databases	HGMD	http://www.hgmd.cf.ac.uk
Single Nucleotide Polymorphism Database	dbSNP	http://www.ncbi.nlm.nih.gov/projects/SNP
NCBI MEDLINE database of life sciences	PubMed	http://www.ncbi.nlm.nih.gov/PubMed
Online Mendelian Inheritance in Man	OMIM	http://www.ncbi.nlm.nih.gov/omim
Mutation database for specific nuclear genes		
mtDNA polymerase gamma	POLG	http://tools.niehs.nih.gov/polg
Optic atrophy gene 1	OPA1	http://lbbma.univ-angers.fr/lbbma.php?id=9
Databases specific for the mitochondrial genome		
	MitoMap	http://www.mitomap.org/MITOMAP
	mtDB	http://www.mtdb.igp.uu.se/
	tRNA	http://mamit-trna.u-strasbg.fr/Summary.asp
Computational algorithms for the prediction of pathogenicity of missense variants		
Sorting Intolerant From Tolerant	SIFT	http://sift.jcvi.org
Polymorphism Phenotyping-2	PolyPhen2	http://genetics.bwh.harvard.edu/pph2
Prediction of pathogenicity	PONP	http://bioinf.uta.fi/PON-P/Pathogenic_or_not_prediction_methods.shtml
Computational algorithms for the prediction of pathogenicity of splice site alterations		
ESE Finder 2.0	ESE2	http://rulai.cshl.edu/tools/ESE2
NetGene2		http://www.cbs.dtu.dk/services/NetGene2
BDGP splice site predictor		http://www.fruitfly.org/seq_tools/other.html

Conclusion

Diagnosis of the complex dual genome mitochondrial disorders requires application of multiple biochemical and molecular approaches. The methods mentioned in this chapter likely will continue to be used in diagnostic laboratories for several years until a more reliable comprehensive whole genome molecular strategy is developed. Although whole exome and whole genome approaches are currently available, these methods have their own limitations. Overall, the quality of MPS tests varies tremendously among different laboratories, and so does the cost of these tests. In addition, it is important that clinicians understand the quality control and quality assurance procedures and compliance to the regulatory agents.

References

1. DiMauro S (2004) The many faces of mitochondrial diseases. Mitochondrion 4(5-6):799-807
2. DiMauro S (2011) A history of mitochondrial diseases. J Inherit Metab Dis 34(2):261-276
3. Smeitink J, van den, Heuvel L DiMauro S (2001) The genetics and pathology of oxidative phosphorylation. Nat Rev Genet 2(5):342-352
4. Spinazzola A, Zeviani M (2007) Disorders of nuclear-mitochondrial intergenomic communication. Biosci Rep 27(1-3):39-51
5. Wong LJ (2010) Molecular genetics of mitochondrial disorders. Dev Disabil Res Rev 16(2):154-162
6. Wong LJ et al (2010) Current molecular diagnostic algorithm for mitochondrial disorders. Mol Genet Metab 100(2):111-117
7. Calvo S et al (2006) Systematic identification of human mitochondrial disease genes through integrative genomics. Nat Genet 38(5):576-582
8. Dimmock DP et al (2008) Clinical and molecular features of mitochondrial DNA depletion due to mutations in deoxyguanosine kinase. Hum Mutat 29(2):330-331
9. El-Hattab AW et al (2010) MPV17-associated hepatocerebral mitochondrial DNA depletion syndrome: new patients and novel mutations. Mol Genet Metab 99(3):300-308
10. Tang S et al (2011) Mitochondrial DNA polymerase {gamma} mutations: an ever expanding molecular and clinical spectrum. J Med Genet 48(10):669-681
11. Wong LJ, Boles RG (2005) Mitochondrial DNA analysis in clinical laboratory diagnostics. Clin Chim Acta 354(1-2):1-20
12. Galbiati S et al (2006) New mutations in TK2 gene associated with mitochondrial DNA depletion. Pediatr Neurol 34(3):177-185
13. Ghezzi D et al (2009) SDHAF1, encoding a LYR complex-II specific assembly factor, is mutated in SDH-defective infantile leukoencephalopathy. Nat Genet 41(6):654-656
14. Hoefs SJ et al (2010) Novel mutations in the NDUFS1 gene cause low residual activities in human complex I deficiencies. Mol Genet Metab 100(3):251-256
15. Nishino I, Spinazzola A, Hirano M (1999) Thymidine phosphorylase gene mutations in MNGIE, a human mitochondrial disorder. Sci 283(5402):689-692
16. Ostergaard E et al (2007) Deficiency of the alpha subunit of succinate-coenzyme A ligase causes fatal infantile lactic acidosis with mitochondrial DNA depletion. Am J Hum Genet 81(2):383-387
17. Quinzii C et al (2006) A mutation in para-hydroxybenzoate-polyprenyl transferase (COQ$_2$) causes primary coenzyme Q$_{10}$ deficiency. Am J Hum Genet 78(2):345-349
18. Randolph LM et al (2011) Fatal infantile lactic acidosis and a novel homozygous mutation in the SUCLG1 gene: a mitochondrial DNA depletion disorder. Mol Genet Metab 102(2):149-152
19. Tang S et al (2012) Mitochondrial neurogastrointestinal encephalomyopathy (MNGIE)-like phenotype: an expanded clinical spectrum of POLG1 mutations. J Neurol 259(5):862-868
20. Trevisson E et al (2011) Coenzyme Q deficiency in muscle. Curr Opin Neurol 24(5):449-456
21. Wong LJ et al (2007) Mutations in the MPV17 gene are responsible for rapidly progressive liver failure in infancy. Hepatol 46(4):1218-1227
22. Scheper GC et al (2007) Mitochondrial aspartyl-tRNA synthetase deficiency causes leukoencephalopathy with brain stem and spinal cord involvement and lactate elevation. Nat Genet 39(4):534-539
23. Spinazzola A et al (2009) Clinical and molecular features of mitochondrial DNA depletion syndromes. J Inherit Metab Dis 32(2):143-158
24. Spinazzola A, Zeviani M (2009) Disorders from perturbations of nuclear-mitochondrial intergenomic cross-talk. J Intern Med 265(2):174-192
25. Milone M et al (2008) Sensory ataxic neuropathy with ophthalmoparesis caused by POLG mutations. Neuromuscul Disord 18(8):626-632
26. Milone M et al (2011) Novel POLG splice site mutation and optic atrophy. Arch Neurol 68(6):806-811

27. Koene S, Smeitink J (2009) Mitochondrial medicine: entering the era of treatment. J Intern Med 265(2):193–209
28. Pagliarini DJ et al (2008) A mitochondrial protein compendium elucidates complex I disease biology. Cell 134(1):112–123
29. Spinazzola A, Zeviani M (2005) Disorders of nuclear-mitochondrial intergenomic signaling. Gene 354:162–168
30. Bykhovskaya Y et al (2004) Missense mutation in pseudouridine synthase 1 (PUS1) causes mitochondrial myopathy and sideroblastic anemia (MLASA). Am J Hum Genet 74(6):1303–1308
31. Edvardson S et al (2007) Deleterious mutation in the mitochondrial arginyl-transfer RNA synthetase gene is associated with pontocerebellar hypoplasia. Am J Hum Genet 81(4):857–862
32. Miller C et al (2004) Defective mitochondrial translation caused by a ribosomal protein (MRPS16) mutation. Ann Neurol 56(5):734–738
33. Smeitink JA et al (2006) Distinct clinical phenotypes associated with a mutation in the mitochondrial translation elongation factor EFTs. Am J Hum Genet 79(5):869–877
34. Valente L et al (2009) The R336Q mutation in human mitochondrial EFTu prevents the formation of an active mt-EFTu.GTP.aa-tRNA ternary complex. Biochim Biophys Acta 1792(8):791–795
35. Valente L et al (2007) Infantile encephalopathy and defective mitochondrial DNA translation in patients with mutations of mitochondrial elongation factors EFG1 and EFTu. Am J Hum Genet 80(1):44–58
36. Antonicka H et al (2003) Mutations in COX15 produce a defect in the mitochondrial heme biosynthetic pathway, causing early-onset fatal hypertrophic cardiomyopathy. Am J Hum Genet 72(1):101–114
37. Hoefs SJ et al (2008) NDUFA2 complex I mutation leads to Leigh disease. Am J Hum Genet 82(6):1306–1315
38. Saada A et al (2009) Mutations in NDUFAF3 (C3ORF60), encoding an NDUFAF4 (C6ORF66)-interacting complex I assembly protein, cause fatal neonatal mitochondrial disease. Am J Hum Genet 84(6):718–727
39. Tay SK et al (2004) Association of mutations in SCO2, a cytochrome c oxidase assembly gene, with early fetal lethality. Arch Neurol 61(6):950–952
40. Tiranti V et al (1998) Mutations of SURF-1 in Leigh disease associated with cytochrome c oxidase deficiency. Am J Hum Genet 63(6):1609–1621
41. Valnot I et al (2000) Mutations of the SCO1 gene in mitochondrial cytochrome c oxidase deficiency with neonatal-onset hepatic failure and encephalopathy. Am J Hum Genet 67(5):1104–1109
42. Willems PH et al (2008) Mitochondrial Ca2+ homeostasis in human NADH:ubiquinone oxidoreductase deficiency. Cell Calcium 44(1):123–133
43. de Lonlay P et al (2001) A mutant mitochondrial respiratory chain assembly protein causes complex III deficiency in patients with tubulopathy, encephalopathy and liver failure. Nat Genet 29(1):57–60
44. Dieteren CE et al (2008) Subunits of mitochondrial complex I exist as part of matrix- and membrane-associated subcomplexes in living cells. J Biol Chem 283(50):34753–34761
45. Bianchi MS, Bianchi NO, Bailliet G (1995) Mitochondrial DNA mutations in normal and tumor tissues from breast cancer patients. Cytogenet Cell Genet 71(1):99–103
46. Corral-Debrinski M et al (1992) Mitochondrial DNA deletions in human brain: regional variability and increase with advanced age. Nat Genet 2(4):324–329
47. Geisler S et al (2010) PINK1/Parkin-mediated mitophagy is dependent on VDAC1 and p62/SQSTM1. Nat Cell Biol 12(2):119–131
48. Walker UA, Collins S, Byrne E (1996) Respiratory chain encephalomyopathies: a diagnostic classification. Eur Neurol 36(5):260–267
49. Bernier FP et al (2002) Diagnostic criteria for respiratory chain disorders in adults and children. Neurol 59(9):1406–1411

50. Scaglia F et al (2006) Molecular bases of hearing loss in multi-systemic mitochondrial cytopathy. Genet Med 8(10):641–652
51. Calvo SE et al (2012) Molecular diagnosis of infantile mitochondrial disease with targeted next-generation sequencing. Sci Transl Med 4(118):118–ra10
52. Tang S, Huang T (2010) Characterization of mitochondrial DNA heteroplasmy using a parallel sequencing system. Biotech 48(4):287–296
53. Vasta V et al (2009) Next generation sequence analysis for mitochondrial disorders. Genome Med 1(10):100
54. Carrozzo R et al (2007) SUCLA2 mutations are associated with mild methylmalonic aciduria, Leigh-like encephalomyopathy, dystonia and deafness. Brain 130(Pt 3):862–874
55. Fernandez-Vizarra E et al (2007) Nonsense mutation in pseudouridylate synthase 1 (PUS1) in two brothers affected by myopathy, lactic acidosis and sideroblastic anaemia (MLASA). J Med Genet 44(3):173–180
56. Lebon S et al (2007) A novel mutation of the NDUFS7 gene leads to activation of a cryptic exon and impaired assembly of mitochondrial complex I in a patient with Leigh syndrome. Mol Genet Metab 92(1–2):104–108
57. Matthews PM et al (1994) Pyruvate dehydrogenase deficiency. Clinical presentation and molecular genetic characterization of five new patients. Brain 117 (Pt 3):435–443
58. Rahman S et al (2001) A SURF1 gene mutation presenting as isolated leukodystrophy. Ann Neurol 49(6):797–800
59. Elpeleg O et al (2005) Deficiency of the ADP-forming succinyl-CoA synthase activity is associated with encephalomyopathy and mitochondrial DNA depletion. Am J Hum Genet 76(6):1081–1086
60. Mandel H et al (2001) The deoxyguanosine kinase gene is mutated in individuals with depleted hepatocerebral mitochondrial DNA. Nat Genet 29(3):337–341
61. Saada A et al (2001) Mutant mitochondrial thymidine kinase in mitochondrial DNA depletion myopathy. Nat Genet 29(3):342–324
62. Debray FG et al (2007) Diagnostic accuracy of blood lactate-to-pyruvate molar ratio in the differential diagnosis of congenital lactic acidosis. Clin Chem 53(5):916–921
63. Wolf NI, Smeitink JA (2002) Mitochondrial disorders: a proposal for consensus diagnostic criteria in infants and children. Neurol 59(9):1402–1405
64. Campos Y et al (1995) Mitochondrial DNA deletion in a patient with mitochondrial myopathy, lactic acidosis, and stroke-like episodes (MELAS) and Fanconi's syndrome. Pediatr Neurol 13(1):69–72
65. Niaudet P et al (1994) Deletion of the mitochondrial DNA in a case of de Toni-Debre-Fanconi syndrome and Pearson syndrome. Pediatr Nephrol 8(2):164–168
66. Barshop BA (2004) Metabolomic approaches to mitochondrial disease: correlation of urine organic acids. Mitochondrion 4(5–6):521–527
67. Gibson KM et al (1991) Phenotypic heterogeneity in the syndromes of 3-methylglutaconic aciduria. J Pediatr 118(6):885–890
68. Sperl W et al (2006) Deficiency of mitochondrial ATP synthase of nuclear genetic origin. Neuromuscul Disord 16(12):821–829
69. Cizkova A et al (2008) TMEM70 mutations cause isolated ATP synthase deficiency and neonatal mitochondrial encephalocardiomyopathy. Nat Genet 40(11):1288–1290
70. Shchelochkov OA et al (2010) Milder clinical course of type IV 3-methylglutaconic aciduria due to a novel mutation in TMEM70. Mol Genet Metab 101(2–3):282–285
71. Ostergaard E et al (2007) Mitochondrial encephalomyopathy with elevated methylmalonic acid is caused by SUCLA2 mutations. Brain 130(Pt 3):853–861
72. Tanji K (2012) Morphological assessment of mitochondrial respiratory chain function on tissue sections. Methods Mol Biol 837:181–194
73. Milone M et al (2009) Mitochondrial disorder with OPA1 mutation lacking optic atrophy. Mitochondrion 9(4):279–281
74. Frazier AE, Thorburn DR (2012) Biochemical analyses of the electron transport chain complexes by spectrophotometry. Methods Mol Biol 837:49–62

75. Grazina MM (2012) Mitochondrial respiratory chain: biochemical analysis and criterion for deficiency in diagnosis. Methods Mol Biol 837:73–91
76. Mollet J et al (2007) Prenyldiphosphate synthase, subunit 1 (PDSS1) and OH-benzoate polyprenyltransferase (COQ$_2$) mutations in ubiquinone deficiency and oxidative phosphorylation disorders. J Clin Invest 117(3):765–772
77. Quinzii CM et al (2005) Coenzyme Q deficiency and cerebellar ataxia associated with an aprataxin mutation. Neurol 64(3):539–541
78. Li Z, Graham BH (2012) Measurement of mitochondrial oxygen consumption using a Clark electrode. Methods Mol Biol 837:63–72
79. Brand MD, Nicholls DG (2011) Assessing mitochondrial dysfunction in cells. Biochem J 435(2):297–312
80. Trounce IA et al (1996) Assessment of mitochondrial oxidative phosphorylation in patient muscle biopsies, lymphoblasts, and transmitochondrial cell lines. Methods Enzymol 264:484–509
81. Leary SC (2012) Blue native polyacrylamide gel electrophoresis: a powerful diagnostic tool for the detection of assembly defects in the enzyme complexes of oxidative phosphorylation. Methods Mol Biol 837:195–206
82. Sasarman F, Shoubridge EA (2012) Radioactive labeling of mitochondrial translation products in cultured cells. Methods Mol Biol 837:207–217
83. Antonicka H et al (2003) Identification and characterization of a common set of complex I assembly intermediates in mitochondria from patients with complex I deficiency. J Biol Chem 278(44):43081–43088
84. Weraarpachai W et al (2012) Mutations in C12orf62, a factor that couples COX I synthesis with cytochrome c oxidase assembly, cause fatal neonatal lactic acidosis. Am J Hum Genet 90(1):142–151
85. Leary SC et al (2004) Human SCO1 and SCO2 have independent, cooperative functions in copper delivery to cytochrome c oxidase. Hum Mol Genet 13(17):1839–1848
86. Diaz F, Barrientos A, Fontanesi F (2009) Evaluation of the mitochondrial respiratory chain and oxidative phosphorylation system using blue native gel electrophoresis. Curr Protoc Hum Genet 19:19.4
87. Zerbetto E, Vergani L, Dabbeni-Sala F (1997) Quantification of muscle mitochondrial oxidative phosphorylation enzymes via histochemical staining of blue native polyacrylamide gels. Electrophoresis 18(11):2059–2064
88. Weraarpachai W et al (2009) Mutation in TACO1, encoding a translational activator of COX I, results in cytochrome c oxidase deficiency and late-onset Leigh syndrome. Nat Genet 41(7):833–837
89. Ogasahara S et al (1986) Treatment of Kearns-Sayre syndrome with coenzyme Q$_{10}$. Neurol 36(1):45–53
90. Horvath R (2012) Update on clinical aspects and treatment of selected vitamin-responsive disorders II (riboflavin and CoQ(10)). J Inherit Metab Dis 35(4):679–687
91. Liang WC et al (2009) ETFDH mutations, CoQ$_{10}$ levels, and respiratory chain activities in patients with riboflavin-responsive multiple acyl-CoA dehydrogenase deficiency. Neuromuscul Disord 19(3):212–216
92. Ohkuma A et al (2009) Clinical and genetic analysis of lipid storage myopathies. Muscle Nerve 39(3):333–342
93. Tang PH, Miles MV (2012) Measurement of oxidized and reduced coenzyme Q in biological fluids, cells, and tissues: an HPLC-EC method. Methods Mol Biol 837:149–168
94. Hahn SH, Kerfoot S, Vasta V (2012) Assay to measure oxidized and reduced forms of CoQ by LC-MS/MS. Methods Mol Biol 837:169–179
95. Ruiz-Jimenez J et al (2007) Determination of the ubiquinol-10 and ubiquinone-10 (coenzyme Q$_{10}$) in human serum by liquid chromatography tandem mass spectrometry to evaluate the oxidative stress. J Chromatogr A 1175(2):242–248
96. Shanske S, Wong LJ (2004) Molecular analysis for mitochondrial DNA disorders. Mitochondrion 4(5–6):403–415

97. Wong LJ, Lam CW (1997) Alternative, noninvasive tissues for quantitative screening of mutant mitochondrial DNA. Clin Chem 43(7):1241–1243
98. Wong LJ, Senadheera D (1997) Direct detection of multiple point mutations in mitochondrial DNA. Clin Chem 43(10):1857–1861
99. Tang S et al (2012) Analysis of common mitochondrial DNA mutations by allele-specific oligonucleotide and Southern blot hybridization. Methods Mol Biol 837:259–279
100. Brautbar A et al (2008) The mitochondrial 13513G>A mutation is associated with Leigh disease phenotypes independent of complex I deficiency in muscle. Mol Genet Metab 94(4):485–490
101. Wang J et al (2009) Two mtDNA mutations 14487T>C (M63V, ND6) and 12297T>C (tRNA Leu) in a Leigh syndrome family. Mol Genet Metab 96(2):59–65
102. Ware SM et al (2009) Infantile cardiomyopathy caused by a mutation in the overlapping region of mitochondrial ATPase 6 and 8 genes. J Med Genet 46(5):308–314
103. Suomalainen A et al (1992) Use of single strand conformation polymorphism analysis to detect point mutations in human mitochondrial DNA. J Neurol Sci 111(2):222–226
104. Wong LJ, Chen TJ, Tan DJ (2004) Detection of mitochondrial DNA mutations using temporal temperature gradient gel electrophoresis. Electrophoresis 25(15):2602–2610
105. Chen TJ, Boles RG, Wong LJ (1999) Detection of mitochondrial DNA mutations by temporal temperature gradient gel electrophoresis. Clin Chem 45(8 Pt 1):1162–1167
106. Wartell RM, Hosseini SH, Moran CP Jr (1990) Detecting base pair substitutions in DNA fragments by temperature-gradient gel electrophoresis. Nucleic Acids Res 18(9):2699–2705
107. Michikawa Y et al (1997) Comprehensive, rapid and sensitive detection of sequence variants of human mitochondrial tRNA genes. Nucleic Acids Res 25(12):2455–2463
108. Sternberg D et al (1998) Exhaustive scanning approach to screen all the mitochondrial tRNA genes for mutations and its application to the investigation of 35 independent patients with mitochondrial disorders. Hum Mol Genet 7(1):33–42
109. Van Den Bosch BJ et al (2000) Mutation analysis of the entire mitochondrial genome using denaturing high performance liquid chromatography. Nucleic Acids Res 28(20):E89
110. Landsverk ML, Cornwell ME, Palculict ME (2012) Sequence analysis of the whole mitochondrial genome and nuclear genes causing mitochondrial disorders. Methods Mol Biol 837:281–300
111. Cui H, Zhang W, Wong LJC (2011) Comprehensive molecular analyses of mitochondrial genome by next-generation sequencing. In: 12th international congress of human genetics/61st annual meeting of the American Society of Human Genetics, Montreal, Canada
112. Dimmock D et al (2010) Quantitative evaluation of the mitochondrial DNA depletion syndrome. Clin Chem 56(7):1119–1127
113. Bai RK, Wong LJ (2004) Detection and quantification of heteroplasmic mutant mitochondrial DNA by real-time amplification refractory mutation system quantitative PCR analysis: a single-step approach. Clin Chem 50(6):996–1001
114. Cox R et al (2011) Leigh syndrome caused by a novel m.4296G>A mutation in mitochondrial tRNA isoleucine. Mitochondrion 12(2):258–261
115. Enns GM et al (2006) Molecular-clinical correlations in a family with variable tissue mitochondrial DNA T8993G mutant load. Mol Genet Metab 88(4):364–371
116. Venegas V, Halberg MC (2012) Quantification of mtDNA mutation heteroplasmy (ARMS qPCR). Methods Mol Biol 837:313–326
117. Lacbawan F et al (2000) Clinical heterogeneity in mitochondrial DNA deletion disorders: a diagnostic challenge of Pearson syndrome. Am J Med Genet 95(3):266–268
118. Sadikovic B et al (2010) Sequence homology at the breakpoint and clinical phenotype of mitochondrial DNA deletion syndromes. PLoS ONE 5(12):e15687
119. Wong LJ (2001) Recognition of mitochondrial DNA deletion syndrome with non-neuromuscular multisystemic manifestation. Genet Med 3(6):399–404
120. Chinault AC et al (2009) Application of dual-genome oligonucleotide array-based comparative genomic hybridization to the molecular diagnosis of mitochondrial DNA deletion and depletion syndromes. Genet Med 11(7):518–526

121. Wong LJ et al (2008) Utility of oligonucleotide array-based comparative genomic hybridization for detection of target gene deletions. Clin Chem 54(7):1141–1148
122. Wang J et al (2012) Targeted array CGH as a valuable molecular diagnostic approach: experience in the diagnosis of mitochondrial and metabolic disorders. Mole Genet Metab 106(2):221–230
123. Wong LJ et al (2003) Compensatory amplification of mtDNA in a patient with a novel deletion/duplication and high mutant load. J Med Genet 40(11):e125
124. Bourdon A et al (2007) Mutation of RRM2B, encoding p53-controlled ribonucleotide reductase (p53R2), causes severe mitochondrial DNA depletion. Nat Genet 39(6):776–780
125. Hakonen AH et al (2007) Recessive Twinkle mutations in early onset encephalopathy with mtDNA depletion. Brain 130(Pt 11):3032–3040
126. Spinazzola A (2011) Mitochondrial DNA mutations and depletion in pediatric medicine. Semin Fetal Neonatal Med 16(4):190–196
127. Bai RK, Wong LJ (2005) Simultaneous detection and quantification of mitochondrial DNA deletion(s), depletion, and over-replication in patients with mitochondrial disease. J Mol Diagn 7(5):613–622
128. Zhang W, Cui H, Wong LJ (2012) Comprehensive one-step molecular analyses of mitochondrial genome by massively parallel sequencing in a clinical diagnostic laboratory. Clin Chem (Epub PMID: 22777720)
129. Zhang W, Cui H, Wong LJC (2011) Next generation sequencing in clinical diagnostic laboratories: implementation of quantitative and qualitative controls in dual genome analysis. In: Association for molecular pathology 2011 annual meeting, Grapevine, TX
130. Shaibani A et al (2009) Mitochondrial neurogastrointestinal encephalopathy due to mutations in RRM2B. Arch Neurol 66(8):1028–1032
131. Compton AG et al (2011) Application of oligonucleotide array CGH in the detection of a large intragenic deletion in POLG associated with Alpers Syndrome. Mitochondrion 11(1):104–107
132. Douglas GV et al (2011) Detection of uniparental isodisomy in autosomal recessive mitochondrial DNA depletion syndrome by high-density SNP array analysis. J Hum Genet 56(12):834–839
133. Lee NC et al (2009) Simultaneous detection of mitochondrial DNA depletion and single-exon deletion in the deoxyguanosine gene using array-based comparative genomic hybridisation. Arch Dis Child 94(1):55–58
134. Zhang S et al (2010) Application of oligonucleotide array CGH to the simultaneous detection of a deletion in the nuclear TK2 gene and mtDNA depletion. Mol Genet Metab 99(1):53–57
135. Wang J, Rakhade M (2012) Utility of array CGH in molecular diagnosis of mitochondrial disorders. Methods Mol Biol 837:301–312
136. Bainbridge MN et al (2011) Whole-genome sequencing for optimized patient management. Sci Transl Med 3(87):87–re3
137. Lupski JR et al (2011) Clan genomics and the complex architecture of human disease. Cell 147(1):32–43
138. Worthey EA et al (2011) Making a definitive diagnosis: successful clinical application of whole exome sequencing in a child with intractable inflammatory bowel disease. Genet Med 13(3):255–262
139. Wang J et al (2012) An integrated approach for classifying mitochondrial DNA variants: one clinical diagnostic laboratory's experience. Genet Med 14(6):620–626
140. Zhang VW, Wang J (2012) Determination of the clinical significance of an unclassified variant. Methods Mol Biol 837:337–348
141. Richards CS et al (2008) ACMG recommendations for standards for interpretation and reporting of sequence variations: revisions 2007. Genet Med 10(4):294–300

Part II
Genes Involved in Mitochondrial DNA Biogenesis and Maintenance of Mitochondrial DNA Integrity

Chapter 3
Mitochondrial Disorders Associated with the Mitochondrial DNA Polymerase γ: A Focus on Intersubunit Interactions

Matthew J. Young and William C. Copeland

Introduction

The circular, multicopied 16.6 kb human mitochondrial genome is found within the mitochondrial matrix of the dynamic cellular mitochondrial network. Discrete regions of protein-DNA interactions occur along the matrix side of the mitochondrial inner membrane known as nucleoids. Recent superresolution microscopy techniques have revealed that each nucleoid harbors 1–3 mitochondrial DNA (mtDNA) molecules [1, 2] and each cell contains ∼ 300–800 nucleoids. The mitochondrion contains only one mtDNA polymerase to repair and replicate these mtDNAs, the heterotrimeric polymerase gamma (pol γ encoded by two nuclear genes: 1. The *POLG* gene located on chromosomal locus 15q25 encoding one 140 kDa catalytic subunit, p140 and 2. The *POLG2* gene located on chromosomal locus 17q23-24 encoding the ∼ 110 kDa homodimeric accessory subunit p55$_2$. Immunocytochemical experiments have confirmed that both p140 and p55$_2$ are translated in the cytoplasm followed by translocation into the mitochondrial matrix via the mitochondrial import machinery. The catalytic subunit plays an essential role in human and mouse mtDNA maintenance and integrity, while the accessory subunit has been suggested to regulate the mtDNA content of nucleoids and to play a role in conferring processivity to the holoenzyme [3–8]. Therefore, it is not surprising that mutations of both *POLG* and *POLG2* are associated with mitochondrial disease.

Human mitochondrial diseases associated with mutations in *POLG*, *POLG*-related disorders, are comprised of early-childhood-onset diseases and late-adulthood-onset diseases. These include Alpers-Huttenlocher syndrome (AHS), Childhood myocerebrohepatopathy spectrum (MCHS), myoclonic epilepsy myopathy sensory ataxia (MEMSA), ataxia neuropathy spectrum (ANS), autosomal-recessive progressive external ophthalmoplegia (arPEO), and autosomal-dominant progressive external ophthalmoplegia (adPEO). Mutations in *POLG* are one of the most common causes

---

W. C. Copeland (✉) · M. J. Young
Laboratory of Molecular Genetics, National Institute of Environmental Health Sciences,
National Institutes of Health, DHHS, Research Triangle Park, NC 27709, USA
e-mail: copelan1@niehs.nih.gov

of mitochondrial disorders and the cumulative frequency of recessive *POLG* mutations (approaching 1 % in some populations) predicts the prevalence of the silent carriers to be approximately 2 % (or 1/50) by Hardy–Weinberg equilibrium calculations [9]. This frequency of recessive carriers would result in a disease frequency of 1:10,000 (1/50 × 1/50 × 1/4 chance of a child inheriting both alleles) in the general population [9–12]. Cohen et al. [11] have recently described the prevalence of mutations, diagnosis, and clinical descriptions of *POLG*-related disorders. The number of coding allelic variants is staggering, currently approaching 200 known *POLG* mutations (http://tools.niehs.nih.gov/polg/ and [12]).

One of the challenges of mitochondrial medicine will require distinguishing pathogenic mutations from the plethora of neutral SNPs in the general population. The Single Nucleotide Polymorphism Database (dbSNP) of Nucleotide Sequence Variation (http://www.ncbi.nlm.nih.gov/SNP/) currently lists a total of 572 SNPs for the *POLG* gene alone (including the 5' near gene region, 5' UTR, exons, introns, and 3' near gene region) and 345 SNPs for *POLG2*. Thirty-seven of the *POLG2* SNPs are a combination of synonymous and nonsynonymous changes in the coding sequence. Not yet included in dbSNP are the recently published nonsynonymous SNPs associated with mitochondrial disease encoding G103S, P205R, R369G, δ V398-K431, and L475DfsX2 p55, three of which have biochemical defects (see below) [13].

The c.2864A>G mutation encoding Y955C p140 was demonstrated to cosegregate with adPEO in the seminal *POLG* genetic linkage study by Van Goethem et al. [14]. Similar to the *POLG* gene two other genes encoding components of the minimal mtDNA replisome, *POLG2* and *c10orf2* (encoding the mitochondrial *twinkle* helicase), have also been implicated in human mitochondrial diseases associated with mtDNA maintenance [15–17]. Unlike *POLG* and the *twinkle* helicase mutations [18–20], *POLG2* mutations have not yet been subject to genetic linkage or cosegregation analysis. In the absence of available family genetics, in vitro biochemical analyses of purified p55$_2$ or p140 variant proteins provide powerful tools to determine enzymatic deficiencies. Other indications of mutation pathogenicity include: (1) mitochondrial electron transport chain studies, (2) clinical evidence suggestive of mitochondrial disease, (3) mtDNA deletion and/or depletion analysis, (4) evolutionary conservation of amino acid residues, (5) amino acid residue position relative to structural and functional domains, (6) recurrence of the mutation in unrelated patients and absence in control subjects, and (7) existence of different variants at the same amino acid residue position [12]. When possible we will discuss these other indicators with specific mutations addressed in this chapter. Many recent reviews have comprehensively examined mutations associated with *POLG*-related disorders [16, 21–23]. Here we focus on the recent findings that several *POLG2* mutations are associated with mitochondrial disease, *POLG2*-related disorders, and we examine *POLG* mutations that have been demonstrated to disrupt the interface between p140 and p55$_2$.

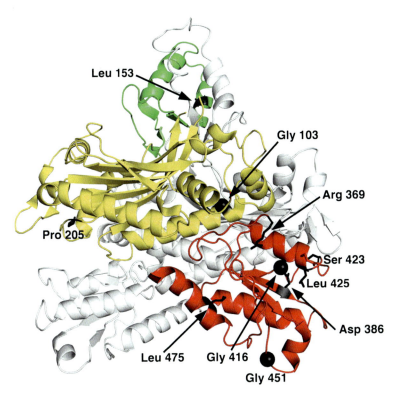

Fig. 3.1 Variant residues associated with *POLG2*-related disorders. One p55 monomer of the homodimer is colored *white* while the other monomer is colored *yellow*, *green*, and *red* to highlight the previously defined domains 1, 2, and 3, respectively [27]. Residues changed by mutation are emphasized in *black* in only the colored monomer. The figure was generated using the program PyMOL and the published crystal structure PDB ID 2G4C [27]

Structure of the Pol γ Homodimeric Accessory Subunit, p55$_2$

The p55$_2$ accessory subunit has been well-characterized biochemically as a DNA-binding protein that confers processivity and physiological salt tolerance on the pol γ catalytic subunit, p140 [8, 24, 25]. Processivity is a measurement of the extent of DNA synthesis during a single enzyme-binding event. When monomers of the murine and human p55 homodimeric crystal structures are superimpositioned they closely resemble each other with a 1.3 Å root-mean-square deviation of the Cα atoms [26]. Each p55 monomer is subdivided into three functional domains (Fig. 3.1). *Domain 1* (amino acid residues 66–131 and 183–353 based on the alignment in Carrodeguas et al. [27]) is homologous to the catalytic domain of aminoacyl-tRNA synthetases and is involved in p55 dimerization as well as binding to p140 [26–29]. This domain lies downstream of the putative 25 amino acid residue N-terminal *m*itochondrial-*t*argeting *s*equence (MTS) required for recognition and direction to the mitochondrial

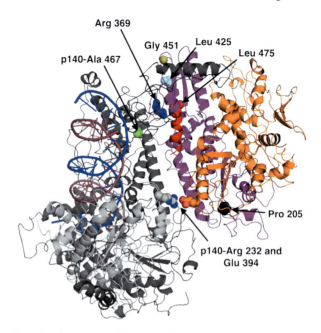

Fig. 3.2 Three-dimensional structure of the p140-p55₂ holoenzyme with DNA modeled into the active site. The catalytic subunit is colored in *dark grey* with the exonuclease domain highlighted in *light gray*. The proximal p55 monomer is colored *purple*, and the distal monomer is *orange*. The primer strand is colored in *light red* and the template strand is colored in *blue*. Residues associated with *POLG2*-related disorders are shown as spheres in the *purple* proximal monomer for clarity. The proximal p55 C-terminal α-helix is colored in *red* highlighting the ten residues truncated in the L475DfsX2 variant. The side chains of p140-Arg232 and the distal p55-Glu394 are shown as spheres. The figure was generated using the program PyMOL and the published holoenzyme model [23]

matrix. *Domain 2* contains a four-helix bundle involved with p55 homodimerization (aa residues 132–182). The accessory subunit has a very large dimer interface of ~ 4000 Å² [30] and analytical ultracentrifugation experiments have shown that the dimer is tightly associated; characterized by an estimated p55₂ intermolecular K_d of < 0.1 nM [31]. The carboxyl-terminal *domain 3* is involved in binding the catalytic subunit (aa residues 354–485) [29] and is homologous to the anticodon-binding domain of type IIa aminoacyl-tRNA synthetases [27]. Biochemical experiments have suggested that in the holoenzyme the proximal p55 subunit (Fig. 3.2, purple) strengthens the interaction with DNA while the distal p55 subunit (orange) is important for accelerating the nucleotide incorporation step [29, 30].

Pol γ accessory subunits from human, mouse, *Drosophila*, and *Xenopus* all share homology with prokaryotic tRNA synthetases suggesting a hypothetical model of p55-mediated loading of pol γ onto primers for initiation of mtDNA replication [6, 7, 28, 32]. The essential role for p55 in mtDNA replication of metazoans has been suggested by the study of two separate null mutations in the *Drosophila melanogaster* pol γ accessory subunit gene, which caused lethality in the early-pupal stage of fly

3 Mitochondrial Disorders Associated with the Mitochondrial DNA Polymerase γ ...

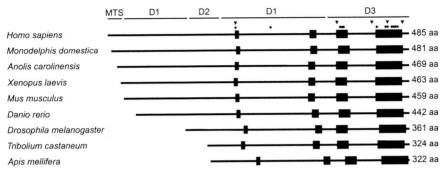

Fig. 3.3 Comparison of vertebrate and invertebrate accessory subunit primary amino acid sequences. The alignment was generated using PRofile ALIgNEment or PRALINE [67]. Among the nine metazoans *black boxes* represent conserved regions where at least five consecutive amino acid residues have PRALINE consistency scores of 6 or greater. In the 10-point consistency scale, zero indicates the least conserved alignment position and 10 indicates the most conserved. From *left* to *right black triangles* represent disease-associated residues Pro205, Arg369, Leu425, Gly451, and Leu475 discussed in the text. From *left* to *right black circles* represent residues conserved among all metazoans Pro205, Trp262, His375, Leu378, Pro380, Asp433, Leu448, Gly451, Arg458, Thr460, Glu464, and His467. MTS, mitochondrial-targeting sequence; D1, D2, and D3 domains 1, 2, and 3 respectively. NCBI accession numbers used in the analysis: NP_009146.2 (Hsap), XP_001379344.1 (Mdom), XP_003224481.1 (Acar), AAI70142.1 (Xlae), NP_056625.2 (Mmus), XP_001922359.2 (Drer), NP_001027262.1 (Dmel), XP_001808672.1 (Tcas), and XP_003250450.1 (Amel)

development [33]. To gain insight into salient features of human pol γ we carried out an alignment on diverse catalytic and accessory subunit sequences from several vertebrates and invertebrates (Figs. 3.3 and 3.4). Select representatives from four orders of fungi were included for the catalytic subunit alignment (Schizosaccharomycetales, Saccharomycetales, Sordariales, and Magnaporthales).

Structure of the Heterotrimeric Pol γ Apoenzyme, p140-p55$_2$

The 3.2 Å resolution structure revealed an approximately 3500 Å^2 unequal subunit contact area between the proximal monomer of the p55 dimer (see Fig. 3.2, purple) and the catalytic subunit [29]. The p55-Δl4 variant used for crystallization lacked the four-helix bundle present in the human and murine crystal structures (residues 147–179 substituted with Gly-Gly) [26, 27]; however, this structure is likely not required for interaction with p140 [29]. A primer-template DNA has recently been modeled into the putative DNA-binding channel (see Fig. 3.2 and [23]). The catalytic subunit contains four known functional domains (*Homo sapiens* Fig. 3.4): (1) *The amino-terminal domain* (p140 amino acid residues 1–170) which harbors the MTS (amino-terminal residues 1–25, in silico prediction) required for mitochondrial import. (2) *The exonuclease domain* (aa 171–440) for 3′→5′ proofreading of mispaired 3′-terminal nucleotides. This domain contains three conserved exonuclease motifs (Exo I, II, and III) and two of the catalytic residues, Asp198 and Glu200, are located

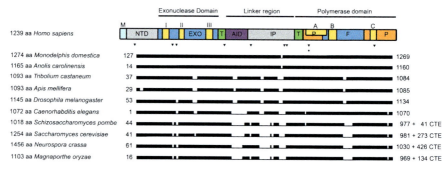

Fig. 3.4 Comparison of vertebrate, invertebrate, and fungal catalytic subunit primary amino acid sequences. The alignment was generated using PRALINE [67]. Among the examined eukaryotes, catalytic subunit sequence identity was relatively high ranging from 29 % (*N. crassa*) to 78 % (*M. domestica* and *A. carolinensis*). Salient features are visualized using a "gap map" for each organism where alignment gaps relative to human pol γ are represented as *white spaces* within *black lines*. Gaps of five amino acid residues or greater are presented. *Mus musculus*, *Xenopus laevis*, and *Danio rerio* sequences were also included in the analysis but are not represented due to the absence of gaps in these sequences. The highly conserved region between human pol γ residues 71–1,230 is displayed for each organism while additional carboxyl-terminal extension (CTE) lengths of fungal sequences are listed on the *right hand side*. From *left* to *right black triangles* represent disease-associated residues Ser64, Arg232, Thr251, Ala467, Pro587, Gly737, Trp748, Gly848, and Asp1143 discussed in the text. Gly848 is conserved among all eukaryotes and is highlighted with a *black circle*. M, mitochondrial-targeting sequence; NTD, amino-terminal domain; I, II, and III exonuclease motifs I, II, and III; A, B, and C polymerase motifs A, B, and C; T, thumb subdomain; P, palm subdomain; F, fingers subdomain; AID, accessory-interacting-determinant subdomain; IP, intrinsic processivity subdomain. NCBI accession numbers used in the analysis: NP_001119603.1 (Hsap), NP_059490.2 (Mmus), XP_003340027.1 (Mdom), NP_001081464.1 (Xlae), XP_001921130.2 (Drer), XP_003227099.1 (Acar), XP_968888.2 (Tcas), XP_395230.4 (Amel), NP_476821.1 (Dmel), CAB60771.1 (Cele), NP_588025.2 (Spom), NP_014975.2 (Scer), XP_957747.2 (Ncra), XP_364068.2 (Mory).

within Exo I [34]. (3) *The spacer/linker domain* (aa 476–785) contains two subdomains, (A) The *i*ntrinsic *p*rocessivity subdomain (*IP*, aa 571–785). In the absence of the accessory subunit, the IP subdomain is predicted to provide a binding site for the primer-template DNA and therefore contribute to the high-intrinsic processivity of the catalytic subunit. (B) The Accessory-*I*nteracting *D*eterminant subdomain (*AID*, aa 476–570) [23] harboring the "L-helix" (p140-V543–L558). The L-helix makes hydrophobic interactions with the C-terminal domain of the proximal p55 monomer (p55-V398–L406, V441–L455) thereby exposing the polylysine-tract region (K-tract, p140–496KQKKAKKVKK505) in a putative upstream DNA-binding channel. Based on two structural models [23, 29] this region of the linker is predicted to interact with the negatively charged phosphodiester backbone of the incoming double stranded DNA. Interestingly, mutating key hydrophobic residues in the L-helix (L549N, L552N, or the entire helix) reduced polymerase activity and processivity in the presence of p55$_2$, likely caused by movement/disruption of the K-tract from the DNA [29]. (4) *The DNA polymerase domain* (aa 441–475; aa 786–1239) for 5′ → 3′ DNA synthesis. This domain harbors three conserved polymerase motifs (Pol A, B, and C). Similar to other DNA polymerases the crystal structure revealed the common resemblance of this domain with a "right-hand" with three subdomains, (A) The thumb (aa 441–475; aa 786–815), (B) The palm (aa 816–910; aa 1096–1239), harboring the "RR loop" (aa 842–856 with R852 and R853 at the tip of the loop), and (C) the fingers (aa 911–1095). The DNA polymerase active site lies within the palm subdomain and contains the catalytic residues harboring the conserved carboxylates (Asp890 located within Pol A and Asp1135, Glu1136 both located within Pol C).

POLG2 Mutations Associated with Mitochondrial Disease, p55$_2$

POLG2 mutations are not as well characterized as compared with the mutations in the gene for the catalytic subunit. However, we recently examined a cohort of 112 patients, exhibiting heterogeneous symptoms similar to *POLG*-related disease but lacking *POLG* gene mutations, and identified novel *POLG2* mutations. In a control subject and the group of 112 patients, eight heterozygous *POLG2* mutations were identified including six novel substitutions [13]. Family genetics were not available for the patients in this study; therefore, a biochemical approach was used to gain insight into these mutations. Analysis of seven variant p55 proteins (G103S, L153V, P205R, R369G, D386E, S423Y, and L475DfsX2) revealed that G103S, L153V, D386E and S423Y functioned similar to wild-type, and the P205R, R369G, and L475DfsX2 variants had biochemical deficiencies. The four variants without detectible biochemical difference probably represent neutral SNPs similar to the earlier findings of the c.1247G>C SNP (encoding p55-G416A, Table 3.1) [35]. In agreement with neutrality, the 1000 Genome catalog of human genetic variation has found that the c.1247G>C SNP has a *minor allele frequency* (MAF) of 0.5 % in the May 2011 data set of 1,094 individuals (Table 3.1 and http://www.1000genomes.org/node/506). In addition, alleles encoding D386E and A169T have MAF values of 0.3 and 25 %, respectively. These results emphasize the importance of following up on predicted disease-associated SNPs with biochemical analyses and with family genetic studies. These results also suggest mutations in the *POLG2* gene may contribute to a portion of *POLG*-like diseases and should be included when screening for these disorders.

Three reports have identified multiple noncoding *POLG2* SNPs (with the exception of c.505G>A, encoding A169T p55) as important markers for risk associated with urinary bladder cancer invasiveness, lung cancer, and head and neck cancers as well as advanced fibrosis associated with chronic hepatitis C (CHC) [36–38]. The contribution of these SNPs to the aforementioned diseases is not clear; however, Ratanajaraya et al. did observe a slight increase of *POLG2* transcription by the intronic allele c.1111–157T>G using an expression quantitative trait loci (eQTL) analysis. These cancer and fibrosis/CHC-associated SNPs are quite frequent in the 1000 Genome sample population, with MAF values ranging between 16.4 and 25.5 % (see Table 3.1).

Table 3.1 Published *POLG2* mutations and SNPs

cDNA mutation	Amino acid change	Gene location	MAF/minor allele count[a]	Reference	Public ID
c.-81T>C	–	5′ UTR	G = 0.247/540	[37]	rs9897606
c.307G>A	p.G103S	Exon 1	–	[13]	–
c.457C>G	p.L153V	Exon 1	N/A	[13]	rs149446102
c.505G>A	p.A169T	Exon 1	T = 0.250/547	[43, 37]	rs1427463
c.562+112T>C	–	Intron 1	G = 0.250/547	[37]	rs3744409
c.614C>G	p.P205R	Exon 2	–	[13]	–
c.796-406T>C	–	Intron 3	G = 0.250/547	[38]	rs6504232
c.969+2149G>A	–	Intron 4	T = 0.164/359	[37]	rs6504231
c.1105A>G	p.R369G	Exon 5	–	[13]	–
c.1110+44T>C	–	Intron 5	G = 0.250/546	[43]	rs7223078
c.1111-157T>G	–	Intron 5	C = 0.255/558	[38, 36]	rs17650301
c.1158T>G	p.D386E	Exon 6	C = 0.003/6	[13]	rs61751983
c.1191+7_1191+8insT	–	Intron 6	N/A	[43]	rs70937066
c.1207_1208ins24	p.δ V398-K431	Minus exon 7 due to exon skipping	–	[43]	–
c.1247G>C	p.G416A	Exon 7	G = 0.005/10	[35, 13]	rs17850455
c.1268C>A	p.S423Y	Exon 7	N/A	[13]	rs181583071
c.1292+146C>T	–	Intron 7	= 0.250/546	[38]	rs9908927
c.1352G>A	p.G451E	Exon 8	N/A	[25]	rs104894632
c.1423_1424delTT	p.L475DfsX2	Exon 8	–	[13]	–

[a]The global minor allele frequency (MAF) represents the second most frequent allele value in the data set. If there are three alleles, with frequencies of 0.50, 0.49, and 0.01, the MAF will be reported as 0.49. The MAF/MinorAlleleCount reports the minor allele (A, T, G, or C), the MAF in the data set, and the number of observations of the minor allele. For example, p.A169T (encoded by c. 505G>**A** or NC_000017.10:g.62492582C>T) has the minor allele 'T', a frequency of 25 % in the 1000 Genome phase population, and 'T' is observed 547 times in the 1,094 sample population, or in 2,188 chromosomes, http://www.ncbi.nlm.nih.gov/projects/SNP/snp_ref.cgi?rs=1427463#Diversity and http://www.1000genomes.org/node/506

G451E p55 (c.1352G>A, Exon 8)

The first heterozygous *POLG2* mutation described (c.1352G>A; p.G451E) was in a 60-year-old female patient with adPEO, multiple mtDNA deletions in muscle, 6 % COX-deficient muscle fibers, and progressive drooping of the eyelids (ptosis) [25]. She was first affected with the disorder at the age of 40. The patient's two sisters were unaffected and contained wild-type *POLG2* sequences while her mother had been similarly affected but was deceased. The mutation was not detected in 144 control individuals from the same geographic region. Likewise the mutation is not seen in the 1000 Genome project.

Biochemically, the p.G451E variant has a 6-fold weaker ability to bind DNA and fails to physically associate with and enhance the processivity of the catalytic subunit [13, 25]. The Gly451 residue found in the homologous aminoacyl-tRNA synthetase anticodon-binding domain 3, is absolutely conserved throughout vertebrates and insects emphasizing the evolutionarily pressure to maintain this residue (see Fig. 3.3). A 3.6 Å van der Waals interaction is made between the side chains of the proximal p55-G451 monomer and the catalytic subunit Leu559 residue in the AID subdomain (see Fig. 3.2). Replacement of the small, uncharged glycine side chain at position 451 with the longer carboxyl containing side chain of glutamate could easily cause a steric clash between the catalytic and accessory subunits and likely disrupt the pol γ intersubunit interactions between these two subunits.

P205R p55 (c.614C>G, Exon 2)

The heterozygous c.614C>G (p.P205R) *POLG2* mutation was found in a 6-month-old male patient with irritability, failure to thrive, lethargy, central hypotonia, liver disease, and refractory seizures. The early presentation of these symptoms is similar to the Alpers syndrome, although this patient only harbored one detectable mutation. The P205R p55 protein behaves as a homodimer via analytical gel filtration analysis and has an approximately 4-fold weaker ability to bind DNA compared with the wild-type protein, similar to G451E p55. This variant maintains a wild-type ability to associate with the catalytic subunit as measured in a functional assay but the ability to coimmunoprecipitate p140 is somewhat reduced. The ability of the holoenzyme containing P205R p55 to stimulate primer-extension to full-length products and to stimulate processivity was compromised at low- and high-salt concentrations [13].

In the alignment of accessory subunit protein sequences, the Pro205 residue is absolutely conserved throughout vertebrates and insects emphasizing the importance of this residue (see Fig. 3.3). Based on the crystal structure of the holoenzyme (see Fig. 3.2) the location of the P205R substitution would not appear to disrupt the interface between p140 and p55$_2$. However, the arginine substitution in this p55 variant is located within a sharp bend between α-helix F and β-strand 5 of a six-stranded β-sheet in domain 1 [27]. The addition of a bulky positively charged side chain in this region could distort this structural feature, and intramolecular p55 misfolding may sufficiently disrupt the function of p55 homodimer or heterodimers to induce the DNA binding and primer-extension defects we observed [13].

R369G p55 (c.1105A>G, Exon 5)

The R369G substitution encoded by the c.1105A>G mutation was first described as a heterozygous substitution in a 19-year-old female subject with neuromuscular and metabolic clinical complications consistent with chronic PEO [13]. Recently, this substitution was also found in a heterozygous 55-year-old male with adPEO, ptosis, ataxia, and peripheral neuropathy symptoms [39]. The patient's asymptomatic father died at the age of 71 and his paternal uncle and nephew were clinically affected but declined genetic testing. Analysis of the patient's skeletal muscle revealed mtDNA deletions, ragged-red fibers, and ~10 % COX-deficient muscle fibers. The mutation was not found in 181 adult control samples and has not been described in the 1000 Genome catalog [39].

The R369G p55 behaves as a homodimer and retains a wild-type ability to bind dsDNA [13]. The holoenzyme containing this variant was resistant to N-ethylmaleimide (NEM), an indicator of the ability of p55$_2$ and p140 to physically associate. Pol γ harboring R369G p55 also had an ~1.7-fold decrease in polymerase-specific activity, decreased processivity, and a high degree of inhibition of primer-extension reactions at physiological salt concentration [13, 39]. Using a functional assay to measure the affinity of variant p55$_2$ subunits for the catalytic subunit, a 4.5-fold decrease relative to wild-type p55$_2$ was observed suggesting a partial defect in subunit association.

The Arg369 residue is conserved among the vertebrates and the red flour beetle (*Tribolium castaneum*, see Fig. 3.3). In the alignment, *Drosophila* harbors an Arg residue one amino acid upstream of human Arg369 that may perform a similar function. Within the homologous anticodon-binding domain 3 of the proximal p55 monomer the Arg369 side chain is located within 4.9 Å of p140 (see Fig. 3.2). The R369G substitution could alter the local van der Waals interactions made between the long side chain of p55-R369 and the thumb subdomain side chain of the p140-L473 residue thus contributing to reduced primer-extension activity and affinity for the catalytic subunit. We predict that the glycine substitution at position 369 decreases the affinity of p55$_2$ for p140 and causes the apparent salt sensitive stalling and decreased polymerase activity due to the modification of a hydrophobic interaction between p140 and the p55 proximal monomer.

L475DfsX2 p55 (c.1423_1424delTT, Exon 8)

The heterozygous c.1423_1424delTT frameshift mutation was identified in an 8-year-old male patient. This mutation deletes two nucleotides in the last exon resulting in a nonsynonymous change at position 475 of the protein and a premature stop codon that truncates ten carboxyl-terminal amino acid residues of p55 domain 3 (L475DfsX2). The location of the frameshift within the last exon of *POLG2* is predicted to escape nonsense mediated decay in vivo due to the inability to form an exon-junction complex downstream of the premature stop [40]. Clinical symptoms included developmental delay, hypotonia, seizures, constipation, elevated transaminases, ketosis, high CSF protein, abnormal EEG, cerebellar atrophy, abnormal MRI, and cortical blindness.

Several similar biochemical deficiencies between the G451E and L475DfsX2 p55 were observed: (1) Using either an end-labeled primer extension assay or a processivity assay these proteins are defective at stimulating p140 at either low or high salt, (2) When using anti-p55-specific antibodies they failed to coimmunoprecipitate p140, and (3) in a functional assay for intersubunit interaction, the affinity for p140 was two orders of magnitude lower than that of wild-type p55. Unlike all other p55 clinical protomers studied till date, the L475DfsX2 variant failed to form a functional homodimer when examined by analytical gel filtration and had an almost complete inability to bind dsDNA. In addition, this variant formed 2-mercaptoethanol reducible crosslinked multimers consistent with protein aggregation due to formation of aberrant disulfide bonds [41]. Therefore, the inability of L475DfsX2 p55 to stimulate DNA synthesis is 3-fold: (1) propensity to aggregate, (2) failure to interact with the catalytic subunit, and (3) failure to bind to DNA. To the best of our knowledge L475DfsX2 p55 is the first case of a clinical variant that is unable to bind DNA and has protein instability.

The Leu475 residue is absolutely conserved among the vertebrates and *Drosophila* while it is chemically conserved as a hydrophobic residue in both the honeybee (*Apis mellifera*, Ile) and the red flour beetle (*T. castaneum*, Val, see Fig. 3.3). Among humans, chimpanzees, orangutans, gibbons, marmosets, pandas, and dogs the Leu475 residue and the C-terminal ten amino acid residues truncated by the mutation are absolutely conserved (data not shown). In vivo the L475DfsX2 p55 variant is likely an unstable protein; however, the status of a heterodimer remains to be determined. Several interesting intersubunit interactions between the C-terminal α-helix truncated in L475DfsX2 and p140 were revealed by the crystal structure. In the proximal monomer, this α-helix makes three close contacts with the AID subdomain of p140: (1) A potential hydrogen bond is made between the side chains of the p140-Cys557 thiol group that is 3.5 Å away from the p55-Tyr478 hydroxyl group; (2) A potential 3.9 Å van der Waals interaction is made between the benzyl ring of the p55-Tyr478 side chain and the p140-Leu576 side chain; and (3) A potential 3.1 Å hydrogen bond is made between the ε-amino group of the p55-Lys477 side chain and the carbonyl group of the p140-Gly571 main chain. We hypothesize that if a L475DfsX2 p55/p55 heterodimer could form in vivo the loss of these three intersubunit contacts would result in impaired pol γ activity. It is conceivable that a heterodimer p55 would realign and allow the wild-type p55 protomer to interact with p140. However, it is tempting to speculate that a variant protomer with DNA-binding defects bound at the distal position could result in defective pol γ mtDNA binding or loading or both. In general, the failure to enhance processivity in the catalytic subunit by the four aforementioned variants would cause the replication complexes to stall during mtDNA replication and is consistent with the accumulation of mtDNA deletions detected in patients with PEO [42].

(A169T-)δV398-K431 p55 (c.505G>A, Exon 1 and c.1207_1208ins24, Exon 7)

A heterozygous 24 bp insertion in exon 7 of *POLG2* (c.1207_1208ins24) that causes missplicing was found in a 62-year-old female patient of German ethnic origin with three other *POLG2* SNPs: c.505G>A encoding A169T, c.1110+44T>C located in intron 5, and c.1191+7_1191+8insT, intron 6 [43]. Skipping of exon 7 and subsequent misligation of exon 6–8 is predicted to result in the deletion of residues Val398 to Lys431. The 24 bp insertion was absent in 100 healthy German control samples as well as 70 German and 100 British patients harboring mtDNA deletions in muscle [43]. All four of the mutations are predicted syntenic (in cis) as the patient's asymptomatic brother also carried all four of the mutations. The patient's unaffected nonconsanguineous parents both died at the age of 83 and 84 and her 22-year-old daughter and two other brothers were reported asymptomatic. Clinical symptoms included late-onset ptosis, which first developed at 30 years of age, myopathy and muscle mtDNA deletions [43].

Although the δ V398-K431 p55 variant has not yet been biochemically characterized, the disruption of the C-terminal domain 3 is predicted to impair interactions with the catalytic subunit. The A169T substitution found in the p55 dimerization domain 2 is likely the result of a neutral SNP as was previously discussed and the c.1110+44T>C has a MAF of 25 % (see Table 3.1). Ala169 is only found in the human, murine, and *Drosophila* sequences of the alignment. The region spanning Val398 to Lys431 is not highly conserved among the select metazoans; however, the Leu425 residue within this region was a noticeable hotspot being only replaced in *Monodelphis domestica* (opossum) by a methionine residue.

Contributions of p55$_2$ to Balancing DNA Polymerase and Exonuclease Activities

Presteady-state exonuclease activity has been measured in the holoenzyme (p140-p55$_2$) and p140 alone with a partial duplex DNA substrate containing four mismatched nucleotides [44]. The amount of mispaired primer-template bound to p140-p55$_2$ subject to excision during the first cycle of the reaction, amplitude for mismatch excision, increased 1.5-fold relative to p140 alone, indicating an increased affinity of p140-p55$_2$ for mismatches. The holoenzyme had a 2.6-fold reduction in the excision rate of mismatched primers compared with p140 alone and excised a correctly paired base 10-fold slower than a mismatch or about twice as slow as the catalytic subunit alone [44]. These experiments suggest that the accessory subunit aids pol γ in distinguishing mismatched from correctly base-paired primer termini and at the same time reduces excision of correctly base-paired nucleotides; therefore, in the holoenzyme p55$_2$ may contribute to balancing DNA synthesis and DNA degradation reactions.

POLG Mutations that Disrupt Interactions with p55$_2$ and are Associated with Mitochondrial Disease

A467T p140 (c.1399G>A, Exon 7)

Disease associated mutations have been found on the catalytic subunit side of the pol γ interface that perturb interactions with p55. A recent estimate of unrelated families with two mutant *POLG* alleles reports the Ala467 to Thr codon substitution is the most common *POLG* disease mutation [12]. The same study reports G848S, W748S, T251I-P587L, and G737R as the second, third, fourth, and eighth, most common *POLG* disease alleles, respectively. A467T is commonly associated with Alpers, PEO, and ataxia-neuropathy. Biochemical studies of the A467T p140 variant demonstrated reduced template binding and lower processivity [22, 45]. DNA polymerase activity of the A467T containing holoenzyme is reduced to 4 % of wild-type and NEM protection and coimmunoprecipitation assays demonstrate compromised $p55_2$ interaction [46]. Although the DNA polymerase activity is drastically compromised, the exonuclease activity is only reduced about 2-fold. Similarly, in a primer extension assay the exonuclease activity of holoenzymes harboring A467T p140 is increased relative to the wild-type as indicated by the complete degradation of the radioactively labeled primer [46]. Also, the accessory subunit partially mitigates the reduced DNA binding and DNA polymerase activity [45]. The Ala467 residue is conserved among the vertebrates, invertebrates, and *Schizosaccharomyces pombe* (see Fig. 3.4) and structurally is located in a hydrophobic center of the thumb subdomain (see Fig. 3.2). The Thr467 hydroxyl group substitution may interrupt the local hydrophobicity of this region as previously suggested [23].

Arg 232 Variants

R232G p140 (c.694C>G, Exon 3)

Mutations of the Arg232 codon associated with mitochondrial disorders have been described in several reports [47–50]. The c.694C>G *POLG* mutation encoding R232G p140 was identified in a 3-month-old female patient that suffered from progressive generalized hypotonia, a state of low muscle tone, and hepatocerebral syndrome, liver and neurological failure (Table 3.2 patient 1) [47]. Analysis of liver homogenate revealed defects of the mtDNA encoded respiratory chain activities relative to age-matched controls and extreme mtDNA depletion (3–5 % of the controls' mean). The patient later died at 6 months of age. Two additional *POLG* mutations in this proband were c.752C>T of exon 3 encoding T251I and c.1760C>T of exon 10 encoding P587L. Sequencing of the *POLG* exons from the patient's three unaffected family members revealed that the father was heterozygous for c.752C>T and c.1760C>T, while the mother was heterozygous for c.694C>G only. The patient's 4-year-old brother was heterozygous for both paternal mutations. Within this family the *POLG* segregation analysis, and the low probability that recombination will occur between closely linked mutations, suggest that the T251I and P587L codon mutations are syntenic (or in cis). In the proband when both mutant alleles (encoding T251I-P587L and R232G) occur in trans the result is a severe mtDNA depletion disorder [47]. The authors suggest this disease could result from a partial dominant-negative effect of the R232G codon mutation in combination with the syntenic pair. A partial dominant-negative effect was also suggested for the K1191R, R853Q, and G848S variants expressed in trans with the cis T251I-P587L mutant codons [51].

Table 3.2 Summary of four patients encoding p140-Arg232 variants. Segregation of *POLG* alleles is provided for first-degree relatives, specifically parents and in half of the cases sibling(s). Affected patients are compound heterozygotes while heterozygous family members are unaffected

Patient[a]	Age of onset/death	*POLG* mutations[b]	Predicted amino acid change	Location in crystal structure	Disease/symptom	Suggested mode of inheritance	Reference
1f	3 months[¹]/6 months	c.694C>G, c.752C>T-c.1760C>T[c]	p.R232G, p.T251I-P587L	Exo, Exo-Link	Progressive generalized hypotonia/infantile hepatocerebral syndrome; extreme liver mtDNA depletion (3–5 % of the controls' mean) by SB[k] and real-time PCR	c.694C>G partially dominant-negative in combination with c.752C>T-c.1760C>T[d]	[47, 68]
F		c.752C>T-c.1760C>T, wild-type	p.T251I-P587L, wild-type	Exo-Link	Unaffected		[47]
M		c.694C>G, wild-type	p.R232G, wild-type	Exo	Unaffected		[47]
B		c.752C>T-c.1760C>T, wild-type	p.T251I-P587L, wild-type	Exo-Link	Unaffected		[47]
2m	5 months/—	c.695G>A, c.2243G>C-c.3428A>G	p.R232H, p.W748S-E1143G[e]	Exo, Link-Pol	Generalized muscular hypotonia/progressive encephalopathy; skeletal muscle mtDNA depletion (12 % relative to mean copy number via quantitative PCR), muscle mitochondria were complex I deficient	Recessive mode of inheritance	[48]

Table 3.2 (continued)

Patient[a]	Age of onset/death	POLG mutations[b]	Predicted amino acid change	Location in crystal structure	Disease/symptom	Suggested mode of inheritance	Reference
F		c.2243G>C-c.3428A>G, (wild-type?)[h]	p.W748S-E1143G, (wild-type?)	Link-Pol	Unaffected		[48]
M		c.695G>A, (wild-type?)	p.R232H, (Wild-type?)	Exo	Unaffected		[48]
3m	22 years (35 years)[f]	c.191C>T-c.695G>A, c.2209G>C	p.S64L-R232H, p.G737R	NTD-Exo, link	ar axonal CMT type 2 g, mosaic defect of cytochrome c oxidase, affecting 6 % of muscle fibers, Longrange PCR of skeletal muscle DNA, revealed multiple deletions of mtDNA	Recessive mode of inheritance	[48, 49]
F[i]	(~ 60 years)	c.2209G>C, wild-type	p.G737R, wild-type	Link	Unaffected		[49]
M	(~ 60 years)	c.191C>T-c.695G>A, wild-type	p.S64L-R232H, Wild-type	NTD-Exo	Unaffected		[49]
B1	(40 years)	c.191C>T-c.695G>A, c.2209G>C	p.S64L-R232H, p.G737R	NTD-Exo	ar axonal CMT type 2	Recessive mode of inheritance	[49]
B2	(41 years)	Wild-type, wild-type	Wild-type, wild-type		Unaffected		[49]
S	9 years (42 years)	N/A[j]	N/A		ar axonal CMT type 2		[49]
4m	6 months/23 months	c.695G>A, c.2542G>A	p.R232H, p.G848S	Exo, Pol	Leigh syndrome; mtDNA depletion in muscle (3 %) and liver (12 %) via SB analysis	Recessive mode of inheritance	[50]

Table 3.2 (continued)

Patient[a]	Age of onset/death	POLG mutations[b]	Predicted amino acid change	Location in crystal structure	Disease/symptom	Suggested mode of inheritance	Reference
M		c.695G>A, (Wild-type?)	p.R232H, (wild-type?)	Exo	Unaffected		[50]
F		c.2542G>A, (wild-type?)	p.G848S, (wild-type?)	Pol	Unaffected		[50]

[a]Proband's of the studies are numbered followed by either lowercase m for male or lowercase f, female; F, father; M, mother; B, brother; S, sister of patient respectively

[b]Syntenic mutations (mutations in cis) are shown on the same line while independently assorting POLG allelic mutations (mutations in trans) are located on a separate line; Wild-type indicates the nonmutant status of the second allele when available

[c]c.752C>T-c.1760C>T has been found in combination with a wild-type allele (heterozygous) in ~1 % of Italian controls [47, 53] in contrast neither mutation was available from 1000 Genome sample population of 1,094 ethnically diverse individuals

[d]It was predicted by the authors that the p.R232G variant may have a partially dominant-negative effect when coexpressed with the second abnormal variant, p.T2511-P587L

[e]E1143G previously reported as a rare SNP that occurred in synteny with W748S in three families [48], in agreement with this statement the c.3428A>G SNP has a MAF/Minor Allele Count of C=0.016/35 while the other mutations in this table were not available from the 1000 Genome sample population

[f]Age during time of study [48]

[g]Autosomal recessive axonal Charcot-Marie-Tooth type 2

[h]Brackets indicate a heterozygous state not explicitly stated in the reference but assumed as these individuals are unaffected

[i]Three half-siblings of the father remain unaffected

[j]Not available

[k]Southern blot

R232H p140 (c.695G>A, Exon 3)

The c.695G>A *POLG* mutation encoding R232H p140 was found in three separate families with patients suffering from mitochondrial disease (see Table 3.2). Similar to the family of patient 1 (see Table 3.2), segregation analysis demonstrated that the mutation encoding the R232H variant occurs in trans (and in one case in trans and in cis) with other nonsynonymous mutations (patient 2 encoding R232H and W748S-E1143G; patient 3, S64L-R232H and G737R; patient 4 R232H and G848S). In the family of patient 3, the mutations also segregate with the disorder in the patient's brother [49].

Catalytic subunits harboring either R232G or R232H substitutions have almost no detectable DNA polymerase or exonuclease defects in comparison to wild-type p140 in the absence of $p55_2$. Analytical gel filtration chromatography of purified R232G p140 or R232H p140 mixed separately with $p55_2$ showed that both complexes chromatographed as single ~ 220-kDa peaks [44]. Densitometric analysis of peak fractions supported the molecular weight determination and therefore it is likely that both of these p140 variants form a wild-type like heterotrimer complex. When analyzed separately in a steady-state $p55_2$-dependant assay, these variants had either an almost complete lack of (R232G) or weakened (R232H) ability to stimulate DNA polymerase activity. In addition, the processivity of R232G and R232H-containing holoenzymes was reduced 7.9- and 2.3-fold, respectively. The histidine substitution presumably retained some positive charge to allow a small response to $p55_2$. In the presence or absence of $p55_2$, the amount of productive protein · DNA complex that can be turned over during the early presteady-state phase of the reaction, the burst amplitude, was similar to wild-type levels for both R232H and R232G p140. This suggests that Arg232 variant-containing holoenzymes bind DNA similar to wild-type complexes.

Presteady-state kinetic examination of exonuclease activity on the holoenzyme containing the R232G p140 variant revealed a 2.5-fold increase in the rate of hydrolysis of a single mismatch relative to the wild-type [44]. The R232G-holoenzyme was also three times faster at hydrolyzing a correctly paired base abut the mismatch. The increased rate of correct base pair excision compared with mismatch excision suggests a reduced ability of the polymerase to discriminate between mismatched and normal base pairs. These results support a dominant negative p140-Arg232 variant model. In this scenario, the aberrant stimulation of polymerase activity and processivity could cause the variant holoenzyme to "stall" or dissociate from the mtDNA template. If the Arg232 variant binds mtDNA similar to wild-type this could cause a "crash" between the variant and the p140 encoded by the second allele. Also, increased exonuclease activity of the R232G p140 variant in combination with decreased polymerase functions could cause excessive $3' \rightarrow 5'$ exonucleolytic degradation of the nascent strand in vivo thereby contributing to mtDNA depletion (see Patient 1, Table 3.2). The imbalance between DNA synthesis and degradation has also been proposed for four orthologous disease mutations in the exonuclease domain of yeast pol γ [52].

Among the select metazoans and yeasts, the p140-Arg232 accessory subunit interacting residue is conserved among the vertebrates and *Drosophila* and is absent in the fungi, nematode (*Caenorhabditis elegans*), honeybee, and red flour beetle (see Fig. 3.4). In the crystal structure of the human holoenzyme, the only interaction between p140 and the distal p55 monomer is a 2.8 Å salt bridge between p140-R232 and p55-E394 emphasizing the importance of this residue (see Fig. 3.2) [29]. The p55-Glu394 residue is absolutely conserved with the exception of *A. mellifera* harboring a chemically conserved Asp residue.

POLG Mutations Associated with Arg232 Variants

The Ser64 residue is located in the amino-terminal domain and is not well conserved. The S64L variant has been documented once in cis with R232H and in trans with G737R codon substitutions associated with axonal Charcot-Marie-Tooth type 2 [49]. The IP subdomain Gly737 residue is mostly conserved among the vertebrates and the insects but there is also a gap around this region and this residue is not conserved among the fungi and the nematode. The G737R substitution has also been associated with Alpers, Parkinsonism, PEO, and seizures but has not yet been biochemically characterized.

T251I and P587L p140 (c.752C > T, Exon 3 and c.1760C > T, Exon 10)

The abundance, complexity, and lack of biochemical experimentation on T251I and P587L p140 variants warrant some discussion. A recent study suggests that the c.752C>T-c.1760C>T allele (encoding T251I-P587L p140) is the fourth most frequent *POLG* mutation associated with *POLG*-mitochondrial disease [12]. These two mutations have also been found in 1 % of Italian control subjects' chromosomes providing a significant reservoir for potential disease-causing alleles [53]. Till date, the majority of the affected individuals harbor this mutation pair with another *POLG* mutation in trans ([12, 53] and http://tools.niehs.nih.gov/polg/). Diseases associated with the syntenic pair include PEO, Alpers, mitochondrial depletion syndrome, and myocerebrohepatopathy spectrum. However, these two mutations exist in other arrangements. The syntenic pair was reported homozygous in four individuals (c.752C>T-c.1760C>T allele 1 and c.752C>T-c.1760C>T allele 2) all of whom presented with PEO [53, 54]. A third case of the homozygous syntenic pair was associated with Alpers [55]. The c.752C>T and c.1760C>T mutations have also been found as compound heterozygous associated with ptosis in one family (four individuals, carrying c.752C>T on the first allele and c.1760C>T on the second allele) [56] and in a separate individual with ataxia/acute disseminated encephalomyelitis [57]. PEO patients with each mutation occurring separately but in compound with a second allelic variant have also been described, c.752C>T (encoding T251I) in trans with c.679C>T (encoding R227W) [58] and c.1760C>T (encoding P587L) in trans with c.2557C>T (encoding R853W) [59]. These nonsyntenic mutations are intriguing as they should occur at a very low frequency due to their close location on chromosome 15, 4544 bp between g.9612C>T (exon 3, encoding T251I) and g.14157C>T (exon 10, encoding P587L). The Pro587 residue is found in the IP subdomain while Thr251 is found in the exonuclease domain. Except for the nematode the Pro587 residue is absolutely conserved among the metazoans while the T251I is not well conserved.

W748S and E1143G p140 (c.2243G > C, Exon 13 and c.3428A > G, Exon 21)

Similar to the T251I-P587L syntenic codon pair the W748S and E1143G codons frequently occur in cis but other arrangements of the codons have also been associated with mitochondrial disease (http://tools.niehs.nih.gov/polg/). Similar to the T251I-P587L pair, the W748S and E1143G codons are located ~ 5 kb apart on the chromosome (g.16370G>C-g.21201A>G). The W748S-E1143G codon pair is associated with a spectrum of disease phenotypes such as Alpers, ANS, and PEO [60]. Analyses of variants harboring W748S, or E1143G, or both has suggested E1143G can modulate the W748S substitution in vitro [60]. Haplotype analysis revealed that the W748S-E1143G syntenic codons likely originated from a common ancient founder in patients from Finland, Norway, the United Kingdom, and Belgium [61]. In contrast, the W748S codon substitution was recently reported without the E1143G substitution in 18 unrelated patients with *POLG* deficiency [12]. Among the vertebrates the Trp748 residue is absolutely conserved while among the insects and three of the fungi it is chemically conserved as a hydrophobic containing side chain. The Trp748 residue is found in the IP subdomain of the linker. Glu1143 of human p140 is conserved or chemically conserved as an aspartic acid in ten of the 14 organisms in the alignment. The other four organisms, *Xenopus laevis*, *Danio rerio*, *Anolis carolinensis*, and *A. mellifera*, contain a serine residue at this position (see Fig. 3.4). Structurally, these conserved features may indicate the importance of a 3.6 Å hydrogen bond between the carboxyl group of the Glu1143 side chain and the amine of the main chain Y884 peptide bond.

G848S p140 (c.2542G > A, Exon 16)

The G848S codon mutation is typically associated with Alpers syndrome. A biochemical comparison of the G848S p140 variant to wild-type revealed ~ 5-fold reduction in DNA-binding affinity and the variant-containing holoenzyme had only 0.03 % polymerase activity [62]. The Gly848 residue is absolutely conserved among metazoans and fungi and is located within the palm subdomain of the crystal structure (see Fig. 3.4). Euro et al. [23] suggested that the G848S variant affects the p140 'RR loop' that likely makes extensive contacts with the first four base pairs of the primer-template DNA. Defects in DNA binding and DNA polymerase activity are predicted to result from this substitution in agreement with the observed biochemical defects [62].

Conservation of Pol γ Amino Acid Residues

Among the select accessory subunit sequences analyzed, several disease-associated residues are either absolutely conserved or chemically conserved (Pro205, Gly451, and Leu475). The Arg369 residue is conserved among vertebrates and one insect while the skipping of exon 7 (c.1207_1208ins24) removes a highly conserved Leu425. The insect accessory subunit sequences are significantly shorter than the other metazoan sequences and do not align well with the vertebrate N-terminal region harboring the human p55 dimerization domain 2 and the first section of domain 1 (see Fig. 3.3). The insect sequences also lack the p55-Asp129 residue that participates in human p55 homodimerization [30]. These deviations from vertebrate accessory subunits support the observation that the *D. melanogaster* pol γ oligomerizes as a heterodimer with a monomeric accessory subunit [63]. We predict that the honeybee and red flour beetle may also contain heterodimeric pol γ's.

Among the catalytic subunit sequences the disease-associated Gly848 residue is strictly conserved across evolution and the Ala467 residue is also highly conserved. Among the fungi and *C. elegans* there is a large gap in the alignment at the p140-AID subdomain region of the linker, including the L-helix (see Fig. 3.4). The gap is less severe for the insects however when compared with the vertebrates the N-terminal region of the AID and the L-helix are not well conserved (data not shown). Another evolutionary difference is the lack of conservation of the distal p55-interacting p140-Arg232 residue among the fungi, the nematode, the honeybee, and the beetle. The lack of this residue is also in agreement with a heterodimeric insect pol γ. Currently, there is no evidence for a gene encoding an accessory subunit in the less complex organisms such as the fungi and the nematode *C. elegans*. In support of this evolutionary difference, two separate groups have demonstrated that Mip1p, the yeast mtDNA polymerase, is a highly processive single subunit enzyme [64, 65]. In addition in vivo, the yeast-specific carboxyl-terminal region of Mip1p was demonstrated to be critical for fidelity of mtDNA replication and for mtDNA maintenance during fermentative growth [66]. We predict that these less complex organisms have evolved separate domains to confer processivity on the enzyme. Based on these evolutionary differences we propose three different subclasses of DNA pol γ's: (1) heterotrimeric complexes found in vertebrates, (2) heterodimeric complex found in insects, and (3) monomeric polymerases found in the less complex eukaryotic organisms.

Conclusion

Pol γ variation may arise by inherited mutant gene transfer, de novo muta-genesis (germline or postzygotic mosaicism), environmental mutagens, or through recombination. Crossover during recombination may generate mutations in cis from compound heterozygotes or in trans from syntenic mutations. Mutations in cis or in trans may influence the pathogenicity of disease and modulate variant enzymatic activity as previously suggested [60]. Monitoring family genetics is essential as it identifies mutations that are vertically transferred as well as mutations arising by other means. Other factors such as gene dose, other mutations in cis or in trans elements, or a combination of these factors may also contribute to mitochondrial disorders. Identifying these elements either by whole genome sequencing or genetic linkage analysis or ideally both is of the utmost importance in addition to biochemically testing uncharacterized variant proteins. These future studies will enhance our understanding of mitochondrial disorders and will ultimately help us assess how mutations additively or synergistically contribute to the pathophysiology of these complex diseases.

Acknowledgments We thank Dr. Rajesh Kasiviswanathan and Margaret Humble for critically reading this manuscript. This research was supported by the Intramural Research Program of the NIH, National Institute of Environmental Health Sciences (ES 065078).

References

1. Kukat C, Wurm CA, Spahr H, Falkenberg M, Larsson NG, Jakobs S (2011) Super-resolution microscopy reveals that mammalian mitochondrial nucleoids have a uniform size and frequently contain a single copy of mtDNA. Proc Natl Acad Sci U S A 108:13534–13539
2. Brown TA, Tkachuk AN, Shtengel G et al (2011) Super-resolution fluorescence imaging of mitochondrial nucleoids reveals their spatial range, limits, and membrane interaction. Mol Cell Biol 31(24):4994–5010
3. Hance N, Ekstrand MI, Trifunovic A (2005) Mitochondrial DNA polymerase gamma is essential for mammalian embryogenesis. Hum Mol Genet 14:1775–1783
4. Spelbrink JN, Toivonen JM, Hakkaart GA et al (2000) In vivo functional analysis of the human mitochondrial DNA polymerase POLG expressed in cultured human cells. J Biol Chem 275:24818–24828
5. Garrido N, Griparic L, Jokitalo E, Wartiovaara J, Van Der Bliek AM, Spelbrink JN (2003) Composition and dynamics of human mitochondrial nucleoids. Mol Biol Cell 14:1583–1596
6. Di Re M, Sembongi H, He J et al (2009) The accessory subunit of mitochondrial DNA polymerase gamma determines the DNA content of mitochondrial nucleoids in human cultured cells. Nucleic Acids Res 37:5701–5713
7. Carrodeguas JA, Kobayashi R, Lim SE, Copeland WC, Bogenhagen DF (1999) The accessory subunit of xenopus laevis mitochondrial DNA polymerase gamma Increases processivity of the catalytic subunit of human DNA polymerase gamma and is related to class II aminoacyl-tRNA synthetases. Mol Cell Biol 19:4039–4046
8. Lim SE, Longley MJ, Copeland WC (1999) The mitochondrial p55 accessory subunit of human DNA polymerase gamma enhances DNA binding, promotes processive DNA synthesis, and confers N-ethylmaleimide resistance. J Biol Chem 274:38197–38203
9. Cohen BH, Naviaux RK (2010) The clinical diagnosis of POLG disease and other mitochondrial DNA depletion disorders. Methods 51:364–373
10. Haas RH, Parikh S, Falk MJ et al (2007) Mitochondrial disease: A practical approach for primary care physicians. Pediatrics 120:1326–1333
11. Cohen BH, Chinnery PF, Copeland WC (2010) POLG-related disorders. Genereviews at genetests: Medical genetics information resource [database online]. Copyright, University of Washington, Seattle, 1997–2010. http://www.genetests.org. Accessed 16 Mar 2010
12. Tang S, Wang J, Lee NC et al (2011) Mitochondrial DNA polymerase {gamma} mutations: an ever expanding molecular and clinical spectrum. J Med Genet 48:669–681

13. Young MJ, Longley MJ, Li FY, Kasiviswanathan R, Wong LJ, Copeland WC (2011) Biochemical analysis of human POLG2 variants associated with mitochondrial disease. Hum Mol Genet 20:3052–3066
14. Van Goethem G, Dermaut B, Lofgren A, Martin JJ, Van Broeckhoven C (2001) Mutation of POLG is associated with progressive external ophthalmoplegia characterized by mtDNA deletions. Nat Genet 28:211–212
15. Longley MJ, Humble MM, Sharief FS, Copeland WC (2010) Disease variants of the human mitochondrial DNA helicase encoded by C10orf2 differentially alter protein stability, nucleotide hydrolysis and helicase activity. J Biol Chem 285:29690–29702
16. Copeland WC (2008) Inherited mitochondrial diseases of DNA replication. Annu Rev Med 59:131–146
17. Farge G, Pham XH, Holmlund T, Khorostov I, Falkenberg M (2007) The accessory subunit B of DNA polymerase {gamma} is required for mitochondrial replisome function. Nucleic Acids Res 35:902–911
18. Li FY, Tariq M, Croxen R et al (1999) Mapping of autosomal dominant progressive external ophthalmoplegia to a 7-cM critical region on 10q24. Neurology 53:1265–1271
19. Spelbrink JN, Li FY, Tiranti V et al (2001) Human mitochondrial DNA deletions associated with mutations in the gene encoding Twinkle, a phage T7 gene 4-like protein localized in mitochondria. Nat Genet 28:223–231
20. Suomalainen A, Kaukonen J, Amati P et al (1995) An autosomal locus predisposing to deletions of mitochondrial DNA. Nat Genet 9:146–151
21. Stumpf JD, Copeland WC (2011) Mitochondrial DNA replication and disease: insights from DNA polymerase gamma mutations. Cell Mol Life Sci 68:219–233
22. Chan SS, Copeland WC (2009) DNA polymerase gamma and mitochondrial disease: understanding the consequence of POLG mutations. Biochim Biophys Acta 1787:312–319
23. Euro L, Farnum GA, Palin E, Suomalainen A, Kaguni LS (2011) Clustering of alpers disease mutations and catalytic defects in biochemical variants reveal new features of molecular mechanism of the human mitochondrial replicase, Pol {gamma}. Nucleic Acids Res 39(21):9072–9084
24. Carrodeguas JA, Pinz KG, Bogenhagen DF (2002) DNA Binding Properties of Human pol gamma B. J Biol Chem 277:50008–50014
25. Longley MJ, Clark S, Yu Wai, Man C et al (2006) Mutant POLG2 Disrupts DNA Polymerase gamma Subunits and Causes Progressive External Ophthalmoplegia. Am J Hum Genet 78:1026–1034
26. Fan L, Kim S, Farr CL et al (2006) A novel processive mechanism for DNA synthesis revealed by structure, modeling and mutagenesis of the accessory subunit of human mitochondrial DNA polymerase. J Mol Biol 358:1229–1243
27. Carrodeguas JA, Theis K, Bogenhagen DF, Kisker C (2001) Crystal structure and deletion analysis show that the accessory subunit of mammalian DNA polymerase gamma, Pol gamma B, functions as a homodimer. Mol Cell 7:43–54
28. Fan L, Sanschagrin PC, Kaguni LS, Kuhn LA (1999) The accessory subunit of mtDNA polymerase shares structural homology with aminoacyl-tRNA synthetases: implications for a dual role as a primer recognition factor and processivity clamp. Proc Natl Acad Sci U S A 96:9527–9532
29. Lee YS, Kennedy WD, Yin YW (2009) Structural insight into processive human mitochondrial DNA synthesis and disease-related polymerase mutations. Cell 139:312–324
30. Lee YS, Lee S, Demeler B, Molineux IJ, Johnson KA, Yin YW (2010) Each monomer of the dimeric accessory protein for human mitochondrial DNA polymerase has a distinct role in conferring processivity. J Biol Chem 285:1490–1499
31. Yakubovshaya E, Chen Z, Carrodeguas JA, Kisker C, Bogenhagen DF (2006) Functional human mitochondrial DNA polymerase gamma forms a heterotrimer. J Biol Chem 281:374–382
32. Carrodeguas JA, Bogenhagen DF (2000) Protein sequences conserved in prokaryotic aminoacyl-tRNA synthetases are important for the activity of the processivity factor of human mitochondrial DNA polymerase. Nucleic Acids Res 28:1237–1244

33. Iyengar B, Luo N, Farr CL, Kaguni LS, Campos AR (2002) The accessory subunit of DNA polymerase gamma is essential for mitochondrial DNA maintenance and development in Drosophila melanogaster. Proc Natl Acad Sci U S A 99:4483–4488
34. Kaguni LS (2004) DNA polymerase gamma, the mitochondrial replicase. Annu Rev Biochem 73:293–320
35. Ferraris S, Clark S, Garelli E et al (2008) Progressive external ophthalmoplegia and vision and hearing loss in a patient with mutations in POLG2 and OPA1. Arch Neurol 65:125–131
36. Ratanajaraya C, Nishiyama H, Takahashi M et al (2011) A polymorphism of the POLG2 gene is genetically associated with the invasiveness of urinary bladder cancer in Japanese males. J Hum Genet 56:572–576
37. Michiels S, Danoy P, Dessen P et al (2007) Polymorphism discovery in 62 DNA repair genes and haplotype associations with risks for lung and head and neck cancers. Carcinogenesis 28:1731–1739
38. Huang H, Shiffman ML, Cheung RC et al (2006) Identification of two gene variants associated with risk of advanced fibrosis in patients with chronic hepatitis C. Gastroenterology 130:1679–1687
39. Craig K, Young MJ, Blakely EL et al (2012) A p.R369G POLG2 mutation associated with adPEO and multiple mtDNA deletions causes decreased affinity between polymerase gamma subunits. Mitochondrion 12(2):313–319
40. Chang YF, Imam JS, Wilkinson MF (2007) The nonsense-mediated decay RNA surveillance pathway. Annu Rev Biochem 76:51–74
41. Baneyx F, Mujacic M (2004) Recombinant protein folding and misfolding in Escherichia coli. Nat Biotechnol 22:1399–1408
42. Krishnan KJ, Reeve AK, Samuels DC et al (2008) What causes mitochondrial DNA deletions in human cells? Nat Genet 40:275–279
43. Walter MC, Czermin B, Muller-Ziermann S et al (2010) Late-onset ptosis and myopathy in a patient with a heterozygous insertion in POLG2. J Neurol 257:1517–1523
44. Lee YS, Molineux IJ, Yin YW (2010) A single mutation in human mitochondrial DNA polymerase pol gammaA affects both polymerization and proofreading activities, but only as a holoenzyme. J Biol Chem 285:28105–28116
45. Luoma PT, Luo N, Loscher WN et al (2005) Functional defects due to spacer-region mutations of human mitochondrial DNA polymerase in a family with an ataxia-myopathy syndrome. Hum Mol Genet 14:1907–1920
46. Chan SSL, Longley MJ, Copeland WC (2005) The common A467T mutation in the human mitochondrial DNA polymerase (POLG) compromises catalytic efficiency and interaction with the accessory subunit. J Biol Chem 280:31341–31346
47. Ferrari G, Lamantea E, Donati A et al (2005) Infantile hepatocerebral syndromes associated with mutations in the mitochondrial DNA polymerase-{gamma}A. Brain 128:723–731
48. Kollberg G, Moslemi AR, Darin N et al (2006) POLG1 mutations associated with progressive encephalopathy in childhood. J Neuropathol Exp Neurol 65:758–768
49. Harrower T, Stewart JD, Hudson G et al (2008) POLG1 mutations manifesting as autosomal recessive axonal Charcot-Marie-Tooth disease. Arch Neurol 65:133–136
50. Taanman JW, Rahman S, Pagnamenta AT et al (2009) Analysis of mutant DNA polymerase gamma in patients with mitochondrial DNA depletion. Hum Mutat 30:248–254
51. Wong LJ, Naviaux RK, Brunetti-Pierri N et al (2008) Molecular and clinical genetics of mitochondrial diseases due to POLG mutations. Hum Mutat 29:E150–172
52. Szczepanowska K, Foury F (2010) A cluster of pathogenic mutations in the 3′-5′ exonuclease domain of DNA polymerase gamma defines a novel module coupling DNA synthesis and degradation. Hum Mol Genet 19:3516–3529
53. Horvath R, Hudson G, Ferrari G et al (2006) Phenotypic spectrum associated with mutations of the mitochondrial polymerase {gamma} gene. Brain 129:1674–1684
54. Stewart JD, Tennant S, Powell H et al (2009) Novel POLG1 mutations associated with neuromuscular and liver phenotypes in adults and children. J Med Genet 46:209–214

55. Stewart JD, Schoeler S, Sitarz KS et al (2011) POLG mutations cause decreased mitochondrial DNA repopulation rates following induced depletion in human fibroblasts. Biochim Biophys Acta 1812:321–325
56. Aitken H, Gorman G, McFarland R, Roberts M, Taylor RW, Turnbull DM (2009) Clinical reasoning: Blurred vision and dancing feet: restless legs syndrome presenting in mitochondrial disease. Neurol 72:e86–90
57. Harris MO, Walsh LE, Hattab EM, Golomb MR (2010) Is it ADEM, POLG, or both? Arch Neurol 67:493–496
58. Agostino A, Valletta L, Chinnery PF et al (2003) Mutations of ANT1, Twinkle, and POLG1 in sporadic progressive external ophthalmoplegia (PEO). Neurol 60:1354–1356
59. Gonzalez-Vioque E, Blazquez A, Fernandez-Moreira D et al (2006) Association of novel POLG mutations and multiple mitochondrial DNA deletions with variable clinical phenotypes in a Spanish population. Arch Neurol 63:107–111
60. Chan SSL, Longley MJ, Copeland WC (2006) Modulation of the W748S mutation in DNA polymerase {gamma} by the E1143G polymorphism in mitochondrial disorders. Hum Mol Genet 15:3473–3483
61. Hakonen AH, Heiskanen S, Juvonen V et al (2005) Mitochondrial DNA polymerase W748S mutation: a common cause of autosomal recessive ataxia with ancient European origin. Am J Hum Genet 77:430–441
62. Kasiviswanathan R, Longley MJ, Chan SS, Copeland WC (2009) Disease mutations in the human mitochondrial DNA polymerase thumb subdomain impart severe defects in mtDNA replication. J Biol Chem 284:19501–19510
63. Wernette CM, Kaguni LS (1986) A mitochondrial DNA polymerase from embryos of Drosophila melanogaster. Purification, subunit structure, and partial characterization. J Biol Chem 261:14764–14770
64. Vanderstraeten S, Van den Brule S, Hu J, Foury F (1998) The role of 3′-5′ exonucleolytic proofreading and mismatch repair in yeast mitochondrial DNA error avoidance. J Biol Chem 273:23690–23697
65. Viikov K, Valjamae P, Sedman J (2011) Yeast mitochondrial DNA polymerase is a highly processive single-subunit enzyme. Mitochondrion 11:119–126
66. Young MJ, Theriault SS, Li M, Court DA (2006) The carboxyl-terminal extension on fungal mitochondrial DNA polymerases: identification of a critical region of the enzyme from Saccharomyces cerevisiae. Yeast 23:101–116
67. Simossis VA, Heringa J (2005) PRALINE: a multiple sequence alignment toolbox that integrates homology-extended and secondary structure information. Nucleic Acids Res 33:W289–294
68. Spinazzola A, Invernizzi F, Carrara F et al (2009) Clinical and molecular features of mitochondrial DNA depletion syndromes. J Inherit Metab Dis 32:143–158

Chapter 4
Alpers–Huttenlocher Syndrome, Polymerase Gamma 1, and Mitochondrial Disease

Russell P. Saneto and Bruce H. Cohen

Introduction

Alpers–Huttenlocher syndrome (AHS, Alpers syndrome) was described in a 4-month-old girl with normal early development who developed intractable seizures in the context of a 1-month illness [1]. Autopsy findings demonstrated a widespread destruction of cerebral cortex and basal ganglia. Although this was not the first reported patient with this illness [2], Alpers' report led to the recognition of this illness and other reports. More than 30 years passed before Wefring and Lamvik [3] proposed that the liver and brain findings were the cause of an inherited metabolic defect. Confirmation of inheritance pattern and involvement of mitochondria occurred a few years later. In a neuropathological study of four children who died of AHS, Sandback and Lerman [4] described abnormal giant mitochondria within neurons and a likely autosomal-recessive inheritance pattern. In 1976, Huttenlocher et al. [5] described the associated hepatic features that occurred in patients with the clinical features of AHS, elevated cerebral spinal fluid protein, as well as confirming the illness followed a pattern of autosomal-recessive inheritance. Over the next several decades, confusion about AHS arose due to patient descriptions using different nomenclatures and incomplete studies. Clarity of the pathological findings coupled with the clinical manifestations of AHS occurred in 1990 when Harding [6] performed a systematic study of 32 patients and noted a distinctive liver and brain pathology, and defined the typical clinical course of the disease. This report outlined the typical early development, an insidious onset of developmental delay, failure to gain weight (thrive), bouts of vomiting, and pronounced hypotonia. Seizure onset would then occur, although seizures could be the heralding event in some children. Once the

R. P. Saneto (✉)
Neurology Department, Division of Pediatric Neurology, Seattle Children's Hospital,
4800 Sand Point Way NE, Seattle, WA 98105, USA
e-mail: russ.saneto@seattlechildrens.org

B. H. Cohen
NeuroDevelopmental Science Center, Neurology Division,
Children's Hospital Medical Center of Akron,
Northeast Ohio Medical University, Akron, OH 44087, USA

seizures began, the clinical course becomes rapidly progressive. Liver involvement was variable and sometimes only occurs at the terminal stages of disease, but in some patients, could be present before or at the time of seizure onset [7]. Postmortem examination of the liver showed characteristic combination of severe microvesicular fatty change, diffuse and haphazard bile duct proliferation/transformation, bridging fibrosis, and disorganization of the normal architecture. The cerebral cortex is variably involved with a consistent and striking involvement of the calcarine cortex. Microscopic changes included spongiosis, neuronal loss, and astrocytosis, which progresses down through the cortical layers.

The molecular elucidation of disorders of mitochondrial DNA (mtDNA) maintenance, in particular AHS, was ushered in with the cloning and characterization of *POLG* in 1996 by Ropp and Copeland [8]. The clinical importance of the *POLG* and associated disease was discovered 3 years later with the first biochemical evidence of mtDNA depletion and altered pol γ enzyme activity in patients with AHS [9]. In 2001, the first series of patients with progressive external ophthalmoplegia were described to have pathogenic mutations in *POLG* [10]. Within 5 years, Naviaux and Nguyen [11] described DNA mutations in *POLG* found to be responsible for the clinical entity of AHS and the full circle of clinical expression, identification of the genetic etiology, and physiological changes within the mitochondria were completed. To date, more than 60 mutations in *POLG* have been associated with AHS (http://tools.niehs.nih.gov/polg/) with a larger cohort of mutations responsible for the broader group of *POLG*-related syndromes and disorders [12–15].

More than 180 mutations in *POLG* are the cause for a broad range of mitochondrial diseases presenting at varying ages ranging from infancy to adulthood. Autosomal-dominant mutations usually involve adult onset syndromes, while autosomal-recessive mutations most often cause syndromes presenting during infancy and childhood. Almost all mutations inducing autosomal-dominant disorders are located in the catalytic residues of the polymerase domain of *POLG* [16]. Bulkier amino acid substitutions, which interfere with both enzyme translocation and binding to an incoming nucleotide, result in decreased affinity of new nucleotide insertion producing increased error rates and/or reduced catalysis. The autosomal-dominant phenotype arises due to the mutant enzyme competing effectively with the wild-type enzyme for binding. This altered nucleotide affinity induces errors in DNA synthesis by stalling of mutant pol γ, which results in mtDNA base substitutions and deletions [17, 18]. Homozygous-dominant mutations are likely embryonic lethal, as these type of mutations have not been described.

Clinical Aspects of Alpers–Huttenlocher Syndrome

Alpers–Huttenlocher syndrome (AHS) and Alpers syndrome are equivalent and synonymous terms, and describe the same disease process. The age of onset, order of symptom presentation, severity of disease, and speed of progression vary with every patient, including siblings with the same genotype. Therefore, incomplete phenocopies that only partially meet full criteria for AHS are common occurrences. Incomplete phenotypes may be due to diagnosis during the early phase of the illness or another *POLG* disease having some, but not all, features of AHS. The patients with incomplete phenotypes are in fact AHS and should not be labeled "Alpers-like," as this confuses physicians and families. However, as with all *POLG* diseases, AHS may have varying degrees of severity and time of onset, with an undefined time from the onset of the first symptom to its progression. Unfortunately, patients with AHS ultimately express progressive organ demise, with death as a universal outcome. The tempo of demise can vary greatly [9, 19, 20]. Thus, it is not always possible to firmly diagnose a patient with AHS early in the course of the illness, even with molecular confirmation with known pathogenic homozygote or compound heterozygote mutations identified. It is the combination of the elements of AHS and not the sequence of presentation, genetic mutation(s), or severity that assist in making the clinical diagnosis.

Clinical POLG Mutation Syndromes

Mutations in *POLG* cause a broad range of mitochondrial diseases with an age of onset ranging from infancy to adulthood [12, 21–28]. More than 180 point mutations in *POLG* have been reported that cause many clinical syndromes, as well as multiple organ system diseases that do not cluster within distinct phenotypic syndromes (http://tools.niehs.nih.gov/polg/) [12, 21, 25]. The constellation of signs and symptoms segregated *POLG* syndromes into myocerebrohepatopathy syndrome, AHS, myoclonus epilepsy myopathy sensory ataxia syndrome, autosomal-dominant progressive ophthalmoplegia, autosomal-recessive progressive ophthalmoplegia, ataxia neuropathy spectrum, and nonsyndromic multisystem cytopathies [13, 25]. Dominant mutations usually cause adult-onset disease, typically progressive ophthalmoplegia with varying degrees of myopathy, Parkinsonism, premature ovarian failure, sensory ataxia, and/or cataract [25, 29–31]. Recessive mutations result in severe adult or juvenile-onset ataxic syndromes, neuropathy, seizures, and progressive ophthalmoplegia [24].

AHS and myocerebrohepatopathy are two *POLG* autosomal-recessive disorders that primarily affect young children and produce mtDNA depletion. These are both noted by early symptom manifestation with a progressive decline leading to early death [1, 6, 9, 25]. Although there is overlap with age of onset, there are clinical clues that these are separate entities. Sometime in the disease course, patients with AHS almost always have medically intractable seizures and an associated devastating encephalopathy, with eventual liver failure if the patient lives long enough [11]. However, liver involvement is variable and in some present early in the course of AHS [6]. In myocerebrohepatopathy, the clinical expression is distinct from AHS [12, 25]. The age of onset is younger and the liver findings are most often the most devastating symptom without the hepatic pathology of AHS (Table 4.1) [20]. Encephalopathy is an earlier symptom and some patients have medically controlled seizures.

Table 4.1 Histological features of Alpers–Huttenlocher liver disease [20]

At least three of the following eight histological findings, once Wilson disease is excluded

Bile ductular proliferation
Bridging fibrosis or cirrhosis
Collapse of liver cell plates
Hepatocyte dropout or focal necrosis with or without portal inflammation
Microvesicular steatosis
Oncoytic change (mitochondrial proliferation associated with intensely eosinophilic cytoplasm) in scattered hepatocytes not affected by steatosis
Parenchymal disarray or disorganization of the normal lobular architecture
Regenerative nodules

Alpers–Huttenlocher Syndrome

The hallmark features of AHS are the triad of clinical findings of refractory seizures, developmental regression, and liver dysfunction. These findings, with the combination of at least 2 of 11 other findings make up the clinical diagnosis of AHS (Table 4.2). During the course of the disease, seizures will appear, become medically refractory and persistent until the end of the disease. When liver features are present, there is a set of histological features constantly found (see Table 4.1).

The age of onset is bimodal, but the typical age of early onset patients is 2–4 years, with a range of 3 months–8 years [6, 26, 32]; and a later second peak between 17 and 24 years of age, with a range of 10–35 years [27, 33–35]. Infants and children with AHS appear healthy until the onset of the disorder. Onset is influenced, in part, by specific mutations within the *POLG* gene, but there are likely modifier (ecogenic single nucleotide variant) mutations in *POLG* or related genes. Together with environmental factors, these changes in pol γ function not only alter age of onset and severity of disease, but also the phenotypic expression of AHS [12, 25]. Currently, the full mechanism(s) for this variability remain uncertain. The onset and stepwise progression of this disorder often occurs in the context of an illness with a previous normal developmental pattern.

A later age of onset was first noted by Harding [6, 33] and later confirmed by other reports [27, 34, 35]. This older age of onset of this group typically clusters between 17 and 24 years of age. In distinction to the younger onset age group, the majority of patients in this group have homozygous-recessive mutations in *POLG* with homozygous p.A467T or p.W748S transitions. Some patients were heterozygous for the p.A467T and p.W748S mutations. However, recently an older onset patient was found to be a compound heterozygote for the p.T851A and the p.R1047W [35]. So, although mutation location and alterations in pol γ activity may represent partial explanations for later onset, clearly there are other modifying factors at play.

Before the onset of severe symptoms and signs, some patients are healthy while others demonstrate nonspecific symptoms such as clumsiness, migraine-like headache, progressive ataxia, or seizures [27, 35, 36]. As with the younger onset group, the older AHS patients have seizures, which can rapidly evolve into status epilepticus and/or epilepsia partialis continua, and become medical refractory to

Table 4.2 Diagnostic criteria for Alpers–Huttenlocher syndrome. (Adapted from [11, 20])

Clinical triad of refractory seizures, psychomotor regression, and hepatopathy
In the absence of either hepatopathy (Table 4.1) or additional findings (see below), the diagnosis can only be confirmed either by *POLG* gene sequencing, liver biopsy, or postmortem examination
Additional clinical findings
Elevated CSF protein (> 100 mg/dL)
Brain proton MR spectroscopy showing reduced N-acetyl aspartate, normal creatine, and elevated lactate
Cerebral volume loss (central > cortical, with ventriculomegaly) on repeat MRI or CT studies
Electroencephalogram showing a multifocal paroxysmal activity with high-amplitude polyspikes (10–100 µV, 12–25 Hz)
Cortical blindness or optic atrophy
Abnormal visual evoked potentials and normal electroretinogram
Quantitative mitochondrial DNA depletion in skeletal muscle or liver (35 % mean)
Deficiency in pol γ enzymatic activity ($\leq$ 10 %) in skeletal muscle or liver
Elevated blood or CSF lactate (3 mM) on at least one occasion in the absence of acute liver failure
Isolated complex IV or a combination I, III, and IV electron transport complex defects ($\leq$ 20 % of normal) upon liver respiratory chain testing
A sibling confirmed to have Alpers syndrome

all treatment. Multiple organ involvement occurs with abnormal eye movements, severe myoclonus, involuntary movements, hemiparesis, pancreatitis, and liver failure (almost all were exposed to valproic acid). Once seizures appear, the tempo of disease progression is rapid, with death occurring within 4 years after onset [27, 33, 34]. In other cases, progression of disease is slower [27, 36]. The etiology for the tempo of disease progression is not known, although the use of valproic acid (divalproex sodium) has influenced the natural course of the disease in many cases, especially the rapidly progressive disease patients.

Seizures are a common initial symptom in about 50 % of AHS patients. The type of seizure is not constant and can vary from patient to patient, and changes as the disease progresses in a particular patient. Seizure semiology of initial events is not well described, but in many is described as hallucination or compatible with occipital lobe seizures. We have seen a range of initial seizures consisting of febrile convulsions, focal seizures presenting as hallucination auras, focal motor seizures, and generalized tonic clonic [36]. Seizures may be easily managed at the onset, but can soon become intractable despite medical management, including the use of high dosages of anticonvulsants and even more aggressive measures including general anesthesia. Unfortunately, eventual medication intractability is almost universal in AHS. The electroencephalogram (EEG) may provide specific findings in AHS. Very often, there is an initial occipital predominance to epileptiform discharges, described as large amplitude spike/polyspike and wave discharges (> 200 mV), which over time become generalized [36–39]. Occipital discharges may be unilateral or bilateral over the posterior head region. Seizures during the early phase of the illness will often have the semiology (clinical manifestations) of occipital lobe seizures, such as hallucinations, nausea and vomiting, dysautonomia, headache, and eye nystagmus. The rare patient will maintain seizure control with medications. As the disease progresses, most patients have repeated episodes of status epilepticus and/or epilepsia partialis continua [9, 11, 26, 27, 36, 40–42]. Prolonged seizures are difficult to treat and death may occur during a status epilepticus event. Myoclonus becomes more prominent, often linked with progression of the epileptic component of the illness. It is often difficult to differentiate myoclonus from myoclonic seizures and both can be medically refractory and devastating. The clinician is tempted to try valproic acid during refractory status epilepticus, epilepsia partialis continua, and frequent bouts of almost continuous myoclonus or myoclonic seizures. However, this medication should be avoided, as it can induce liver dysfunction that results in irreversible liver failure [11, 19, 36, 43–45]. Although the use of high-dose benzodiazepine and other anticonvulsants may lessen the seizure burden or improve the myoclonus, the side effects, which include sedation, depressed respiratory drive, and bulbar dysfunction must be balanced against the benefit of the medication. The current medical management is less than satisfactory for the seizures and myoclonus.

Organs involved in AHS are primarily those that require large amounts of energy and prone to oxidative damage, which include the brain, peripheral nervous system, gastrointestinal tract, and liver. Onset of symptomatic hypoglycemia as early as 1 year of age, with a normal liver biopsy, has been seen in this illness [36]. Older patients have been described with fasting hypoglycemia [46]. For unclear reasons, the hypoglycemic events may resolve over time. The exact etiology is unclear but likely reflects impaired gluconeogenesis. Another early symptom is headaches, usually migrainous, which can be associated with visual hallucination auras due to occipital lobe involvement [27, 41].

Ataxia, often described as "clumsiness" by caregivers, can be an early symptom, but is universally seen during the course of the disease due to cerebellar, sensory nerve, ascending sensory pathways, and/or corticospinal tract involvement [36, 39, 42, 47, 48]. Within the brain, AHS is primarily a gray matter disease involving cortex, as well as cerebellar function [6]. Clinically, choreoathetosis, myoclonus, and Parkinsonism tremors have been described reflecting gray matter involvement [29, 44]. Peripheral nervous system is also involved with neuropathy occurring in most, manifested by loss of muscle stretch reflexes (deep tendon reflexes), distal weakness, and sensory neuropathy. Detecting and evaluating sensory involvement in young patients is difficult and often escapes identification until muscle stretch reflexes are lost. In some series neuropathy was universal, suggesting that this is a common finding [27]. In the older patient, loss of vibratory sensation, pain, and temperature loss are usually later findings. However, sensory neuropathy has been reported as an early finding demonstrating patient heterogeneity [36].

The predominance of neuronal loss in the calcarine and striate cortex presents clinically as blindness. Visual loss may be initially transitory but usually becomes complete as the disease progresses. Not all patients have visual loss, and this may depend on the length of the illness and degree of brain involvement. Visual evoked potentials are usually abnormal, but retinopathy is usually absent [29, 41, 48], which differentiates AHS from other common mitochondrial disorders with similar clinical symptoms including Kearns–Sayre syndrome and the Neuropathy-Ataxia-Retinitis-Pigmentosa disorder. Hallucination aura in patients with migraine or just isolated hallucinations may occur and this is thought to be a reflection of the occipital region neuronal dysfunction. An early finding is the delay in smooth pursuit saccades acquisition, which demonstrated the involvement of cortex networks involving visual cortex, cerebellum, and ocular muscle control centers [36].

Many patients develop swallowing dysfunction, delayed gastric emptying, and gastrointestinal dysmotility, which become pronounced as the disease progresses. These children often require gastric tube for nutrition in the mid-stage of the disorder. Because of defects in gluconeogenesis and poor gastric motility, treatment with continuous feeding via gastrointestinal tube with additional nutritional manipulation is often required. The dysmotility may result in a total inability to digest food and many children at end stage of the disease require total parental nutrition. In our personal series of two patients that have undergone autopsy examination (one with childhood onset and the other with juvenile onset of AHS), the complete loss of the longitudinal muscularis propria external muscle layer in the gastrointestinal tract was observed (Saneto, unpublished data). This is a similar pathology to another disorder leading to mtDNA depletion, mitochondrial neurogastrointestinal encephalomyopathy, or MNGIE disease [49]. The reason(s) for the specific nature of muscle degeneration is not known; however, the mtDNA depletion aspect of both disorders may be responsible. Pancreatitis is also a common finding and is often seen in those that have received valproic acid, suggesting a co-morbid role in the induction pancreatic and liver dysfunction.

Liver dysfunction occurring after exposure to valproic acid has been one of the defining features of AHS, beginning with the introduction of valproic acid use for seizure control [11, 36, 37, 43, 45]. However, as Harding [6] appropriately noted, liver failure can occur without valproic acid exposure. Clearly, liver failure has been described in most early reports, before valproic acid was available [1]. Liver failure with valproic acid usually occurs within 2–3 months of exposure. The liver failure is often heralded by hypoglycemia, decreased albumin synthesis, reduced synthesis of coagulation factors, and mild transaminase elevation. If captured early, within a week of liver manifestations, the use of intravenous levocarnitine may provide some benefit in the course of valproate-induced hepatic failure [50]. In some rare cases, liver dysfunction with exposure to valproate can be reversed by drug withdrawal [37, 51]. When the liver involvement is present, there is a specific pattern of liver histochemical features (see Table 4.1). If the patient lives long enough, most patients eventually develop liver involvement and subsequent failure.

The tempo of neurodegeneration and other organ deterioration varies and periods of stabilization can occur. Death due to infections, seizures, respiratory issues related to overall nutritional weakness and liver failure result in variability in life span. Genetic factors influence the severity of disease. These factors alter life expectancy and perhaps organ involvement, with the onset of first symptoms to death ranging from 3 months to about 12 years.

Diagnosis of Alpers–Huttenlocher Syndrome

When the clinical suspicion of AHS is present, complete *POLG* sequencing should be considered. With greater than 60 mutations in *POLG* reported associated with

AHS, the range and combination of mutations makes full sequencing the most sensitive and specific method in confirming the diagnosis. There is some evidence that 3-methylglutaconic aciduria may be a clue for more severely affected patients having AHS and electron transport chain (ETC) involvement [52]. There are no specific or sensitive biochemical markers for *POLG* disease. Muscle tissue may show a mosaic pattern with cytochrome c oxidase histochemical staining and ragged-red fibers (trichrome histochemical staining) may be present, but neither of these findings is diagnostic for AHS.

Enzymatic activities of the ETC complexes in muscle or liver specimens are not specific for AHS and may be an insensitive method for diagnosis. *POLG* mutations are not necessarily correlated with measurable defects in ETC enzyme activity or abnormal oxidative phosphorylation. Although reports demonstrated biochemical enzymatic abnormalities, other studies demonstrate such a variable pattern, including normal activity, single electron transport chain enzyme defects, or multiple complex deficiencies [32, 43]; therefore, this methodology should not be used for the purpose of screening for or diagnosing AHS. The variability of ETC complex enzyme activity findings is likely related to stage of disease. As the disease progresses, the abnormalities in activity become more pronounced as the catalytic protein subunits encoded by mtDNA that comprise part of the ETC result in abnormal enzyme function.

A recent report suggests that there is clustering of AHS disease mutations into distinct functional regions that are located throughout *POLG* [53]. The mtDNA copy number in AHS patients is 3–40 % of normal [11, 23, 28, 54]. The use of quantitative mtDNA content may seem useful for diagnosis, but it is not sensitive or specific. Early in the course of AHS, mtDNA content may be normal. Some mutations do not affect mtDNA content but fidelity of the mtDNA synthesis, and therefore the content may be normal, but the errors in the genome may render the mtDNA useless. Furthermore, muscle and liver mtDNA depletion (defined as less than 50 % and 20 % of age-matched control mtDNA content, respectively) may lag behind disease symptoms [55]. If mtDNA content is measured, only muscle or liver tissue should be used, as mutations in the *POLG* gene do not always induce mtDNA depletion in blood cells [55]. With disease progression, almost all AHS patients will demonstrate mtDNA depletion. Two patients have been described with AHS and mtDNA deletions without depletions [56]. The finding of mtDNA depletion may be helpful in disease diagnosis, but its absence cannot be used to exclude AHS.

A clue to the presence of AHS is the EEG findings early in the disease. Explosive seizures with occipital lobe predominance of epileptiform discharges that evolve into status epilepticus and/or epilepsia partialis continua, combined with psychomotor development should signal the possibility of Alpers syndrome [36–38]. Seizure semiology will reflect the occipital lobe involvement.

There is a wide variety of findings on computerized tomography (CT) and magnetic resonance (MR) imaging. CT or MR imaging may be normal early in the disease course, but usually as the disease progresses atrophy is noted on CT. MR imaging demonstrates atrophy and other changes using various sequence acquisitions [56]. These radiographic findings represent gliosis following neuronal cell loss. The

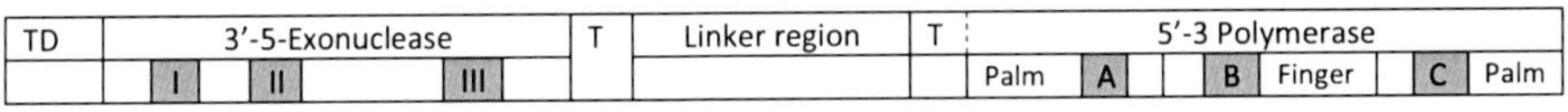

TD	3′-5-Exonuclease		T	Linker region	T	5′-3 Polymerase				
	I	II	III				Palm	A	B	Finger	C	Palm

Fig. 4.1 Functional domains of the pol γ protein. The NTD region is the mitochondrial targeting sequence and is located at the N-terminal of the protein. The T regions are the thumb subdomain of the protein and are found between the exonuclease and linker region as well as within the polymerase region [53]. The exonuclease domain contains essential motifs I, II, and III for its activity. The polymerase domain contains subdomains comprising the thumb, palm, and finger, which contain motifs A, B, and C, respectively. The motifs located within the polymerase domain are critical for polymerase activity [16, 58]

occipital location of EEG abnormalities can sometimes be seen on MR, as hyperintensity on T2/FLAIR sequences in the occipital regions suggest site of mitochondrial dysfunction involvement in EEG changes [57]. Often, when the seizures become uncontrolled, T2/FLAIR hyperintensities are prominent in the thalami and basal ganglia [37]. Interestingly, there may be transient resolution of MR changes with normalization of T2/FLAIR changes [57]. However, as the relentless neurological involvement progresses, the MR shows cortical atrophy. As with other clinical symptoms, MR changes are somewhat variable, but most show a progression of cortical involvement with later basal ganglia and brainstem involvement. Some patients have cerebellar involvement, which would correspond to prominent Purkinje cell drop out seen in autopsy cases [6].

The diagnosis of AHS in a patient takes a great deal of clinical acumen with the knowledge that clinical signs and symptoms vary in onset, intensity, severity, and timing. Clinical suspicion is required early in the diagnosis as the sequence of symptoms may be distinctive for each patient. However, eventually the constellation of symptoms will reveal itself and the identification of *POLG* mutations will confirm the clinical diagnosis.

Genetics of Alpers–Huttenlocher Syndrome: *POLG* Function, Structure, and Mutations

The mtDNA polymerase, pol γ, is the only known DNA polymerase in mammalian mitochondria and is responsible for mtDNA replication and repair [16, 58]. The pol γ protein is synthesized in the cell nucleus and transported into the mitochondria across the outer and inner mitochondrial membranes to the inner face of the inner mitochondrial membrane, where it associates with other proteins that make up the mtDNA replisome and nucleoid. The pol γ protein is composed of three DNA activities, 5′ -> 3′DNA polymerase, 3′ -> 5′ exonuclease and a 5′-deoxyribose phosphate lyase activity. The exonuclease activity (catalytic domain) is located in the N-terminal region and is connected by the linker region to the C-terminal domain, which contains the 5′ -> 3′polymerase activity (Fig. 4.1). The polymerase region has three subdomains termed the palm, fingers, and thumb. The 5′-deoxyribose phosphate lyase activity is also located in the C-terminal domain, 5′ -> 3 polymerase region, but the exact location of the active site is not known [59]. There are motifs found in the exonuclease region, I, II, and III, and polymerase region A, B, and C, which are essential for full activity [16, 58].

Pol γ protein is part of a heterotrimer, one molecule of pol γ associated with two molecules of POLG2 [8, 60]. The 55-kDA accessory subunit, POLG2 or p55, enhances the processivity (the average number of nucleotides added by the enzyme per association–disassociation with the template DNA) of the holoenzyme [18]. The association site of one POLG2 subunit with pol γ is in the AID subdomain of the catalytic subunit [61]. This interaction increases DNA-binding affinity of the holoenzyme. The second POLG2 protein makes very limited contact with pol γ and is designated as the distal accessory subunit [18]. This distal interaction with the second POLG2 protein subunit enhances the polymerization rate of the holoenzyme [61]. The replisome complex is also associated with a single-stranded DNA-binding protein, mtSSB, mtDNA helicase PEO1 (Twinkle), and a number of accessory proteins and transcription factors [16].

Both copies of *POLG* are expressed in mammalian cells and mono-allelic expression of wild-type *POLG* is sufficient to compensate and avoid disease [62]. Although recessive mutations causing human illness are distributed along the length of the *POLG* gene, they tend to cluster within distinct regions of the tertiary structure within the catalytic domain [53]. Recessive mutations in the polymerase region of *POLG* decrease polymerase activity, alter DNA-binding affinity, and lower catalytic efficiency [53]. Mutations in the linker region, AID subunit, mediate interactions with POLG2 and destabilize the pol γ-DNA complex, and decrease processivity [63]. With rare exceptions, linker region mutations are expressed as recessive traits as the reduced processivity can be partially corrected by the wild allele. Mutations in the exonuclease region, specifically the fingers subdomain, confer partitioning of the DNA substrate between the polymerase and exonuclease sites, and diminish the fidelity of the polymerase [53]. Mutations in these regions are expressed also as recessive traits, as the transcript of the wild allele can compensate.

Recessive *POLG* mutations when present in the homozygote or compound heterozygote state result in severe adult or juvenile-onset syndromes and the catastrophic early-childhood syndromes of AHS and myocerebrohepatopathy [11, 25]. In general, compound heterozygous mutations usually induce the most severe disease, while homozygous recessive mutations are most often associated with milder phenotypes [16, 43]. However, for unclear reasons the same recessive mutation, either compound heterozygous or homozygous recessive, can be found in the most severe early-onset phenotype as well as in patients with the milder late-onset disease. The reasons for such genotype/phenotype variation within a single syndrome, AHS and between the larger groups of *POLG* syndromes are not completely clear.

The phenotypic variability within AHS can be explained in part by the location of the mutation within the gene. As previously mentioned, homozygote of a known dominant *POLG* mutation or a known dominant *POLG* mutation *in trans* with another recessive mutation have not been described, as these would be expected to be lethal in utero. Therefore, *POLG* nucleotide variants leading to AHS are strictly expressed as recessive mutations. Compound heterozygous-recessive mutations are usually associated with early-onset or childhood disease, while homozygous-recessive mutations cause latter-onset or juvenile disease. Compound heterozygous mutations giving rise to AHS are hypothesized to cause more severe disease because combinations of *POLG* defects in distinct domains/motifs may induce compounding enzyme abnormalities severely compromising pol γ activity [53]. The most common example is seen with the *in trans* compound heterozygous mutations within the linker region, p.A467T/p.W748S, which is known to have decreased survival and increased incidence of liver failure compared with homozygous p.A467T and p.W748S patients [21, 27, 64]. This attractive hypothesis cannot fully explain why these same homozygous-recessive mutations can also give rise to early-onset AHS or the same compound heterozygous mutation can give rise to other distinct *POLG* syndromes with a wide range of disease onset and severity [22, 27]. However, combinations of mutations from distinct regions likely do, in part, determine the AHS phenotype.

Another mechanism for phenotypic variation is silent polymorphisms that are only expressed in complex epigenetic and environmental situations. In these patients, a particular *POLG* genotype would remain silent unless specific conditions are present that can either lead to disease or modify phenotype. The polymorphic nature of *POLG* suggests that many of these ecogenic single nucleotide variants (ESNV) exist [12]. Patterns are beginning to emerge; two ESNVs have been identified within the Northern European population, p.E1143G and p.Q1236H. The p.E1143G has a frequency of approximately 4.5 % and the p.Q1236H approximately 8.6 % of the northern European population, but is almost nonexistent in the Asian, sub-Saharan African or African-American populations (http://ww.ncbi/htm.nih.gov/SNP_ref.cgi?rs=23-7441). The ESNV p.E1134G results in an increased catalytic rate for incoming nucleotide as well as increased intrinsic stability in *in-vitro* studies [63]. When present in cis with certain pathological mutations, especially the p.W748S, p.E1143G can modify the disease phenotype by enhancing enzyme activity [26, 64, 65]. This same ESNV can also increase organ sensitivity to an environmental toxin such as the medication valproic acid. Thus, p.E1143G can be disease modifying either by enhancing enzyme activity or compromising liver function, depending on the situation and environmental exposure. Independently when the p.E1143G or the p.Q1236H ESNV is present, there is a > 20-fold increase of valproic acid toxicity [66]. A significant question arises, whether either ESNV is wholly or partially responsible for the induced liver failure by valproic acid exposure in AHS. We have preliminary data suggesting that p.Q1236H is independent from other AHS mutations that induce liver toxicity; six previously reported patients did not have the p.Q1236H ESNV, yet had liver failure with valproic acid exposure [36, Wong, personal communication]. Full understanding of liver failure due to valproic acid exposure in AHS remains unclear, but clearly there are likely multiple factors involved.

Alterations in other genes that are part of the mtDNA replisome can cause AHS. Mutations in the mtDNA helicase PEO1 (Twinkle) has been reported in two patients [67]. Thus far, other possible genes involved in the AHS phenotype have not been described.

There are some clues that a phenotype/genotype correlation exists in AHS. There is some evidence that distinct combination of mutations may influence phenotype. For example, there are multiple occurrences of specific *in trans* combinations of *POLG* mutations regions found in AHS while other combinations are absent [53]. For instance, the combination of mutations within the linker region and other specific domain/subdomains produce severe early-onset disease. However, this does not explain why the same heterozygous mutations can result in both early and late-onset AHS. In addition, the same mutations can cause a completely distinct *POLG* disease [68]. The finding of an ESNV within *POLG* that can modify phenotype(s) in the Northern European population suggests that other ESNVs in *POLG* may exist as well. The presence of ESNVs in other ethnic groups may help explain the variation in phenotype. More intensive investigations need to be performed to help answer possible phenotype/genotype correlations.

Epidemiology of *POLG* Mutations in Alpers–Huttenlocher Syndrome

The true prevalence of AHS is unknown. Due to the high mortality, epidemiological parameters of incidence and prevalence are not informative. It is estimated that overall, the minimum birth frequency of children who will develop mitochondrial cytopathies is about 1 in 5,000 [69]. Of this population, up to 25 % will develop *POLG* disease [24]. Crude estimations from published reports estimate about 33 % of those patients with recessive *POLG* mutations have AHS, however, true figures are not known. More study is needed, both in trying to find more reliable prevalence figures as well as numbers of patients with *POLG* mutations that develop AHS.

The most common *POLG* mutation causing AHS and associated *POLG*-spectrum disease is the p.A467T mutation. The p.W748S mutation is also frequently found in *POLG* disease and AHS. Studies have shown that the p.A467T and p.W748S mutations arose from a common ancestor, frequencies varying within various populations of European decent [71]. The carrier frequency of p.W748S is estimated to be 1:125 in Finland and the p.A467T mutation as high as 0.6 % in the Belgium and 1 % in Norway populations [41, 42, 70, 71]. However, other founder gene mutations have not been described in AHS in other ethnic groups. Whether a founder effect is exclusive to the Northern European population is not clear, as most of the early mutational research has concentrated on the northern European population. Recent evidence suggests that *POLG* mutations in AHS are found in many other ethnic groups [36, 72, 73]. The multiethnic distribution of AHS allows the possibility of more mutational "hot spots."

There does not seem to be a gender preference in AHS [12, 19, 25, 27]. However, there may be a slight and nonsignificant female predominance in those AHS with onset in the childhood and adolescent years [12].

Conclusion

AHS is a mitochondrial cytopathy caused by mutations in *POLG* that decrease mtDNA replication or fidelity of the transcript. The result of pol γ dysfunction is depletion of the wild mtDNA with evolving organ dysfunction that leads to early demise. This mitochondrial disorder is exclusively an autosomal-recessive disorder caused by homozygous and compound heterozygous mutations. There is no precise phenotype to genotype correlation, although there are clues that location of mutations within specific regions of *POLG* may play a role in the phenotypic expression of the disorder. Most patients will have a normal early clinical history with onset and progression of disease that varies in presentation, severity, sequence of organ involvement, and tempo of progression. However, the constellation of clinical symptoms is progressive dysfunction mainly within the brain, muscle, and liver. Many questions remain, specifically the natural history of this disorder and how genetic and external factors affect the disease course. In addition, treatment is only supportive at present. The fatal progression of this disorder makes AHS a challenging disease to treat and emotionally wrenching disease for families. There is a need to understand the natural history of this disorder to enable clinicians and researcher to intervene and halt the disease's relentless progression. There is a novel possible treatment medication that is making its way through the clinical trial pathways that may offer hope to those patients with AHS syndrome [74]. Since the molecular elucidation of AHS over a decade ago, there has been a large body of information concerning genetic and environmental elements enhancing our understanding of this disorder. Hopefully, through research and astute clinical judgement we will be able to offer effective therapy to patients and their families in the not so distant future.

Summary

Alpers–Huttenlocher syndrome (AHS, Alpers syndrome) is a rare disorder most often associated with mutations in the mtDNA replicase, polymerase gamma 1 (*POLG*; [4, 11]). These mutations function as autosomal-recessive mutations and may occur either as homozygote or compound heterozygote pairs *in trans*. The polymerase gamma (pol γ) protein contains a polymerase domain responsible for the replication process as well as an endonuclease domain responsible for proofreading and repair. Alterations in pol γ activity result in reduced levels of mtDNA or deletions within the mtDNA, which in turn causes impairment of mitochondrial function and ultimately lead to the expression of this disorder. AHS is characterized by the clinical triad; seizures, liver degeneration, and progressive developmental regression [6]. There is a bimodal range of disease onset; children develop normally over the first few months to years. The typical age of disease onset is 2–4 years, with a range of 3 months–8 years; with the second cluster of patients having an age of onset between 17 and 24 years, with a range of 10–35 years. Some children may manifest variable degrees of "developmental delay" before the onset of other clinical symptoms. Seizures usually herald the onset of a progressive neurodegeneration and other organ system failure. The psychomotor regression is generally episodic and stepwise, and usually triggered by intercurrent infection. The seizures vary and can begin as subtle focal events or as a status epilepticus event, but most, if not all, seizures eventually become refractory to medications. If the patient lives long enough, liver dysfunction and ultimately liver failure ensues, resulting in multiple organ system failure. In rare situations, liver failure may precede the seizures or be an early manifestation of the illness. The tempo of disease progression varies from patient to patient, which is only partially explained by the pathogenic effects of a given specific genetic mutation(s) causing the illness. Diagnosis is made by the constellation of organ involvement, not the sequence of symptoms. Unfortunately, the disorder is uniformly fatal.

References

1. Alpers BJ (1931) Diffuse progressive degeneration of the gray matter of the cerebrum. Arch Neurol Psychiatry 25:469–505
2. Bullard WM (1890) Diffuse cortical sclerosis of the brain in children. J Nerv Ment Dis 15:699–709
3. Wefring KW, Lanvik JO (1967) Familial progressive poliodystrophy with cirrhosis of the liver. Acta Paediatr Scand 56:295–300
4. Sandback U, Lerman P (1972) Progressive cerebral poliodystrophy-Alpers' disease: disorganized giant neuronal mitochondria on electron microscopy. J Neruol Neurosurg Psychiatry 35:749–755
5. Huttenlocher PR, Solitare GB, Adams G (1976) Infantile diffuse cerebral degeneration with hepatic cirrhosis. Arch Neurol 33:186–192
6. Harding BN (1990) Progressive neuronal degeneration of childhood with liver disease (Alpers-Huttenlocher syndrome): a personal review. J Child Neurol 5:273–287
7. Eggar J, Harding BN, Boyd SG, Wilson J, Erdohazi M (1987) Progressive neuronal degeneration of childhood (PNDC) with liver disease. Clin Pediatr 26:167–173
8. Ropp PA, Copeland WC (1996) Cloning and characterization of the human mitochondrial DNA polymerase gamma. Genomics 35:449–458
9. Naviaux RK, Nyhan WJ, Barshop BA et al (1999) Mitochondrial DNA polymerase gamma deficiency and mtDNA depletion in a child with Alpers syndrome. Ann Neurol 25:54–58
10. Van Goethem G, Dermaut B, Lofgren A, Martin JJ, Van Broeckhoven C (2001) Mutation of POLG is associated with progressive external ophthalmoplegia characterization by mtDNA deletions. Nat Genet 28:211–212
11. Naviaux RK, Nguyen KV (2004) POLG mutations associated with Alpers syndrome and mitochondrial DNA depletion. Ann Neurol 55:706–712
12. Saneto RP, Naviaux RK (2010) Polymerase gamma disease through the ages. Dev Disabil Res Rev 16:163–174
13. Cohen BH, Naviaux RK (2010) The clinical diagnosis of POLG disease and other mitochondrial DNA depletion disorders. Methods 51:364–373
14. Copeland WC (2008) Inherited mitochondrial disease of DNA replication. Ann Rev Med 59:131–146
15. Cohen BH, Chinnery PF, Copeland WC (2010) GeneReviews at Gene Tests: Medical Genetics Information Resource, Copyright, University of Washington, Seattle. http://www.genetests.org Accessed 05 Feb 2012
16. Graziewicz MA, Longley MJ, Copeland WC (2006) DNA polymerase gamma in mitochondrial DNA replication and repair. Chem Rev 106:383–405

17. Atanassova N, Fuste JM, Wanrooij S et al (2011) Sequence-specific stalling of DNA polymerase (gamma) and the effects of mutation causing progressive ophthalmoplegia. Hum Mol Genet 20:1212–1223
18. Lee Y-S, Kennedy WD, Yin YW (2009) Structural insight into processive human mitochondrial DNA synthesis and disease-related polymerase mutations. Cell 139:312–324
19. Davidzon G, Mancuso M, Ferraris S et al (2005) POLG mutations and Alpers syndrome. Ann Neurol 57:921–923
20. Nguyen KV, Shariel FS, Chan SSL, Copeland WC, Naviaux RK (2006) Molecular diagnosis of Alpers syndrome. J Hepatol 45:108–116
21. Tang S, Wang J, Lee N-C et al (2011) Mitochondrial DNA polymerase γ mutations: an ever expanding molecular and clinical spectrum. J Med Genet 48:669–681
22. Blok MJ, van den Bosch BJ, Jongen E et al (2009) The unfolding clinical spectrum of POLG mutations. J Med Genet 46:776–785
23. Spinazzola A, Zeviani M (2009) Disorders from perturbations of nuclear-mitochondrial intergenomic cross-talk. J Intern Med 265:174–192
24. Chinnery PF, Zeviani M (2008) 135th ENMC workshop: polymerase gamma and disorders of mitochondrial DNA synthesis, 21–23 September 2007. Naarden, The Netherlands. Neuromuscul Disord 18:257–267
25. Wong L-JC, Naviaux RK, Brunetti-Pierri N et al (2008) Molecular and clinical genetics of mitochondrial diseases due to POLG mutations. Hum Mut 29:E150-E172
26. Horvath R, Hudson G, Ferrari G et al (2006) Phenotypic spectrum associated with mutations of the mitochondrial polymerase γ gene. Brain 129:1674–1684
27. Tzoulis C, Engelsen BA, Telstad W et al (2006) The spectrum of clinical disease caused by the A467T and W748S POLG mutations: a study of 26 cases. Brain 129:1685–1692
28. Ferrari G, Lamanea E, Donati A et al (2005) Infantile hepatocerebral syndromes associated with mutations in the mitochondrial DNA polymerase-γ A. Brain 128:723–731
29. Luoma P, Melberg A, Rinne JO et al (2004) Parkinsonism, premature menopause, and mitochondrial DNA polymerase gamma mutations: clinical and molecular genetic study. Lancet 364:875–882
30. Pagnamenta A, Taanman J-W, Wilson CJ et al (2006) Dominant inheritance of premature ovarian failure associated with mutant mitochondrial DNA polymerase gamma. Hum Reprod 21:2467–2473
31. Suomalainen A, Majander P, Wallin M et al (1997) Autosomal dominant progressive external ophthalmoplegia with multiple deletions of mtDNA: clinical, biochemical, and molecular genetic features of the 10q-linked disease. Neurol 48:1244–1253
32. De Vries MC, Rodenburg RJ, Morava E et al (2007) Multiple oxidative phosphorylation deficiencies in severe childhood multi-system disorders due to polymerase gamma (POLG1) mutations. Eur J Pediatr 166:229–234
33. Harding BN (1995) Progressive neurological degeneration of childhood with liver disease (Alpers' disease) presenting in young adults. J Neurol Neurosurg Psych 58:320–325
34. Uusimaa J, Hinttala R, Rantala H et al (2008) Homozygous W748S mutation in the POLG1 gene in patients with juvenile-onset Alpers syndrome. Epilepsia 49:1038–1045
35. Wiltshire E, Davidzon G, DiMauro S et al (2008) Juvenile Alpers disease. Arch Neurol 65:121–124
36. Saneto RP, Lee I-C, Koenig MK et al (2010) POLG DNA testing as an emerging standard of care before instituting valproic acid therapy for pediatric seizure disorders. Seizure 19:140–146
37. Wolf NI, Rahman S, Schmitt B et al (2009) Status epilepticus in children with Alpers' disease caused by POLG1 mutations: EEG and MRI findings. Epilepsia 50:1596–1607
38. Engelsen BA, Tzoulis C, Karlsen B et al (2008) POLG1 mutations cause a syndromic epilepsy with occipital lobe predilection. Brain 131:818–828
39. Tulinius MH, Hagne I (1991) EEG findings in children and adolescents with mitochondrial encephalomyopathies: a study of 25 cases. Brain Dev 13:167–173
40. Gauthier-Villars M, Landrieu P, Cormier-Daire V et al (2001) Respiratory chain deficiency in Alpers syndrome. Neuropediatrics 32:150–152

41. Hakonen AH, Heiskanen S, Juvonen V et al (2005) Mitochondrial DNA polymerase W748S mutation: a common cause of autosomal recessive ataxia with ancient European origin. Am J Hum Genet 77:430–441
42. Winterhun S, Ferrara G, He L et al (2005) Autosomal recessive mitochondrial ataxia syndrome due to mitochondrial polymerase gamma mutation. Neurology 64:1204–1208
43. Nguyen KV, Ostergaard E, Ravin SH et al (2005) POLG mutations in Alpers syndrome. Neurology 65:1493–1495
44. Mancuso M, Filosto M, Bellan M et al (2004) POLG mutations causing ophthalmoplegia, sensorimotor polyneuropathy, at axia, and deafness. Neurology 27:316–318
45. Bicknese AR, May W, Hickey WR, Dodson WE (1992) Early childhood hepatocerebral degeneration misdiagnosed as valproate hepatotoxicity. Ann Neurol 32:767–775
46. Mochel F, Slama A, Touati G et al (2005) Respiratory chain defects may present only with hypoglycemia. J Clin Endocrinol Metab 90:3780–3785
47. Van Goethem G, Luoma P, Rantamaki M et al (2004) POLG mutations in neurodegenerative disorders with ataxia but no muscle involvement. Neurology 63:1251–1257
48. Di Fonzo A, Bordoni A, Crimi M et al (2003) POLG mutations in sporadic mitochondrial disorders with multiple mtDNA deletions. Hum Mutat 22:498–499
49. Giordano C, Sebastiani M, De Giorgio R et al (2008) Gastrointestinal dysmotility in mitochondrial neurogastrointestinal encephalomyopathy is caused by mitochondrial DNA depletion. Am J Pathol 173:1120–1128
50. Bohan TP, Helton E, McDonald I et al (2008) Effect of L-carnitine treatment for valproate-induced hepatotoxicity. Neurology 56:1405–1409
51. McFarland R, Hudson G, Taylor RW et al (2008) Reversible valproate hepatotoxicity due to mutations in mitochondrial DNA polymerase gamma (POLG1). Arch Dis Child 93:151–153
52. Wortmann SB, Rodenburg RJT, Jonckheere A et al (2009) Biochemical and genetic analysis of 3-mtehylglutaconic aciduria type IV: a diagnostic strategy. Brain 132:136–146
53. Euro L, Farnum GA, Palin E, Suomalainen A, Kaguni LS (2011) Clustering of Alpers disease mutations and catalytic defects in biochemical variants reveal new features of molecular mechanism of the human mitochondrial replicase, Pol γ. Nucleic Acids Res 39:9072–9084
54. Taanman J-W, Rahman S, Pagnamenta AT et al (2009) Analysis of mutant DNA polymerase γ in patients with mitochondrial DNA depletion. Hum Mut 30:248–254
55. Dimmock D, Tang L-Y, Schmitt ES, Wong LJ (2010) Quantitative evaluation of the mitochondrial DNA depletion syndrome. Clin Chem 56:1119–1127
56. Kollberg G, Moslemi A-R, Darin N et al (2006) POLG1 mutations associated with progressive encephalopathy in childhood. J Neuropathol Exp Neurol 65:463–469
57. Saneto RP, Friedman S, Shaw DWW (2008) Neuroimaging in mitochondrial disease. Mitochondrion 8:396–413
58. Kaguni LS (2004) DNA polymerase gamma, the mitochondrial replicase. Ann Rev Biochem 73:293–320
59. Longley MJ, Prasad R, Srivastava DK, Wilson SH, Copeland WC (1998) Identification of 5'deoxyribose phosphate lyase activity to human DNA polymerase gamma and its role in mitochondrial base excision repair in vitro. Proc Natl Acad Sci U S A 95:12244–12248
60. Lecrenier N, Van Der Gruggen P, Poury F (1997) Mitochondrial DNA polymerase from yeast to man: a new family of polymerases. Gene 185:147–152
61. Lee Y-S, Molineux IJ, Yin YW (2010) A single mutation in human mitochondrial DNA polymerase pol gamma A affects both polymerization and proofreading activities, but only as a holoenzyme. J Biol Chem 285:28105–28116
62. Chan SSL, Longley MJ, Copeland WC (2005) The common A467T mutation in the human mitochondrial DNA polymerase (POLG) compromises catalytic efficiency and interaction with the accessory subunit. J Biol Chem 36:31341–31346
63. Chan SSL, Longley M, Copeland WC (2006) Modulation of the W748S mutation in DNA polymerase gamma by the E1143G polymorphism in mitochondrial disorders. Hum Mol Genet 15:3473–3483

64. Chan SS, Copeland WC (2009) DNA polymerase gamma and mitochondrial disease: understanding the consequence of POLG mutations. Biochim Biophys Acta 1787:312–319
65. Hisama FM, Mancuso M, Filosto M, DiMauro S (2005) Progressive external ophthalmoplegia: a new family with tremor and peripheral neuropathy. Am J Med Genet A 135:217–219
66. Stewart JD, Horvath R, Baruffini E et al (2010) Polymerase γ gene POLG determines the risk of sodium valproate-induced liver toxicity. Hepatology 52:1791–1796
67. Hakonen AH, Isohanni P, Paetau A, Herva R, Suomalainen A, Lonnqvist T (2007) Recessive Twinkle mutations in early onset encephalopathy with mtDNA depletion. Brain 130:3032–3040
68. Weiss MD, Saneto RP (2010) Sensory ataxic neuropathy with dysarthria and ophthalmoparesis (SANDO) in late life due to compound heterozygous POLG mutation. Muscle Nerve 41:882–885
69. Smeitink JA, Zeviani M, Turnbull DM, Jacobs HT (2006) Mitochondrial medicine: a metabolic perspective on the pathology of oxidative phosphorylation disorders. Cell Metabol 3:9–13
70. Hakonen AH, Davidzon G, Salemi R et al (2007) Abundance of the POLG disease mutations in Europe, Australia, New Zealand, and the United States explained by single ancient European origin. Eur J Hum Genet 15:779–783
71. Van Goethem G, Martin JJ, Dermaut B et al (2003) Recessive POLG mutations presenting with sensory and ataxic neuropathy in compound heterozygote patients with progressive external ophthalmoplegia. Neuromuscul Disord 13:133–142
72. Gonzalez-Vioque E, Bazquez A, Ferandex-Moreira D et al (2006) Association of novel POLG mutations and multiple mitochondrial DNA deletions with variable clinical phenotypes in a Spanish population. Arch Neurol 63:107–111
73. Mohamed K, FathAllah W, Ahmed E (2011) Gender variability in presentation with Alpers' syndrome: a report of eight patients for the UAE. J Inherit Metab Dis 34:439–441
74. Enns GM, Kinsman SL, Perlman SL et al (2011) Initial experience in the treatment of inherited mitochondrial disease with EPI-743. Mol Genet Metab. doi:10.1016/j.ymgme.2011.10.009

Chapter 5
Deoxyguanosine Kinase

David Paul Dimmock

Protein Product

Deoxyguanosine kinase (DGUOK) (EC 2.7.1.113) is the nuclear-encoded mitochondrial enzyme responsible for the recycling of deoxyadenosine and deoxyguanosine into their respective monophosphates, enabling their use for incorporation into the mitochondrial DNA (mtDNA) when the cell is senescent.

$$ATP + deoxyguanosine = ADP + dGMP$$
$$ATP + deoxyadenosine = ADP + dAMP$$

The gene, *DGUOK,* is found on chromosome 2 with five predicted transcript variants. The mature protein is localized to the mitochondria [1] and is ubiquitously expressed with highest levels seen in tissues with a high demand for ATP including the liver, muscle, and brain [2]. The reference nucleotide sequence for the gene is NM_080916.1 (CCDS 1931.1). The 1.3 kb mRNA is located more than 33 kb of genomic DNA and consists of seven coding exons. The protein product is a 277 amino acid protein (NP_550438). This 32 kDa protein forms an active homodimer [1–5]. Reduced enzyme activity causes a reduction in the availability of dAMP and dGMP leading to a failure of mtDNA replication or repair. As the mitochondria depend heavily on the salvage pathway for the supply of deoxynucleotides, DGUOK deficiency results in mtDNA depletion [6]. This can be rescued in vitro by supplementation of the deoxynucleotides [7–9].

D. P. Dimmock (✉)
Department of Pediatrics, Children's Hospital of Wisconsin,
Medical College of Wisconsin, 9000 W. Wisconsin Ave., MS716,
Milwaukee, WI 53226, USA
e-mail: ddimmock@mcw.edu

Regulation and Expression

In mammalian cells, there are two enzymes which can phophorylate deoxyguanosine and deoxyadenosine: deoxycytidine kinase (DCK) which is located in the cytosol and DGUOK which is located in the mitochondria. Although DCK is also ubiquitously expressed throughout the body, it is typically not expressed in senescent cells, but is upregulated in association with DNA replication during cellular division. Consequently, when the cell is in the replicative phase, the cytosolic enzyme deoxcytidine kinase is the main source of deoxypurine phosphorylation [1]. This alternate enzyme enables mitochondrial DNA maintenance in dividing cells such as fibroblasts and hepatocytes with induced or genetic DGUOK deficiency [10, 11]. Conversely, DGUOK is expressed throughout the cell cycle [1, 12] and therefore mtDNA depletion can be detected in cell lines that are induced into senescence [8]. These differences in expression relative to cell cycle have significant implications for the measurement of enzyme activity or functional depletion [1]. The significance of the cross functionality of these two enzymes in human disease is uncertain at this point in time.

In silico analysis of the promoter region of the *DGUOK* gene shows putative common binding sites for Sterol Regulatory Element-Binding Protein-1(SREBP-1), in addition to a peroxisome proliferator-activated receptor (PPARG) site along with three putative cAMP response element binding (CREB), sites and a CCAAT/enhancer-binding protein alpha (C/EBPalpha) site. However, limited data are available to demonstrate in vivo effects of regulation of these genes on *DGUOK* protein expression. Nonetheless, these sites may provide potential avenues for modulation of gene expression.

Significance in Human Health

Role in Drug Metabolism

The most significant clinical implication for deoxyguanosine kinase in human drug therapy is its role in the key metabolic pathway for phosphorylation of several deoxyribonucleosides and certain nucleoside analogs employed as antiviral or chemotherapeutic agents [13–20]. However, there is no clear data that single nucleotide variants in this gene alter metabolism of currently studied or prescribed drugs (PharmGKB).

Dominant Mutations/Carriers

In many nuclear-inherited mitochondrial disorders, individuals with a single mutation will frequently have a distinct later onset disease, for example progressive external

5 Deoxyguanosine Kinase

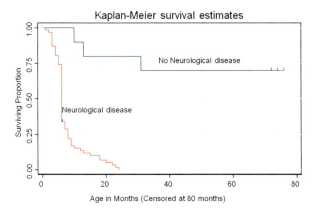

Fig. 5.1 Kaplan–Meier survival graph comparing patients with hypotonia, developmental delay, or nystagmus (neurological disease) and those without them (no neurological disease) censored at 80 months. Statistical analyses were performed with STATA 9 for Windows. Survival analysis was performed by Cox regression analysis with the Breslow method for ties and reanalyzed with a robust method

ophthalmoplegia that is seen in carriers of *POLG* or *TWINKLE (C10orf2)* mutations, whereas liver failure and ataxia is seen in individuals with mutations in both alleles (see Chaps. 1, 3, 4, and 9).

In contrast, close clinical evaluation of several obligate carrier parents (Dimmock, unpublished data) has not revealed overt clinical symptomatology including no external ophthalmoplegia. This is consistent with other enzymatic defects such as TK2 in which there appears to be no overt disease in carriers (Chap. 7). Consequently DGUOK in particular, and likely other enzymatic defects that lead to mtDNA depletion should be considered as distinct from the defects of mtDNA maintenance caused by mutations in the helicase or polymerase and accessory subunit in not confirming a dominant phenotype.

Autosomal-Recessive Deoxyguanosine Kinase Deficiency

Clinical Presentation

Individuals with recessive mutations in DGUOK may present with a spectrum of clinical presentations. However, the long-term prospects for those with liver disease can be divided into two cohorts based on the presence or absence of nytstagmus, severe ypotonia, and delayed milestones (Fig. 5.1).

Most affected infants had lactic acidosis and hypoglycemia in the first week of life. Within weeks of birth, all infants with the severe class of DGUOK deficiency have hepatic disease and neurologic dysfunction. However, such liver failure typically responds to supportive care and may show spontaneous improvement [21] even in children who will subsequently show neurological progression [22].

Table 5.1 Elevations in serum concentrations of several commonly tested chemistry parameters

Test	Range in patients	Normal range
Alanine aminotransferase (ALT)	<50–1000 IU/L	<50 IU/L
Aspartate aminotransferase (AST)	<45–1000 IU/L	<45 IU/L
Gammaglutamyltransferase (GGT) ranging	<51–100 IU/L	<51 IU/L
Conjugated bilirubin	3–12 μmol/L	<2 μmol/L
Alpha fetoprotein (AFP)	Normal-18000	Age dependent
Ferritin	<150–934 ng/ml	<150 ng/ml

Such individuals may be ascertained through newborn screening. Specifically, elevated serum concentrations of tyrosine or phenylalanine are nonspecific abnormalities observed on newborn screening in the majority of infants with the multisystem disease form of DGUOK deficiency diagnosed in the USA in the past 10 years [22] (unpublished data). When transient, elevation of serum concentration of tyrosine may be falsely attributed to transient tyrosinemia of the newborn [23]. However, succinylacetone is not detected.

Severe myopathy, developmental regression, and typical rotary nystagmus developing into opsoclonus are also seen. In particular, nystagmus and opsoclonus are associated with poor long-term outcome. A minority of currently described affected individuals initially presents in infancy or childhood with isolated hepatic disease. Liver disease may present this at birth [21] or following a viral illness. Long-term follow-up suggests that these individuals may subsequently develop mild hypotonia but otherwise have a good outcome without severe neurologic involvement [24].

Cholestasis is prominent early in the clinical course [22]. Liver involvement may cause neonatal- or infantile-onset liver failure. Hepatic dysfunction is progressive to liver failure in the majority of individuals with both forms of DGUOK deficiency. The clinical course is typically dominated by ascites, edema, and hemorrhage due to the decrease of clotting factors. Liver failure is the most common cause of death. However, hepatic disease has undergone reversal with supportive care in one individual with isolated liver disease and currently this individual continues to thrive with supportive care [21, 24]. Significantly, one individual with isolated liver disease subsequently developed hepatocellular carcinoma. An abdominal mass was detected in another person with multisystemic disease prior to death [24]. These findings suggest caution when considering a conservative approach to management of liver failure in such patients.

Individuals with DGUOK deficiency may present with nonspecific abnormalities in a constellation of routine clinical laboratory tests (Table 5.1) [25]. Some individuals with DGUOK deficiency may also have renal involvement manifest as proteinuria and aminoaciduria. No correlation is observed between the presence of this renal tubulopathy and long-term outcome [24].

The majority of affected individuals, described to date have had multisystem illness with hepatic disease (cholestasis) and neurologic dysfunction evident within weeks of birth. Individuals with this severe form of the disease subsequently manifest severe hypotonia, developmental regression, and typical rotary nystagmus that evolve into opsoclonus [22, 24]. However, a subset of individuals with neonatal presentation may show resolution of liver disease with supportive care, or demonstrate long-term survival without neurological defects into teenage years. Consequently, age at presentation does not predict outcome [24].

Similar to other mitochondrial disorders caused by mutations in nuclear-encoded genes [26], individuals with DGUOK deficiency may present in early childhood with liver failure associated with concurrent viral infections [22]. More recently a single individual with two milder mutations in DGUOK has been described with the onset of a myopathic form of the disease in his late teenage years [9]. As this is a new presentation for this disorder, and historically individuals with such myopathic disease have not been tested for DGUOK deficiency, it is unclear how frequent this presentation of DGUOK deficiency is in the population.

Findings frequent in other forms of mtDNA depletion are not observed in DGUOK deficiency. These include seizures; elevated serum CK concentration; or abnormalities on brain imaging [25]. One individual has been described to date with a myopathic form of DGUOK deficiency without elevated CK levels but with reduced left ventricular function [9].

Consistent with other hepatic forms of mtDNA depletion [27], a significant proportion of neonates with DGUOK deficiency have findings that overlap with neonatal hemochromatosis specifically, elevated ferritin and abnormal hepatic imaging, but without accumulation in the salivary glands [22, 28, 29].

Biochemical Testing

Electron Transport Chain (ETC) Activity

Liver typically shows a combined deficiency of ETC complexes I, III, and IV [30–34], but may exhibit isolated deficiency of ETC complex IV [28].

In skeletal muscle, activity of the ETC complexes is normal in the majority of individuals tested [22, 30, 31, 35–37]; however, partially reduced activities of ETC complexes I, III, and IV have been detected in a few cases [24, 32]. Several individuals with DGUOK deficiency have had ETC complex II deficiency [22, 32] suggesting that measurement of ETC activity in muscle is not helpful in distinguishing this nuclear-encoded mtDNA depletion disorder.

Evaluation of ETC activity in brain has seldom been conducted, however in one publication; partial reduction of ETC complexes I, III, and IV activity was detected postmortem [32].

DGUOK Enzyme Activity

DGUOK enzyme activity has been used to establish the pathogenicity of mutations in skin fibroblasts and hepatic tissue [21]; however, pitfalls have been described [38].

More recently several authors have demonstrated that the use of dAMP and dGMP supplementation provides mtDNA rescue to serum starved myoblasts and fibroblasts from individuals with DGUOK deficiency but not those with other nuclear defects [7, 9].

Pathology

In isolated hepatic disease, liver biopsy may be necessary to reduce the number of alternative diagnoses under consideration [39]. At the time of biopsy, liver histology typically reveals microvesicular cholestasis, but may show bridging fibrosis, giant cell hepatitis, or cirrhosis. In several individuals, the histopathologic picture may be more consistent with neonatal hemochromatosis [22, 23, 28, 29, 31, 35]. Liver electron microscopy may reveal an increase in the number of mitochondria and is commonly associated with abnormal cristae. These findings are nonspecific and may be seen in all of the mtDNA depletion syndromes that affect the liver [25, 26].

Skeletal muscle histology is usually normal [28, 35, 36]; however, multisystem disease may be associated with abnormal muscle histology. In one case, increased mitochondrial accumulation in several fibers with subsarcolemmal accumulation of mitochondria, cytochrome *c* oxidase-negative ragged red fibers, and lipid accumulation in ragged red fibers were observed [30].

Neuropathologic examination may be normal or may reveal focal losses of Purkinje cells with Bergmann gliosis [40].

Molecular Testing

Hepatic mtDNA content can be readily tested using real-time quantitative PCR (qPCR) [10] or oligonucleotide array comparative genomic hybridization (oligo aCGH) [23]. To date, hepatic mtDNA depletion has been seen in all individuals with DGUOK deficiency, regardless of severity [23] [unpublished data]. Similarly, while modestly less sensitive or specific, mtDNA depletion has been seen in muscle of all tested patients. Earlier studies suggesting that whole blood mtDNA content may be a useful screen [22], have not been borne out in larger scale studies [10]. Nonetheless, reduction of mtDNA content in affected tissues is not specific for DGUOK deficiency, but is seen in all mtDNA depletion disorders caused by defects in nuclear-encoded genes. The measurement of mtDNA depletion in liver requires an invasive biopsy that may be risky in the face of coagulapothy. Consequently, a definitive diagnosis requires either functional proof of DGUOK enzyme deficiency [9, 21] or the identification of proven deleterious mutations in the DGUOK gene [41–43]. Point mutations are detected by Sanger sequencing in 97 % of disease alleles, with two different intragenic gene deletions accounting for two cases in approximately 120 families diagnosed to date [23, 44]. The ready availability of nuclear gene panels coupled with the cost savings of next generation sequencing will likely lead to such testing being the first line approach for diagnosis of mtDNA depletion syndrome in the near future [27].

Genotype–Phenotype Correlations

To date, all individuals with two null mutations have had multisystem disease and a more severe clinical phenotype [24]. Conversely, no clear genotype–phenotype correlation is evident among individuals with missense mutations. Specifically, affected sibs with the same missense mutations have exhibited divergent long-term outcomes [45]. Similarly, diverse long-term outcomes have been observed in affected individuals from unrelated families harboring the same missense mutations. These findings suggest that in individuals with missense mutations, the genotype and/or family history may not be helpful in predicting long-term outcome or guiding therapeutic decisions [24].

Prevalence

No large population-based studies have evaluated the prevalence of mtDNA depletion in general or DGUOK deficiency specifically. However, review of 10,000 exomes deposited in the NHLBI Grand opportunity Exome Sequencing Project repository demonstrates that two of approximately 5,000 cases have a stop codon in the coding regions captured (both were heterozygous for p.W65X). Significantly, three other known pathogenic missense mutations were seen in 1–2 carriers each in the cohort. The p. Q170R variant was seen in 171 of 10,587 (1.6 %) alleles with four individuals apparently homozygous. Aside from this variant, there are 35 variants predicted to be pathogenic based on Polyphen, Grantham and conservation (GERP++ and Phastcon) scores. These have a combined population frequency suggesting that between 15 per million and approximately 1 per 1,000 individuals would have two recessive missense mutations. Till date, approximately 120 families with DGUOK deficiency have had a confirmed molecular diagnosis. The majority of which have two null mutations [22, 24], suggesting that missense mutations may have a significantly reduced penetrance, individuals may have a different clinical presentation [22], or they may not be recognized as having DGUOK deficiency and consequently not tested.

Nonetheless, when considering liver failure, in one recent study, 50 of 100 children with defects in multiple electron transport chain complexes had mtDNA depletion with an mtDNA copy number less than 35 % of matched normal controls [46]. Nine of those with mtDNA depletion had DGUOK mutations and nine had POLG mutations. Among those with mtDNA depletion and hepatic dysfunction, DGUOK deficiency was the most common single cause [46].

Prognosis

There is a clear prognostic demarcation between multisystem disease in neonates with features of nystagmus, hypotonia, and developmental delay and those with isolated hepatic disease (Fig. 5.1). Affected sibs sharing the same mutations have exhibited multisystemic and isolated hepatic disease with divergent long-term outcomes [45]. Similarly, diverse long-term outcomes have been observed in affected individuals from unrelated families harboring the same mutations [25]. Consequently, it is challenging to provide prognostic information based on molecular testing results, but rather such counseling should be in the light of molecular results and clinical features.

Management

To date, there are no therapeutic interventions that have been proven effective in the setting of rigorous clinical trials to improve the outcome of children with DGUOK deficiency. However, case reports have demonstrated reversal of liver disease parameters following supportive care of liver failure [21, 47]. Consequently, it is recommended that an individual diagnosed with DGUOK deficiency is evaluated by a physician familiar with the care of children with liver failure. Initial testing should include measurement of serum concentrations of AST, ALT, GGT, albumin and prealbumin, and coagulation profile (including PT and PTT). Management of liver disease should be guided by the results of this initial evaluation in consultation with a hepatologist. Given anecdotal evidence of the benefit of supportive care, and data that children with cholestatic liver disease or renal disease are at significant risk of nutritional insufficiency [48]; a dietician with experience in managing children with hepatic and renal failure should be involved in their care. Formulas with an enriched medium-chain-triglyceride content may provide better nutritional support for infants with cholestasis than formulas with predominantly long-chain triglycerides [48]. Cornstarch may reduce symptomatic hypoglycemia in individuals with isolated hepatic disease [22]. Fractional meals and enteral nutrition during the night can result in good nutritional control [47]. Nutritional deficiencies such as essential fatty acid deficiency and fat-soluble vitamin deficiency need to be prevented. Therefore, affected individuals may need to be supplemented with fat-soluble vitamins and essential fatty acids [48].

For children with multisystem illness, orthotopic liver transplantation provides no survival benefit [24]. Several children with isolated hepatic or hepatorenal disease have had excellent 10-year survival with orthotopic liver transplantation and, thus, it is a potential therapeutic option. However, this option warrants careful discussion with parents because at least one child with isolated liver disease prior to transplant, developed neurologic features after liver transplantation [24]. Beyond the issues surrounding hepatic dysfunction, liver transplantation may be indicated in this disease because of the potential for malignant transformation to hepatocellular carcinoma

[22, 24]. The decision regarding orthotopic liver transplantation must be weighed against the known long-term risks of liver transplantation.

AFP is a sensitive, but not specific, marker used to differentiate hepatocellular carcinoma from nonmalignant liver disease. Although the value of a highly elevated serum concentration of AFP in the detection of hepatocellular carcinoma in DGUOK deficiency is not known, the possibility of hepatocellular carcinoma should be considered in individuals with a solid tumor detected by abdominal ultrasound examination and a highly increased serum AFP concentration [35].

Given the risks of renal disease, and its implication for dietary management urine analysis and evaluation for urine amino acids are recommended [24].

Even in long-term survivors, hypotonia may be a significant source of impaired quality-of-life. Therefore, ongoing monitoring of gross motor development and skills with the intervention of appropriate therapies is appropriate in these children [24].

Although it is used in other cholestatic disorders [48], ursodeoxycholic acid has no proven efficacy in this disease. Specifically, one infant with multisystemic disease did not benefit from treatment with this medication [23].

As routine immunizations have not been associated with clinical decompensation in persons with DGUOK deficiency but intercurrent infections have been associated with acute decompensation in this and other mtDNA deletion disorders [22, 26], routine immunizations, including influenza vaccine, are recommended at this time for all individuals with DGUOK deficiency and their household contacts.

Future

The recent development of a rat model of DGUOK deficiency produced by Zinc figure exonuclease technology [49] and the continued study of potential therapies in vitro [7] may provide further insights into disease pathogenicity and enforce the hope of new treatment.

References

1. Gower WR Jr, Carr MC, Ives DH (1979) Deoxyguanosine kinase. Distinct molecular forms in mitochondria and cytosol. J Biol Chem 254(7):2180–2183
2. Wang L, Hellman U, Eriksson S (1996) Cloning and expression of human mitochondrial deoxyguanosine kinase cDNA. FEBS Lett 390(1):39–43
3. Johansson M, Karlsson A (1996) Cloning and expression of human deoxyguanosine kinase cDNA. Proc Natl Acad Sci U S A 93(14): 7258–7262
4. Park I, Ives DH (1988) Properties of a highly purified mitochondrial deoxyguanosine kinase. Arch Biochem Biophys 266(1):51–60
5. Wang L, Karlsson A, Arner ES, Eriksson S (1993) Substrate specificity of mitochondrial 2′-deoxyguanosine kinase. Efficient phosphorylation of 2-chlorodeoxyadenosine. J Biol Chem 268(30):22847–22852

6. Ashley N, Adams S, Slama A, Zeviani M, Suomalainen A, Andreu AL, Naviaux RK, Poulton J (2007) Defects in maintenance of mitochondrial DNA are associated with intramitochondrial nucleotide imbalances. Hum Mol Genet 16(12)1400–1411
7. Bulst S, Abicht A, Holinski-Feder E, Muller-Ziermann S, Koehler U, Thirion C, Walter MC, Stewart JD, Chinnery PF, Lochmuller H, Horvath R (2009) In vitro supplementation with dAMP/dGMP leads to partial restoration of mtDNA levels in mitochondrial depletion syndromes. Hum Mol Genet 18(9):1590–1599
8. Taanman JW, Kateeb I, Muntau AC, Jaksch M, Cohen N, Mandel H (2002) A novel mutation in the deoxyguanosine kinase gene causing depletion of mitochondrial DNA. Ann Neurol 52(2): 237–239
9. Buchaklian A, Helbling D, Ware S, Dimmock D (2012) Recessive deoxyguanosine kinase deficiency causes juvenile onset mitochondrial myopathy. Mol Genet Metab (in press)
10. Dimmock D, Tang LY, Schmitt E, Wong LJ (2010) A quantitative evaluation of the mitochondrial DNA depletion syndrome. Clin Chem 56(7):1119–1127
11. Martinie R, Helbling D, Buchaklian A, Dimmock D (2010) Assessing HepG2 cells as a model for DGUOK deficiency pathology. In: ACMG. Albuquerque, NM
12. Fyrberg A, Mirzaee S, Lotfi K (2006) Cell cycle dependent regulation of deoxycytidine kinase, deoxyguanosine kinase, and cytosolic 5′-nucleotidase I activity in MOLT-4 cells. Nucleosides Nucleotides Nucleic Acids 25(9–11):1201–1204
13. Bennett LL Jr, Allan PW, Arnett G, Shealy YF, Shewach DS, Mason WS, Fourel I, Parker WB (1998) Metabolism in human cells of the D and L enantiomers of the carbocyclic analog of 2′-deoxyguanosine: substrate activity with deoxycytidine kinase, mitochondrial deoxyguanosine kinase, and 5′-nucleotidase. Antimicrob Agents Chemother 42(5):1045–1051
14. Bennett LL Jr, Parker WB, Allan PW, Rose LM, Shealy YF, Secrist JA 3rd, Montgomery JA, Arnett G, Kirkman RL, Shannon WM (1993) Phosphorylation of the enantiomers of the carbocyclic analog of 2′-deoxyguanosine in cells infected with herpes simplex virus type 1 and in uninfected cells. Lack of enantiomeric selectivity with the viral thymidine kinase. Mol Pharmacol 44(6):1258–1266
15. Gaubert G, Gosselin G, Boudou V, Imbach JL, Eriksson S, Maury G (1999) Low enantioselectivities of human deoxycytidine kinase and human deoxyguanosine kinase with respect to 2′-deoxyadenosine, 2′-deoxyguanosine and their analogs. Biochimie 81(11):1041–1047
16. Herrstrom Sjoberg A, Wang L, Eriksson S (2001) Antiviral guanosine analogs as substrates for deoxyguanosine kinase: implications for chemotherapy. Antimicrob Agents Chemother 45(3):739–742
17. Lotfi K, Mansson E, Peterson C, Eriksson S, Albertioni F (2002) Low level of mitochondrial deoxyguanosine kinase is the dominant factor in acquired resistance to 9-beta-D-arabinofuranosylguanine cytotoxicity. Biochem Biophys Res Commun 293(5):1489–1496
18. Rodriguez CO Jr, Mitchell BS, Ayres M, Eriksson S, Gandhi V (2002) Arabinosylguanine is phosphorylated by both cytoplasmic deoxycytidine kinase and mitochondrial deoxyguanosine kinase. Cancer Res 62(11):3100–3105
19. Zhu C, Johansson M, Permert J, Karlsson A (1998) Phosphorylation of anticancer nucleoside analogs by human mitochondrial deoxyguanosine kinase. Biochem Pharmacol 56(8):1035–1040
20. Zhu C, Johansson M, Permert J, Karlsson A (1998) Enhanced cytotoxicity of nucleoside analogs by overexpression of mitochondrial deoxyguanosine kinase in cancer cell lines. J Biol Chem 273(24):14707–14711
21. Mousson de Camaret B, Taanman JW, Padet S, Chassagne M, Mayencon M, Clerc-Renaud P, Mandon G, Zabot MT, Lachaux A, Bozon D (2007) Kinetic properties of mutant deoxyguanosine kinase in a case of reversible hepatic mtDNA depletion. Biochem J 402(2):377–385
22. Dimmock DP, Zhang Q, Dionisi-Vici C, Carrozzo R, Shieh J, Tang LY, Truong C, Schmitt E, Sifry-Platt M, Lucioli S, Santorelli FM, Ficicioglu CH, Rodriguez M, Wierenga K, Enns GM, Longo N, Lipson MH, Vallance H, Craigen WJ, Scaglia F, Wong LJ (2008) Clinical and molecular features of mitochondrial DNA depletion due to mutations in deoxyguanosine kinase. Hum Mutat 29(2):330–331

23. Lee NC, Dimmock D, Hwu WL, Tang LY, Huang WC, Chinault AC, Wong LJ (2009) Simultaneous detection of mitochondrial DNA depletion and single-exon deletion in the deoxyguanosine gene using array-based comparative genomic hybridisation. Arch Dis Child 94(1):55–58
24. Dimmock DP, Dunn JK, Feigenbaum A, Rupar A, Horvath R, Freisinger P, Mousson de Camaret B, Wong LJ, Scaglia F (2008) Abnormal neurological features predict poor survival and should preclude liver transplantation in patients with deoxyguanosine kinase deficiency. Liver Transpl 14(10):1480–1485
25. Scaglia F, Dimmock D, Wong L-J (2009) DGUOK-related mitochondrial DNA depletion syndrome, hepatocerebral form. In: GeneReviews at GeneTests: Medical Genetics Information Resource (database online). Copyright, University of Washington, Seattle. 1997–2010. http://www.genetests.org. Accessed 9 Jan 2010
26. Lutz RE, Dimmock D, Schmitt E, Zhang Q, Tang L-Y, Reyes C, Truemper E, McComb RD, Hernandez A, Basinger A, Wong LJ (2009) De novo mutations in POLG presenting with acute liver failure or encephalopathy. J Pediatr Gastroenterol Nutr 49(1):126–129
27. Goh V, Helbling D, Biank V, Jarzembowski J, Dimmock D (2012) Next generation sequencing facilitates the diagnosis in a child with Twinkle mutations causing cholestatic liver failure. J Pediatr Gastroenterol Nutr 54(2):291–294
28. Labarthe F, Dobbelaere D, Devisme L, De Muret A, Jardel C, Taanman JW, Gottrand F, Lombes A (2005) Clinical, biochemical and morphological features of hepatocerebral syndrome with mitochondrial DNA depletion due to deoxyguanosine kinase deficiency. J Hepatol 43(2):333–341
29. Hanchard NA, Shchelochkov OA, Roy A, Wiszniewska J, Wang J, Popek EJ, Karpen S, Wong LJ, Scaglia F (2011) Deoxyguanosine kinase deficiency presenting as neonatal hemochromatosis. Mol Genet Metab 103(3):262–267
30. Mancuso M, Ferraris S, Pancrudo J, Feigenbaum A, Raiman J, Christodoulou J, Thorburn DR, DiMauro S (2005) New DGK gene mutations in the hepatocerebral form of mitochondrial DNA depletion syndrome. Arch Neurol 62(5):745–747
31. Mandel H, Hartman C, Berkowitz D, Elpeleg ON, Manov I, Iancu TC (2001) The hepatic mitochondrial DNA depletion syndrome: ultrastructural changes in liver biopsies. Hepatology 34(4 Pt1):776–784
32. Salviati L, Sacconi S, Mancuso M, Otaegui D, Camano P, Marina A, Rabinowitz S, Shiffman R, Thompson K, Wilson CM, Feigenbaum A, Naini AB, Hirano M, Bonilla E, DiMauro S, Vu TH (2002) Mitochondrial DNA depletion and dGK gene mutations. Ann Neurol 52(3):311–317
33. Slama A, Giurgea I, Debrey D, Bridoux D, de Lonlay P, Levy P, Chretien D, Brivet M, Legrand A, Rustin P, Munnich A, Rotig A (2005) Deoxyguanosine kinase mutations and combined deficiencies of the mitochondrial respiratory chain in patients with hepatic involvement. Mol Genet Metab 86(4):462–465
34. Wang L, Limongelli A, Vila MR, Carrara F, Zeviani M, Eriksson S (2005) Molecular insight into mitochondrial DNA depletion syndrome in two patients with novel mutations in the deoxyguanosine kinase and thymidine kinase 2 genes. Mol Genet Metab 84(1):75–82
35. Freisinger P, Futterer N, Lankes E, Gempel K, Berger TM, Spalinger J, Hoerbe A, Schwantes C, Lindner M, Santer R, Burdelski M, Schaefer H, Setzer B, Walker UA, Horvath R (2006) Hepatocerebral mitochondrial DNA depletion syndrome caused by deoxyguanosine kinase (DGUOK) mutations. Arch Neurol 63(8): 1129–1134
36. Mandel H, Szargel R, Labay V, Elpeleg O, Saada A, Shalata A, Anbinder Y, Berkowitz D, Hartman C, Barak M, Eriksson S, Cohen N (2001) The deoxyguanosine kinase gene is mutated in individuals with depleted hepatocerebral mitochondrial DNA. Nat Genet 29(3):337–341
37. Taanman JW, Muddle JR, Muntau AC (2003) Mitochondrial DNA depletion can be prevented by dGMP and dAMP supplementation in a resting culture of deoxyguanosine kinase-deficient fibroblasts. Hum Mol Genet 12(15):1839–1845
38. Arner ES, Spasokoukotskaja T, Eriksson S (1992) Selective assays for thymidine kinase 1 and 2 and deoxycytidine kinase and their activities in extracts from human cells and tissues. Biochem Biophys Res Commun 188(2):712–718

39. Suchy FJ (2007) Approach to the infant with cholestasis. In: Sokol RJ, Suchy FJ, Balistreri WF (eds) Liver disease in children. Cambridge University Press, New York, pp 190–231
40. Filosto M, Mancuso M, Tomelleri G, Rizzuto N, Dalla Bernardina B, DiMauro S, Simonati A (2004) Hepato-cerebral syndrome: genetic and pathological studies in an infant with a dGK mutation. Acta Neuropathol (Berl) 108(2):168–171
41. Ji JQ, Dimmock D, Tang LY, Descartes M, Gomez R, Rutledge SL, Schmitt ES, Wong LJ (2010) A novel c.592-4_c.592-3delTT mutation in DGUOK gene causes exon skipping. Mitochondrion 10(2):188–191
42. Maddalena A, Bale S, Das S, Grody W, Richards S (2005) Technical standards and guidelines: molecular genetic testing for ultra-rare disorders. Genet Med 7(8):571–583
43. Richards CS, Bale S, Bellissimo DB, Das S, Grody WW, Hegde MR, Lyon E, Ward BE (2008) ACMG recommendations for standards for interpretation and reporting of sequence variations: revisions 2007. Genet Med 10(4):294–300
44. Mudd SH, Wagner C, Luka Z, Stabler SP, Allen RH, Schroer R, Wood T, Wang J, Wong LJ (2012) Two patients with hepatic mtDNA depletion syndromes and marked elevations of S-adenosylmethionine and methionine. Mol Genet Metab 105(2):228–236
45. Tadiboyina VT, Rupar A, Atkison P, Feigenbaum A, Kronick J, Wang J, Hegele RA (2005) Novel mutation in DGUOK in hepatocerebral mitochondrial DNA depletion syndrome associated with cystathioninuria. Am J Med Genet A 135(3):289–291
46. Sarzi E, Bourdon A, Chretien D, Zarhrate M, Corcos J, Slama A, Cormier-Daire V, de Lonlay P, Munnich A, Rotig A (2007) Mitochondrial DNA depletion is a prevalent cause of multiple respiratory chain deficiency in childhood. J Pediatr 150(5):531–534, 534 e1–534 e6
47. Ducluzeau PH, Lachaux A, Bouvier R, Duborjal H, Stepien G, Bozon D, Mousson de Camaret B (2002) Progressive reversion of clinical and molecular phenotype in a child with liver mitochondrial DNA depletion. J Hepatol 36(5):698–703
48. Feranchak AP, Sokol RJ (2007) Medical and nutrional managment of cholestasis in infants and children. In: Sokol RJ, Suchy FJ, Balistreri WF (eds) Liver disease in children. Cambridge University Press, New York, pp 190–231
49. Geurts AM, Cost GJ, Freyvert Y, Zeitler B, Miller JC, Choi VM, Jenkins SS, Wood A, Cui X, Meng X, Vincent A, Lam S, Michalkiewicz M, Schilling R, Foeckler J, Kalloway S, Weiler H, Menoret S, Anegon I, Davis GD, Zhang L, Rebar EJ, Gregory PD, Urnov FD, Jacob HJ, Buelow R (2009) Knockout rats via embryo microinjection of zinc-finger nucleases. Science 325(5939):433

Chapter 6
MPV17-Associated Hepatocerebral Mitochondrial DNA Depletion Syndrome

Ayman W. El-Hattab

Introduction

Mitochondrial diseases are a clinically heterogeneous group of disorders caused by defects in mitochondrial DNA (mtDNA) or nuclear genes involved in mitochondrial biogenesis and function [1]. Defects in nuclear genes involved in the maintenance of mtDNA integrity can be associated with large-scale rearrangements of mtDNA (mtDNA multiple deletion syndromes) or with reduction in the mtDNA content (mtDNA depletion syndromes, MDS). MDS are autosomal-recessive disorders that primarily affect infants or children and are characterized by a severe decrease of mtDNA content leading to organ dysfunction due to insufficient amount of respiratory chain components [2]. MDS are phenotypically heterogeneous and may affect either a specific organ or a combination of organs, including muscle, liver, brain, and kidney. Clinically, MDS are classified as one of three forms: a myopathic form (OMIM # 609560) associated with mutations in the thymidine kinase 2 (*TK2*) gene or the p53-induced ribonucleotide reductase B subunit (*RRM2B*) gene [3–5], an encephalomyopathic form (OMIM # 612073) associated with mutations in the ATP-dependant succinyl CoA synthase gene (*SUCLA2*) or the GTP-dependant succinyl CoA synthase gene (*SUCLG1*) [6, 7], and a hepatocerebral form (OMIM # 251880) associated with mutations in the Twinkle (*PEO1*) gene, the polymerase gamma (*POLG1*) gene, the deoxyguanosine kinase (*DGUOK*) gene [8–10], or more recently the *MPV17* gene [11–19].

Mutations in the *MPV17* gene were initially identified in infants who presented in the first year of life with liver failure, failure-to-thrive and neurological symptoms [11]. Subsequently, a homozygous p.R50Q mutation was identified in subjects with Navajo neurohepatopathy, an autosomal-recessive multisystem disorder prevalent in the Native American Navajo population that manifests as liver disease, sensory and motor neuropathy, corneal anesthesia and scarring, cerebral leukoencephalopathy,

A. W. El-Hattab (✉)
Medical Genetics Section, Department of Pediatrics, The Children's Hospital at
King Fahad Medical City and King Saud bin Abdulaziz University for Health Science,
P. O. Box 59046, Riyadh 11525, Kingdom of Saudi Arabia
e-mail: elhattabaw@yahoo.com

Table 6.1 Clinical manifestations of *MPV17*-associated hepatocerebral mitochondrial DNA depletion syndrome

Clinical manifestations	
Hepatic	*31/31 (100 %)*
Liver dysfunction	31/31 (100 %)
Liver failure	28/31 (90 %)
Hepatomegaly	14/29 (48 %)
Cirrhosis	7/29 (24 %)
Hepatocellular cancer	2/31 (6 %)
Neurologic	*26/29 (90 %)*
Developmental delay	24/29 (83 %)
Hypotonia, muscle weakness	19/29 (66 %)
White matter abnormalities in brain MRI	10/29 (34 %)
Peripheral neuropathy	8/29 (28 %)
Seizures	3/29 (10 %)
Microcephaly	2/29 (7 %)
Ataxia	2/29 (7 %)
Failure to thrive	*26/29 (90 %)*
Metabolic	*23/27 (85 %)*
Lactic acidosis	20/27 (74 %)
Hypoglycemia	13/27 (48 %)
Other manifestations	
Tubulopathy	3/29 (10 %)
Gastroesophageal reflux	3/29 (10 %)
Vomiting, diarrhea, or gastrointestinal dysmotility	3/29 (10 %)
Hypoparathyroidism	2/29 (7 %)
Corneal anesthesia and ulcers	2/29 (7 %)

failure-to-thrive, and metabolic acidosis [13]. Therefore, *MPV17*-associated hepatocerebral MDS syndrome can present with a spectrum of combination of hepatic, neurologic, and metabolic manifestations. Till date, molecularly confirmed *MPV17*-associated hepatocerebral MDS has been reported in 31 individuals with 20 different mutations [11–19].

Clinical Manifestations

The clinical manifestations of *MPV17*-associated hepatocerebral MDS are summarized in Table 6.1 and based on the 31 individuals reported in literature with molecularly confirmed diagnosis [11–19].

All affected individuals came to medical attention in infancy with liver dysfunction with elevated transaminases, jaundice, and/or cholestasis. Approximately half of those individuals were reported to have hepatomegaly. Liver disease progressed to liver failure in 90 % of affected individuals. About a quarter were reported with liver cirrhosis and two individuals developed hepatocellular carcinoma [13, 18].

The vast majority of the affected individuals exhibited neurological manifestations, including developmental delay ($\sim$ 85 %), hypotonia and muscle weakness

(~ 65 %), and motor and sensory peripheral neuropathy (~ 30 %). Brain MRI abnormalities were reported in about a third of affected individuals with white matter abnormalities (leukodystrophy) being the most common. Brainstem lesions have been reported in two individuals. Other less frequent neurological manifestations include seizures, ataxia, dystonia, microcephaly, cerenrovascular infarction, and subdural hematoma.

Failure-to-thrive is one of the common manifestations, although some patients have normal growth, especially early in the course of the disease. Metabolic derangements occur in the vast majority with lactic acidosis occurring in ~ 75 % and hypoglycemia in ~ 50 %.

Less frequent manifestations include renal tubulopathy, hypoparathyroidism, and gastrointestinal dysmotility manifested as gastroesophageal reflux, cyclic vomiting, or diarrhea. Corneal anesthesia and ulcers were reported in individuals homozygous for p.R50Q.

Diagnosis

The diagnosis is suspected clinically in infants with a combination of neurological, hepatic, and metabolic manifestations. In *MPV17*-associated hepatocerebral MDS, liver involvement appears early in the course of the disease, in contrast to the other multisystem mitochondrial disorders with prominent neuromuscular involvement in which liver complications are more commonly a late feature. On the other hand, unlike other MDS, neurological involvement in *MPV17*-associated hepatocerebral MDS is generally mild at the onset of disease [17]. Therefore, *MPV17* sequencing should be considered in all infants with liver dysfunction associated with mtDNA depletion even in the absence of neurological manifestations. As a part of unexplained infantile liver dysfunction workup, if a liver biopsy is performed for pathological examination, mtDNA content in liver DNA should be measured. The identification of severely reduced mtDNA copy number should prompt further evaluation that includes sequencing analysis of *MPV17*.

Mitochondrial DNA Content (mtDNA Copy Number)

Consistent with the predominant liver involvement in *MPV17*-associated hepatocerebral MDS, mtDNA copy number is severely reduced in liver compared to muscle, with mtDNA content being 3–30 % and 8–76 % of tissue and age-matched control in liver and muscle, respectively. However, mtDNA copy number values in blood are not a reliable indicator of mtDNA depletion [18, 20].

Electron Transport Chain (ETC) Activity

ETC activity assays can show decreased activity of different complexes with complex I or I+III being the most affected components. Similar to the findings of mtDNA content, the liver ETC activities are typically more reduced than that of muscle tissue [18].

MPV17 Gene Sequencing

Sequencing of all coding exons and the flanking intronic regions of the *MPV17* gene has identified homozygosity or compound heterozygosity in the vast majority of affected individuals [18].

Management and Prognosis

The treatment of liver disease in mitochondrial hepatopathies remains unsatisfactory. Present treatment involves the use of various vitamins and cofactors, none of which have been proven to be effective. Liver transplantation in mitochondrial hepatopathy is controversial, largely because of the multisystemic involvement of this disorder [14]. Liver disease typically progressed to liver failure in *MPV17*-associated hepatocerebral MDS. Liver transplant remains the only option as liver failure develops. Liver transplant was performed in about one-third of affected individuals; however, the outcome was not very satisfactory with half of the transplanted individuals died posttransplantation because of multiorgan failure and/or sepsis. Approximately half of the reported affected individuals did not receive liver transplant and died because of progressive liver failure. The majority of death occurred during infancy. Only four affected individuals were reported to survive without liver transplant with the oldest being 13 years. Two out of those four are p.R50Q homozygous including the 13-year-old affected individual indicating a better prognosis associated with p.R50Q homozygosity as discussed later. Two affected individuals developed hepatocellular carcinoma (Table 6.2) [11–19].

Prevalence

Mitochondrial respiratory chain disorders of all types are estimated to affect approximately one in 20,000 children under 16 years of age [21]. About 10 % of children with respiratory chain defects have liver dysfunction, with the onset of liver disease in the neonatal period in about half of cases [22]. Given the clinical heterogeneity and the difficulties with establishing the diagnosis, the prevalence of primary

Table 6.2 Outcome of individuals affected with *MPV17*-associated hepatocerebral mitochondrial DNA depletion syndrome

Outcome	
Death without liver transplant	17/31 (55 %)
Death in infancy	14/31 (45 %)
Death beyond infancy	3/31 (10 %)
Alive without liver transplant	4/31 (13 %)
Underwent liver transplant	10/31 (32 %)
Death after liver transplant	5/31 (16 %)
Survived after liver transplant	5/31 (16 %)

mitochondrial hepatopathies is likely to be underestimated [23]. The prevalence of *MPV17*-associated hepatocerebral MDS is unknown, but likely to be rare with only 31 cases being reported to date [11–19].

Molecular Aspects

The *MPV17* gene maps to chromosome 2p23.3, spans 13.6 kb, and contains eight exons. The MPV17 protein, which is an inner mitochondrial membrane protein, is predicted to contain four transmembrane (TM) segments (http://pir.uniprot.org). A putative protein kinase C phosphorylation site is predicted to be in the region between TM2 and TM3 [17]. The function of the MPV17 protein and its role in the pathogenesis of MDS are still unknown. However, it has been suggested that it plays a role in controlling mtDNA maintenance and oxidative phosphorylation activity in mammals and yeast [11]. The *sym1* gene, the ortholog of *MPV17* gene in *Saccharomyces cerevisiae*, has a suggested role in the cellular response to metabolic stress and in maintaining oxidative phosphorylation, mitochondrial morphology, and mtDNA integrity and stability [24, 25]. *Mpv17*-deficient mice have been reported to develop focal segmental glomerulosclerosis with massive proteinuria and renal failure, age-dependent hearing loss, and early graying of their coat. MtDNA depletion was demonstrated in renal and liver tissues. In spite of severe mtDNA depletion, only a moderate decrease in respiratory chain enzymatic activities and mild cytoarchitectural alterations were observed in liver; neither cirrhosis nor liver failure occurred at any age [26–29].

Till date, 20 different mutations have been reported in *MPV17* gene in individuals with hepatocerebral MDS [11–19]. About half of those mutations are missense and the remaining half includes nonsense, small deletions and insertions, and splice site mutations. A large deletion spanning exon 8 has also been reported. The most common mutation is p.R50Q which has been identified in seven families. The mutations p.R50W, p.K88del, p.L91del, and 1.57 kb deletion spanning the last exon, each has been observed in two families. All other mutations are private mutations. The p.R50Q (c.149G>A) mutation occurs at a CpG dinucleotides and has been observed in both Navajo and Italian patients. It has been estimated that up to 25 % of all disease-causing human mutations occur at CpG sites [30]. This fact may explain the observation that p.R50Q is the most common mutation identified in the *MPV17* gene. However, the

Table 6.3 *MPV17* gene mutations

MPV17 mutations	Frequency	
	Homozygous	Heterozygous
Missense		
c.149G>A (p.R50Q)	10[a]	0
c.148C>T (p.R50W)	1	1
c.498C>A (p.N166K)	1	0
c.70G>T (p.G24W)	0	1
c.509C>T (p.S170F)	0	2[b]
c.262A>G (p.K88E)	1	0
c.280G>C (p.G94R)	0	1
c.293C>T (p.P98L)	0	1
c.485C>A (p.A162D)	0	1
Nonsense		
c.359G>A (p.W120X)	2[b]	0
c.206G>A (p.W69X)	2[b]	0
In-frame deletion		
c.263_265del3 (p.K88del)	0	2
c.271_273del3 (p.L91del)	0	2
c.234_242del9 (p.G79_T81del)	0	1
Frame shift deletions		
c.116-141del25	0	1
Splicing site		
c.70+5G>A	1	0
c.186+2T>C	0	1
Insertion		
c.451insC	0	2[b]
c.22_23insC	0	1
Large deletion		
1.57 kb deletion-spanning exon 8	1	1
Total number of affected individuals	19	12[c]

[a] Seven unrelated families
[b] Members of the same family
[c] Only one mutation was identified in two affected individuals [18] and the mutations were not specified in two other reported siblings [19]

infrequent missense mutations c.148C>T (p.R50W) and c.293C>T (p.P98L) are also located at CpG dinucleotides (Table 6.3). There is clustering of mutations in the region of the putative protein kinase C phosphorylation site. Whether this reflects some important functional aspect for this domain of the protein or simply sites of common mutations remains unknown [18]. The p.R50Q mutation is located between TM1 and TM2. Since homozygosity for p.R50Q causes a less severe phenotype, this suggests that that this substitution better preserves function or that this region is less vital than other regions of the protein (Fig. 6.1) [18].

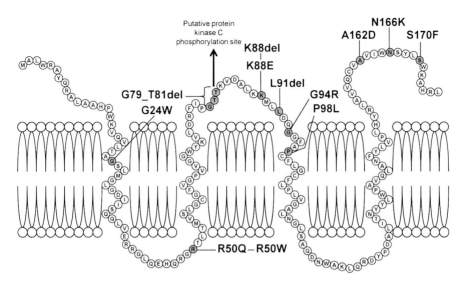

Fig. 6.1 The MPV17 protein, composed of 176 amino acids, is predicted to contain four transmembrane (TM) spans: TM1 from amino acid 18–38, TM2 from 53–73, TM3 from 94–114, and TM4 from 131–151, with short flanking hydrophilic intermembrane and matrix regions (http://pir.uniprot.org). A putative protein kinase C phosphorylation site predicted to be in the region between TM2 and TM3 [17]. The missense mutations and in-frame deletions are demonstrated

Genotype–Phenotype Correlation

Till date, the p.R50Q mutation has only been reported in the homozygous state. In contrast to most *MPV17* gene mutations that are associated with death in infancy or early childhood, the p.R50Q mutation is associated with longer survival, suggesting that this is a hypomorphic mutation. Homozygous p.R50Q mutation has been reported in ten individuals half of whom died usually in their second decade of life. However, liver failure and death during infancy has been reported in an infant homozygous for p.R50Q whose mtDNA copy number was very low. On the other hand, carriers for homozygous p.R50Q who survived longer have a relatively higher mtDNA copy number, suggesting that the degree of mtDNA depletion correlates with the outcome for individuals with p.R50Q mutations. The other *MPV17* mutations have been associated with a lower mtDNA copy number and death in infancy or early childhood if not treated by liver transplantation [18].

Liver transplant was performed in ten affected individuals with only five survived. Out of the five individuals who survived posttransplantation, three carry the homozygous p.R50Q mutation indicating a relatively better outcome in individuals homozygous for p.R50Q posttransplant [18].

Conclusion

MPV17-associated hepatocerebral MDS presents in infancy with liver dysfunction that progresses to liver failure in most of the affected individuals. Cholestasis, hepatomegaly, liver cirrhosis, and hepatocellular carcinoma can develop. The vast majority of the affected individuals exhibited neurological manifestations, including developmental delay, hypotonia, muscle weakness, motor and sensory peripheral neuropathy, and leukodystrophy. Other common manifestations include failure-to-thrive, lactic acidosis, and hypoglycemia. The diagnosis is based on clinical presentation, demonstration of liver mtDNA depletion, and identification of *MPV17* gene mutations. The prognosis is unfavorable with half of affected individuals dying in infancy. Liver transplant remains the only option with about half of the transplanted individuals not surviving posttransplantation. About half of the *MPV17* mutations are missense and there is clustering of mutations in the region of the putative protein kinase C phosphorylation site. The p.R50Q mutation, which occurs in a CpG dinucleotide, is the most common *MPV17* mutation and has only been found in the homozygous state. In contrast to most *MPV17* gene mutations that are associated with death in infancy or early childhood, the p.R50Q mutation is associated with longer survival and better posttransplant outcomes.

References

1. Wallace DC (1999) Mitochondrial diseases in man and mouse. Science 283:1482–1488
2. Spinazzola A, Invernizzi F, Carrara F, Lamantea E, Donati A, Dirocco M, Giordano I, Meznaric-Petrusa M, Baruffini E, Ferrero I, Zeviani M (2009) Clinical and molecular features of mitochondrial DNA depletion syndromes. J Inherit Metab Dis 32:143–158
3. Saada A, Shaag A, Mandel H, Nevo Y, Eriksson S, Elpeleg O (2001) Mutant mitochondrial thymidine kinase in mitochondrial DNA depletion myopathy. Nat Genet 29:342–344
4. Bourdon A, Minai L, Serre V, Jais JP, Sarzi E, Aubert S, Chrétien D, de Lonlay P, Paquis-Flucklinger V, Arakawa H, Nakamura Y, Munnich A, Rötig A (2007) Mutation of RRM2B, encoding p53-controlled ribonucleotide reductase (p53R2), causes severe mitochondrial DNA depletion. Nat Genet 39:776–780
5. Bornstein B, Area E, Flanigan KM, Ganesh J, Jayakar P, Swoboda KJ, Coku J, Naini A, Shanske S, Tanji K, Hirano M, DiMauro S (2008) Mitochondrial DNA depletion syndrome due to mutations in the RRM2B gene. Neuromuscul Disord 18:453–459
6. Elpeleg O, Miller C, Hershkovitz E, Bitner-Glindzicz M, Bondi-Rubinstein G, Rahman S, Pagnamenta A, Eshhar S, Saada A (2005) Deficiency of the ADP-forming succinyl-CoA synthase activity is associated with encephalomyopathy and mitochondrial DNA depletion. Am J Hum Genet 76:1081–1086
7. Ostergaard E, Christensen E, Kristensen E, Mogensen B, Duno M, Shoubridge EA, Wibrand F (2007) Deficiency of the alpha subunit of succinate-coenzyme A ligase causes fatal infantile lactic acidosis with mitochondrial DNA depletion. Am J Hum Genet 81:383–387
8. Hakonen AH, Isohanni P, Paetau A, Herva R, Suomalainen A, Lönnqvist T (2007) Recessive Twinkle mutations in early onset encephalopathy with mtDNA depletion. Brain 130:3032–3040
9. Ferrari G, Lamantea E, Donati A, Filosto M, Briem E, Carrara F, Parini R, Simonati A, Santer R, Zeviani M (2005) Infantile hepatocerebral syndromes associated with mutations in the mitochondrial DNA polymerase-gammaA. Brain 128:723–731

10. Mandel H, Szargel R, Labay V, Elpeleg O, Saada A, Shalata A, Anbinder Y, Berkowitz D, Hartman C, Barak M, Eriksson S, Cohen N (2001) The deoxyguanosine kinase gene is mutated in individuals with depleted hepatocerebral mitochondrial DNA. Nat Genet 29:337–341
11. Spinazzola A, Viscomi C, Fernandez-Vizarra E, Carrara F, D'Adamo P, Calvo S, Marsano RM, Donnini C, Weiher H, Strisciuglio P, Parini R, Sarzi E, Chan A, DiMauro S, Rötig A, Gasparini P, Ferrero I, Mootha VK, Tiranti V, Zeviani M (2006) MPV17 encodes an inner mitochondrial membrane protein and is mutated in infantile hepatic mitochondrial DNA depletion. Nat Genet 38:570–555
12. Parini R, Furlan F, Notarangelo L, Spinazzola A, Uziel G, Strisciuglio P, Concolino D, Corbetta C, Nebbia G, Menni F, Rossi G, Maggioni M, Zeviani M (2009) Glucose metabolism and diet-based prevention of liver dysfunction in MPV17 mutant patients. J Hepatol 50:215–221
13. Karadimas CL, Vu TH, Holve SA, Chronopoulou P, Quinzii C, Johnsen SD, Kurth J, Eggers E, Palenzuela L, Tanji K, Bonilla E, De Vivo DC, DiMauro S, Hirano M (2006) Navajo neurohepatopathy is caused by a mutation in the MPV17 gene. Am J Hum Genet 79:544–548
14. Spinazzola A, Santer R, Akman OH, Tsiakas K, Schaefer H, Ding X, Karadimas CL, Shanske S, Ganesh J, Di Mauro S, Zeviani M (2008) Hepatocerebral form of mitochondrial DNA depletion syndrome: novel MPV17 mutations. Arch Neurol 65:1108–1113
15. Navarro-Sastre A, Martín-Hernández E, Campos Y, Quintana E, Medina E, de Las Heras RS, Lluch M, Muñoz A, del Hoyo P, Martín R, Gort L, Briones P, Ribes A (2008) Lethal hepatopathy and leukodystrophy caused by a novel mutation in MPV17 gene: description of an alternative MPV17 spliced form. Mol Genet Metab 94:234–239
16. Kaji S, Murayama K, Nagata I, Nagasaka H, Takayanagi M, Ohtake A, Iwasa H, Nishiyama M, Okazaki Y, Harashima H, Eitoku T, Yamamoto M, Matsushita H, Kitamoto K, Sakata S, Katayama T, Sugimoto S, Fujimoto Y, Murakami J, Kanzaki S, Shiraki K (2009) Fluctuating liver functions in siblings with MPV17 mutations and possible improvement associated with dietary and pharmaceutical treatments targeting respiratory chain complex II. Mol Genet Metab 97:292–296
17. Wong LJ, Brunetti-Pierri N, Zhang Q, Yazigi N, Bove KE, Dahms BB, Puchowicz MA, Gonzalez-Gomez I, Schmitt ES, Truong CK, Hoppel CL, Chou PC, Wang J, Baldwin EE, Adams D, Leslie N, Boles RG, Kerr DS, Craigen WJ (2007) Mutations in the MPV17 gene are responsible for rapidly progressive liver failure in infancy. Hepatology 46:1218–1227
18. El-Hattab AW, Li FY, Schmitt E, Zhang S, Craigen WJ, Wong LJ (2010) MPV17-associated hepatocerebral mitochondrial DNA depletion syndrome: new patients and novel mutations. Mol Genet Metab 99:300–308
19. Merkle AN, Nascene DR, McKinney AM (2012) MR imaging findings in the reticular formation in siblings with MPV17-related mitochondrial depletion syndrome. AJNR Am J Neuroradiol 33(3):E34–5
20. Dimmock D, Tang LY, Schmitt ES, Wong LJ (2010) Quantitative evaluation of the mitochondrial DNA depletion syndrome. Clin Chem 56:1119–1127
21. Skladal D, Halliday J, Thorburn DR (2003) Minimum birth prevalence of mitochondrial respiratory chain disorders in children. Brain 126:1905–1912
22. Cormier-Daire V, Chretien D, Rustin P, Rötig A, Dubuisson C, Jacquemin E, Hadchouel M, Bernard O, Munnich A (1997) Neonatal and delayed-onset liver involvement in disorders of oxidative phosphorylation. J Pediatr 130:817–822
23. Lee WS, Sokol RJ (2007) Mitochondrial hepatopathies: advances in genetics and pathogenesis. Hepatology 45:1555–1565
24. Trott A, Morano KA (2004) SYM1 is the stress-induced Saccharomyces cerevisiae ortholog of the mammalian kidney disease gene Mpv17 and is required for ethanol metabolism and tolerance during heat shock. Eukaryot Cell 3:620–631
25. Dallabona C, Marsano RM, Arzuffi P, Ghezzi D, Mancini P, Zeviani M, Ferrero I, Donnini C (2010) Sym1, the yeast ortholog of the MPV17 human disease protein, is a stress-induced bioenergetic and morphogenetic mitochondrial modulator. Hum Mol Genet 19:1098–1107
26. Weiher H, Noda T, Gray DA, Sharpe AH, Jaenisch R (1990) Transgenic mouse model of kidney disease: insertional inactivation of ubiquitously expressed gene leads to nephrotic syndrome. Cell 62:425–434

27. Meyer zum Gottesberge AM, Reuter A, Weiher H (1996) Inner ear defect similar to Alport's syndrome in the glomerulosclerosis mouse model Mpv17. Eur Arch Otorhinolaryngol 253:470–474
28. Meyer zum Gottesberge AM, Felix H (2005) Abnormal basement membrane in the inner ear and the kidney of the Mpv17-/- mouse strain: ultrastructural and immunohistochemical investigations. Histochem Cell Biol 124:507–516
29. Viscomi C, Spinazzola A, Maggioni M, Fernandez-Vizarra E, Massa V, Pagano C, Vettor R, Mora M, Zeviani M (2009) Early-onset liver mtDNA depletion and late-onset proteinuric nephropathy in Mpv17 knockout mice. Hum Mol Genet 18:12–26
30. Krawczak M, Ball EV, Cooper DN (1998) Neighboring-nucleotide effects on the rates of germ-line single-base-pair substitution in human genes. Am J Hum Genet 63:474–488

Chapter 7
Mitochondrial DNA Depletion due to Mutations in the *TK2* Gene

Fernando Scaglia

Introduction

Mitochondrial DNA (mtDNA) depletion syndromes are a group of clinically and genetically heterogeneous disorders that were first reported in 1991 [1] that are associated with a severe reduction in mtDNA copy number in affected tissues. The most aggressively affected tissues are those that are postmitotic and these syndromes can present as myopathic, encephalomyopathic, or hepatocerebral [2]. Mutations in the gene encoding the mitochondrial kinase of pyrimidine nucleosides, thymidine kinase 2 (TK2), were the first ascertained genetic cause of the myopathic form of mtDNA depletion syndrome [3]. Mitochondrial DNA replication is not linked to nuclear DNA replication, occurring throughout the cell cycle in a constitutive way with the requirement of a constant supply of deoxynucleotide triphosphates (dNTPs) [4]. In post-mitotic cells, the mitochondrial salvage pathway plays a major role as a source of dNTPs in the maintenance of mtDNA. TK2 is a pyrimidine deoxyribonucleoside kinase and the first and rate limiting step in the phosphorylation of deoxythymidine and deoxycytidine that takes place in the mitochondrial matrix [5].

It has been estimated that *TK2* mutations may account for approximaterly 20 % of the cases of myopathic mtDNA depletion syndrome [6], although this could be due to the fact that *TK2* was the first gene associated with the disorder. The first reported patients with mutations in the *TK2* gene exhibited severe infantile myopathy with motor regression and early demise from respiratory compromise [3, 7]. Additional cases expanded the clinical spectrum with a spinal muscular atrophy type 3-like presentation [8, 9] and milder forms with prolonged survival [10]. Other clinical presentations such as rigid spine syndrome and subacute myopathy without motor regression and with longer survival have been described [11]. Although there could

---

F. Scaglia (✉)
Department of Molecular and Human Genetics, Baylor College of Medicine
and Texas Children's Hospital, Clinical Care Center, Suite 1560, 6701 Fannin Street,
Mail Code CC1560, Houston, TX 77030, USA
e-mail: FSCAGLIA@bcm.edu

be a broad spectrum of phenotypes associated with TK2 deficiency, many of the affected subjects tend to present with common signs and symptoms. The age of onset of symptoms is usually within the first year of life and most patients die in the fourth year of life with respiratory failure the major cause of death in these patients. Proximal muscle weakness and hypotonia are common to all patients. These features tend to worsen progressively and eventually affect the function of the respiratory muscles. Less common clinical features include ptosis, facial diplegia, ophthalmoplegia, exercise intolerance, and feeding difficulties [11]. Patients present with increase in serum creatine kinase levels and this marked increase is a useful diagnostic clue [12]. Other laboratory and histological features of mitochondrial dysfunction in most patients include lactic acidosis, ragged red or ragged blue- and cytochrome-c oxidase-negative fibers on skeletal muscle, and features of myopathy on electromyogram.

TK2 Deficiency Generally Involves Skeletal Muscle

Mitocondrial DNA depletion is typically observed on skeletal muscle in cases of TK2 deficiency [11]. However, it is not entirely clear why TK2 deficiency is associated with a predominant skeletal muscle involvement. The observed tissue specificity and predominant skeletal muscle involvement could be due to the low basal activity in skeletal muscle of the TK2 enzyme and the high enzymatic requirement for mitochondrial-encode proteins and TK2 enzymatic activity in muscle [13]. Moreover, it is possible that other cell types may compensate for the TK2 deficiency by adapting the enzyme network that regulates the dTTP synthesis outside S-phase. For example, normal fibroblasts may have more TK2 activity than required to maintain dTTP during quiescence, which could explain why TK2-mutated fibroblasts do not exhibit mtDNA depletion despite reduced TK2 activity [14].

TK2 deficiency has also been associated with a progressive myofiber loss and a histological analysis suggestive of late stage muscular dystrophy [15]. In that report, the affected patient had congenital muscle weakness, a progressive mitochondrial myopathy and an mtDNA content of 24 % in skeletal muscle, but not in blood. Two novel heterozygous mutations, p.K85NfsX9 and p.R172Q, were detected in a *trans* configuration in the *TK2* gene. Also, the patient's electrocardiogram showed prolonged QT and an incomplete right bundle branch block. This additional feature demonstrates that cardiac rhythm abnormalities can also be part of the spectrum of TK2 deficiency. In this patient, respiratory chain complex activities were not strikingly reduced and this phenomenon which has been previously observed associated with slow clinical progression of disease [16], could be due to compensatory mechanisms. This clinical report underlined the importance of considering TK2 deficiency in patients with dystrophic-like myofiber changes and no evidence of substantial respiratory chain complex deficiency on skeletal muscle.

Another report describes a 12-year-old patient with TK2 deficiency who was serially assessed more than 10 years [10]. Similar to other subjects with long survival and late-onset myopathic mtDNA depletion syndrome, this patient had mtDNA depletion confined to skeletal muscle [17–19]. Progressive muscle atrophy with selective loss of type 2 muscle fibers was also observed. Two mutations, p.T77M and p.R161K, were detected in the *TK2* gene, and the *in vitro* activity of TK2 was markedly decreased [10]. Despite the observed mtDNA depletion, normal activities of the mitochondrial respiratory chain complexes and levels of COX II mitochondrial protein in the remaining muscle fibers were observed. One possible explanation for this phenomenon is that upregulation of mitochondrial transcription and mtDNA level in surviving myofibers may ameliorate the clinical phenotype of patients with mtDNA depletion syndrome [17]. The dissection of the compensatory mechanisms for mtDNA depletion may shed light on new therapeutic strategies for this group of disorders.

Mutations in TK2 Reveal a Spectrum of Clinical Presentations

On the milder end of the spectrum, *TK2* mutations may cause isolated mitochondrial myopathy without evidence of mtDNA depletion. A patient reported to have an isolated mitochondrial myopathy and homozygosity for a *TK2* mutation, p.H90N mutation, did not demonstrate mtDNA depletion [20]. The patient's clinical evolution was relatively mild. A muscle biopsy demonstrated ragged red fibers with a mild decrease in complex I and increases in complexes II and IV activities. This report highlights the fact that even patients with normal mtDNA content should be screened for *TK2* mutations in the context of an isolated mitochondrial myopathy.

On the more severe end of the clinical spectrum, TK2 deficiency can present with an early-onset fatal encephalomyopathy [21]. Lesko et al. [21] reported a patient who became symptomatic before 3 weeks of age who developed a severe epileptic phenotype not typically associated with TK2 deficiency, although it had been previously reported [22]. Her seizures consisted of early myoclonic encephalopathy refractory to treatment with anticonvulsants. In addition, she exhibited immature myelination on brain MRI. She had deceleration of head growth with ensuing microcephaly, strabismus and nystagmus. Initially, her transaminases were elevated. Later in the disease, gamma-glutamyl-transpeptidase was increased although it could have been a consequence of the use of phenobarbital and phenytoin. The liver was noted to be enlarged further expanding the clinical spectrum of TK2 deficiency. She had a rapidly progressive course dying at 3 months of age due to respiratory failure. The early-onset fatal encephalomyopathy may have been caused by the observed severe mtDNA depletion with less than 5 % of age-matched controls. This patient carried a 219insCG that generated a stop codon and a missense mutation, p.R130W. In particular, the second mutation when expressed *in vitro* caused a severe reduction of TK2 enzymatic activity (< 3 %). Thus, these two mutations precipitated a severe clinical phenotype with predominant central nervous system clinical features that are not typically observed in TK2 deficiency and the more severe clinical presentation may have been caused by an almost abolished TK2 enzymatic activity.

TK2 mutations can also cause a rapidly progressive fatal infantile mitochondrial syndrome [22]. All of the seven patients reported by Gotz et al. [22] had rapidly progressive myopathy or encephalomyopathy that led to respiratory failure within the first 3 years of life. The phenotype was accompanied by high creatine kinase values and dystrophic changes on skeletal muscle with cytochrome c oxidase negative fibers. Moreover, three subjects had a less common clinical feature associated with TK2 deficiency, seizures and two among them terminal-phase seizures. One of these patients presented epilepsia partialis continua and one had cortical laminar necrosis. Although all of these patients exhibited a severe dystrophic myopathy reminiscent of an SMA-like phenotype accompanied by mtDNA depletion, two manifested multi-tissue mtDNA depletion on skeletal muscle, brain, and liver. Two different mutations were identified in all the patients either in a homozygous or compound heterozygous state, p.R172W and p.R225W. Both *TK2* mutations changed a highly conserved hydrophilic arginine to hydrophobic tryptophan in the TK2 protein. These mutations originated from ancient founders, with a Finnish origin for the p.R172W mutation and possibly Scandinavian origin for the p.R225W mutation.

The clinical phenotype associated with the p.R172W mutation is among the most severe reported in TK2 deficiency. The onset of disease was observed in the first months of life, with rapid loss of motor skills and death ensuing within few months. The fact that mtDNA depletion was observed in the brain in two children harboring the p.R172W mutation suggests that the observed central nervous system features are due to mtDNA depletion. Homozygosity for the p.R225W mutation presented during the second year of life with myopathy and severe mtDNA depletion restricted to the skeletal muscle. This patient presented later and lived longer than those patients who were homozygous or compound heterozygous for the p.R172W mutation.

All the reported subjects in this study had 5–10 % residual mtDNA amount in skeletal muscle when compared to age-matched controls. Counterintuitively, the observed low mtDNA content was able to maintain 20–30 % of residual activities of the mitochondrial complexes of the respiratory chain. This phenomenon suggested compensatory mechanisms involving upregulated mitochondrial transcription or protein expression [16]. In addition, a downregulation of complex II was also detected. This finding could be due to a secondary effect caused by the severe dystrophic changes on muscle.

Other Phenotypes Associated with TK2 Deficiency

Further evidence that liver involvement could be found in association with TK2 deficiency was presented in an additional report [23]. Two, among three, affected siblings from a nonconsanguineous Jordanian family had elevated transaminases expanding the clinical and laboratory spectrum associated with TK2 mutations. Analysis of liver and skeletal muscle specimens from one of the deceased infants revealed a striking reduction in the mtDNA content with 10 and 20 % of age-matched controls, respectively. This report by Zhang et al. suggested that mtDNA depletion in liver may be observed in cases of TK2 deficiency. Of interest, a novel intragenic 5.8-kb deletion was reported in one of the *TK2* mutant alleles [23] with a p.I254N missense mutation in the other allele. Although most of the reported *TK2* mutations thus far have been point mutations, small insertions, and deletions; intragenic deletions had not been previously reported. This finding expanded the molecular spectrum of TK2 deficiency. Analysis of liver and muscle specimens from one of the deceased infants revealed compound heterozygosity for this maternal intragenic deletion and a paternal point mutation.

Historically, sensorineural hearing loss has been associated with mtDNA depletion syndromes [24] but not with TK2 deficiency. However, recently, sensorineural hearing loss has been reported as an additional clinical feature of TK2 deficiency, further expanding the clinical spectrum [25]. This novel feature underlined the fact that TK2 deficiency may compromise other organs besides skeletal muscle. In this report, the patient harbored a previously reported deletion, p.K244del, and a novel nucleotide duplication in exon 2 that generated a frameshift and a premature stop codon. It was remarked that this patient survived until the age of 8.5 years whereas a substantial number of patients would usually expire prior to the age of 5 years, mostly due to respiratory insufficiency. The prolonged survival in this patient seems to provide further evidence that these subjects could have a variable life span.

TK2 mutations have also been reported associated with the novel phenotype of adult-onset autosomal recessive progressive external ophthalmoplegia [26]. It is characterized by a late-onset slowly progressive myopathy with no evidence of peripheral nervous system involvement. Both reported patients also manifested evidence of late-onset blepharoptosis, with slowly progressive ophthalmoplegia and proximal muscle weakness and atrophy. One patient developed dysarthria and both had evidence of dysphagia. Purkinje cell loss was observed when autopsy was performed in one of the patients, suggesting that late-onset CNS involvement cannot be ruled out. There was no evidence of mtDNA depletion when mtDNA copy number was measured in one subject and the muscle sample of the other subject was not available for analysis. However, mtDNA instability with multiple mtDNA deletions was observed. By exome sequencing, two compound heterozygous mutations, p.R225W and p.T230A were observed. The *in vitro* activity of recombinant mutant proteins and the TK2 activity in patients' cells demonstrated defective TK2 activity. The p.R225W mutation has been previously identified in Finnish pediatric patients with mtDNA depletion syndrome as a homozygous mutation and the recombinant mutant protein is associated with low TK2 activity [22]. Therefore, TK2-autosomal recessive progressive external ophthalmoplegia is among the latest onset of PEO disorders with multiple mtDNA deletions. This finding suggested that mtDNA depletion-associated genes should be evaluated when looking for the molecular causes of PEO disorders.

Tk2 Mouse Models

A Tk2-deficient mouse model has been generated [27]. The mice develop normally for the first week of life. From the first week on, the mice begin to exhibit growth retardation and a high rate of early mortality. Several animals die by the second week of life with no animals surviving more than 4 weeks of life. Progressive loss of mtDNA was found in skeletal muscle, heart, liver, and spleen although mtDNA levels were normal at birth in several tissues. These findings suggest that mtDNA replication is normal during fetal development implying that dNTP supply to mitochondrial is likely to be derived from *de novo* cytosolic dNTP synthesis. No secondary mtDNA mutations or deletions were observed suggesting that Tk2 deficiency in mice causes a severe intramitochondrial deficiency of pyrimidine dNTPs leading to a complete impairment of mtDNA replication due to lack of dNTP substrate. There were no changes on histology in skeletal muscle, however, disorganized muscle fibers were observed in the cardiac muscle suggesting major damage to the integrity of cardiac muscle cells. On electron microscopy of heart mitochondria, the cristae looked abnormal. The mice demonstrated shivering and severe hypothermia compared to the wild-type mice. These phenomena could be explained by defective non shivering thermogenesis caused by loss of hypodermal fat and abnormal brown adipose tissue. However, it is entirely possible that the shivering may be due to the presence of a neurological disorder in these mice. The observed mtDNA depletion in all organs investigated in the $Tk2^{-/-}$ mice is different from the phenotype associated with TK2 deficiency in humans. One explanation for this difference may be the fact that most humans with TK2 deficiency have point mutations in the *TK2* gene leading to a certain degree of residual activity.

Using the knock-out mouse model, the role of TK2 in neuronal homeostasis has been addressed [28]. This study has been useful as it has become evident that neurological phenotypes have been found in patients harboring *TK2* mutations. In the mice, *in vivo* loss of TK2 activity causes an ataxic phenotype after 10–12 days of life with impaired motor coordination and gait. Reduced mtDNA copy number with ~ 70 % reduction was observed in all brain regions by P12. This mtDNA depletion was associated with reduced expression of mtDNA-encoded electron transport chain subunits in the cerebellum, hippocampus, and cortex by P7 corresponding to the onset of neurological impairment. In Tk2-deficient cerebellar neurons, these abnormalities are associated with impaired mitochondrial bioenergetics function, abnormal mitochondrial ultrastructure, and degeneration of selected neuronal types. There are large numbers of Purkinje cells in the P12 $Tk2^{-/-}$ mice that lack expression of the mtDNA complex IV (COX) subunit I. This finding has been associated with deficient COX activity *in vivo,* and reduced respiration capacity and cellular oxygen consumption rate. P12 $Tk2^{-/-}$ Purkinje cells exhibit a reduction in the number of dendrites and decreased dendritic arborization. These findings underscore the fact that lack of TK2 activity hampers the development of correctly arborized neuronal networks leading to neuronal death. These findings underline the necessity of considering a neurological phenotype within the broadening clinical spectrum of TK2 deficiency.

In order to study the pathogenesis of TK2 deficiency and to dissect the tissue-specific effects of TK2 deficiency, mice harboring the p.H126N *Tk2* mutation were generated [29]. This mutation is analogous to the human p.H121N mutation which has been identified in a child with congenital myopathy who died at the age of 4 years [3, 30] and in compound heterozygosity with a p.T113M mutation in two siblings with rapidly progressive myopathy beginning at 12 months [9]. Homozygous *Tk2*-mutant mice exhibited rapidly progressive weakness after age of 10 days and died between the ages of 2 and 3 weeks. Compared with patients with p.H121N *TK2* mutation, Tk2$^{-/-}$ mice exhibited more severely reduced Tk2 activity in tested tissues. These mice manifested an imbalance in deoxynucleotide triphosphate pools, and mtDNA depletion and respiratory chain defects that were prevalent in the central nervous system. Histopathological analysis demonstrated an encephalomyelopathy with vacuolar changes in the anterior horn of the spinal cord. One of the explanations for the selective vulnerability of the central nervous system in these mice may reside in the fact that in wild-type mice, Tk2 may account for ~ 60 % of total brain activity. This finding may imply that the CNS in mice is more dependent on the mitochondrial pyrimidine salvage pathway that in other tissues. It is possible that the fact that most mature neurons are in a postmitotic status may contribute to this severe CNS pathology. Cells that are in the G$_0$ phase seem to be more dependent on the Tk2-mediated mitochondrial pyrimidine salvage pathway. Despite partial mtDNA depletion observed in other tissues, the activities of mitochondrial enzymes and the levels of mtDNA-encoded polypeptides were maintained at similar levels to controls indicating a compensatory effect for the mtDNA defect. This finding has already been observed in the skeletal muscle of patients carrying *TK2* mutations [16].

TK2 in vitro Model

An *in vitro* model to study mtDNA depletion caused by *TK2* mutations has been developed [31]. Small interfering RNA-targeting *TK2* mRNA has been used to decrease the expression of the *TK2* gene in the established human tumor cell line OST TK1$^-$ which has no endogenous thymidine kinase 1. This cell line represents the first *in vitro* model of TK2 deficiency. TK2-deficient cell lines revealed a drastic reduction of TK2 levels (close to 80 %). Moreover, TK2-deficient cell lines revealed severe mtDNA depletion. Nevertheless, TK2-deficient cell lines exhibited increased cytochrome c oxidase activity, higher cytochrome c oxidase subunit I transcript levels and higher cytochrome c oxidase subunit II protein expression. No alterations in the dNTP pools were found but reductions in the *SLC29A1*, *POLG*, *TFAM*, and *TYMP* transcript levels were observed. It seems that in addition to TK2 deficiency, other factors are required in order to lead to mtDNA depletion-associated impairment of respiratory activity. These findings stressed the importance of cellular compensatory mechanisms that may facilitate the expression of respiratory subunits to ensure respiratory activity. Understanding the nature of these compensatory mechanisms will be fundamental in order to design novel therapeutic strategies for TK2 deficiency and other mtDNA depletion syndromes.

Summary

In summary, the molecular and clinical features associated with TK2 deficiency have broadened considerably since it was ascertained as the first cause of myopathic mtDNA depletion syndrome. TK2 deficiency should be suspected in cases of multiorgan involvement even in the absence of mitochondrial respiratory chain defects or mtDNA depletion. The application of novel technologies such as exome sequencing will undoubtedly continue to expand the molecular and clinical spectrum of this disease.

References

1. Moraes CT, Shanske S, Tritschler HJ, Aprille JR, Andreetta F, Bonilla E, Schon EA, DiMauro S (1991) mtDNA depletion with variable tissue expression: A novel genetic abnormality in mitochondrial diseases. Am J Hum Genet 48:492–501
2. Suomalainen A, Isohanni P Mitochondrial DNA depletion syndromes–many genes, common mechanisms, Neuromuscul. Disord 20:429–437
3. Saada A, Shaag A, Mandel H, Nevo Y, Eriksson S, Elpeleg O (2001) Mutant mitochondrial thymidine kinase in mitochondrial DNA depletion myopathy. Nat Genet 29:342–344
4. Bogenhagen D, Clayton DA (1977) Mouse L cell mitochondrial DNA molecules are selected randomly for replication throughout the cell cycle. Cell 11:719–727
5. Rampazzo C, Fabris S, Franzolin E, Crovatto K, Frangini M, Bianchi V (2007) Mitochondrial thymidine kinase and the enzymatic network regulating thymidine triphosphate pools in cultured human cells. J Biol Chem 282:34758–34769
6. Alberio S, Mineri R, Tiranti V, Zeviani M (2007) Depletion of mtDNA: syndromes and genes. Mitochondrion 7:6–12
7. Mancuso M, Filosto M, Bonilla E, Hirano M, Shanske S, Vu TH, DiMauro S (2003) Mitochondrial myopathy of childhood associated with mitochondrial DNA depletion and a homozygous mutation (T77M) in the TK2 gene. Arch Neurol 60:1007–1009
8. Pons R, Andreetta F, Wang CH, Vu TH, Bonilla E, DiMauro S, De Vivo DC (1996) Mitochondrial myopathy simulating spinal muscular atrophy. Pediatr Neurol 15:153–158
9. Mancuso M, Salviati L, Sacconi S, Otaegui D, Camano P, Marina A, Bacman S, Moraes CT, Carlo JR, Garcia M, Garcia-Alvarez M, Monzon L, Naini AB, Hirano M, Bonilla E, Taratuto AL, DiMauro S, Vu TH (2002) Mitochondrial DNA depletion: mutations in thymidine kinase gene with myopathy and SMA. Neurology 59:1197–1202
10. Wang L, Limongelli A, Vila MR, Carrara F, Zeviani M, Eriksson S (2005) Molecular insight into mitochondrial DNA depletion syndrome in two patients with novel mutations in the deoxyguanosine kinase and thymidine kinase 2 genes. Mol Genet Metab 84:75–82
11. Oskoui M, Davidzon G, Pascual J, Erazo R, Gurgel-Giannetti J, Krishna S, Bonilla E, De Vivo DC, Shanske S, DiMauro S (2006) Clinical spectrum of mitochondrial DNA depletion due to mutations in the thymidine kinase 2 gene. Arch Neurol 63:1122–1126
12. Galbiati S, Bordoni A, Papadimitriou D, Toscano A, Rodolico C, Katsarou E, Sciacco M, Garufi A, Prelle A, Aguennouz M, Bonsignore M, Crimi M, Martinuzzi A, Bresolin N, Papadimitriou A, Comi GP (2006) New mutations in TK2 gene associated with mitochondrial DNA depletion. Pediatr Neurol 34:177–185
13. Saada A, Shaag A, Elpeleg O (2003) mtDNA depletion myopathy: elucidation of the tissue specificity in the mitochondrial thymidine kinase (TK2) deficiency. Mol Genet Metab 79:1–5
14. Frangini M, Rampazzo C, Franzolin E, Lara MC, Vila MR, Marti R, Bianchi V (2009) Unchanged thymidine triphosphate pools and thymidine metabolism in two lines of thymidine kinase 2-mutated fibroblasts. FEBS J 276:1104–1113

15. Collins J, Bove KE, Dimmock D, Morehart P, Wong LJ, Wong B (2009) Progressive myofiber loss with extensive fibro-fatty replacement in a child with mitochondrial DNA depletion syndrome and novel thymidine kinase 2 gene mutations. Neuromuscul Disord 19:784–787
16. Vila MR, Villarroya J, Garcia-Arumi E, Castellote A, Meseguer A, Hirano M, Roig M (2008) Selective muscle fiber loss and molecular compensation in mitochondrial myopathy due to TK2 deficiency. J Neurol Sci 267:137–141
17. Vila MR, Segovia-Silvestre T, Gamez J, Marina A, Naini AB, Meseguer A, Lombes A, Bonilla E, DiMauro S, Hirano M, Andreu AL (2003) Reversion of mtDNA depletion in a patient with TK2 deficiency. Neurology 60:1203–1205
18. Vu TH, Tanji K, Valsamis H, DiMauro S, Bonilla E (1998) Mitochondrial DNA depletion in a patient with long survival. Neurology 51:1190–1193
19. Barthelemy C, Ogier de Baulny H, Diaz J, Cheval MA, Frachon P, Romero N, Goutieres F, Fardeau M, Lombes A (2001) Late-onset mitochondrial DNA depletion: DNA copy number, multiple deletions, and compensation. Ann Neurol 49:607–617
20. Leshinsky-Silver E, Michelson M, Cohen S, Ginsberg M, Sadeh M, Barash V, Lerman-Sagie T, Lev D (2008) A defect in the thymidine kinase 2 gene causing isolated mitochondrial myopathy without mtDNA depletion. Eur J Paediatr Neurol 12:309–313
21. Lesko N, Naess K, Wibom R, Solaroli N, Nennesmo I, von Dobeln U, Karlsson A, Larsson NG (2010) Two novel mutations in thymidine kinase-2 cause early onset fatal encephalomyopathy and severe mtDNA depletion. Neuromuscul Disord 20:198–203
22. Gotz A, Isohanni P, Pihko H, Paetau A, Herva R, Saarenpaa-Heikkila O, Valanne L, Marjavaara, S, Suomalainen A (2008) Thymidine kinase 2 defects can cause multi-tissue mtDNA depletion syndrome. Brain 131:2841–2850
23. Zhang S, Li FY, Bass HN, Pursley A, Schmitt ES, Brown BL, Brundage EK, Mardach R, Wong LJ (2010) Application of oligonucleotide array CGH to the simultaneous detection of a deletion in the nuclear TK2 gene and mtDNA depletion. Mol Genet Metab 99:53–57
24. Elpeleg O, Miller C, Hershkovitz E, Bitner-Glindzicz M, Bondi-Rubinstein G, Rahman S, Pagnamenta A, Eshhar S, Saada A (2005) Deficiency of the ADP-forming succinyl-CoA synthase activity is associated with encephalomyopathy and mitochondrial DNA depletion. Am J Hum Genet 76:1081–1086
25. Marti R, Nascimento A, Colomer J, Lara MC, Lopez-Gallardo E, Ruiz-Pesini E, Montoya J, Andreu AL, Briones P, Pineda M (2010) Hearing loss in a patient with the myopathic form of mitochondrial DNA depletion syndrome and a novel mutation in the TK2 gene. Pediatr Res 68:151–154
26. Tyynismaa H, Sun R, Ahola-Erkkila S, Almusa H, Poyhonen R, Korpela M, Honkaniemi J, Isohanni P, Paetau A, Wang L, Suomalainen A (2012) Thymidine kinase 2 mutations in autosomal recessive progressive external ophthalmoplegia with multiple mitochondrial DNA deletions. Hum Mol Genet 21:66–75
27. Zhou X, Solaroli N, Bjerke M, Stewart JB, Rozell B, Johansson M, Karlsson A (2008) Progressive loss of mitochondrial DNA in thymidine kinase 2-deficient mice. Hum Mol Genet 17:2329–2335
28. Bartesaghi S, Betts-Henderson J, Cain K, Dinsdale D, Zhou X, Karlsson A, Salomoni P, Nicotera P (2010) Loss of thymidine kinase 2 alters neuronal bioenergetics and leads to neurodegeneration. Hum Mol Genet 19:1669–1677
29. Akman HO, Dorado B, Lopez LC, Garcia-Cazorla A, Vila MR, Tanabe LM, Dauer WT, Bonilla E, Tanji K, Hirano M (2008) Thymidine kinase 2 (H126N) knockin mice show the essential role of balanced deoxynucleotide pools for mitochondrial DNA maintenance. Hum Mol Genet 17:2433–2440
30. Nevo Y, Soffer D, Kutai M, Zelnik N, Saada A, Jossiphov J, Messer G, Shaag A, Shahar E, Harel S, Elpeleg O (2002) Clinical characteristics and muscle pathology in myopathic mitochondrial DNA depletion. J Child Neurol 17:499–504
31. Villarroya J, Lara MC, Dorado B, Garrido M, Garcia-Arumi E, Meseguer A, Hirano M, Vila MR (2011) Targeted impairment of thymidine kinase 2 expression in cells induces mitochondrial DNA depletion and reveals molecular mechanisms of compensation of mitochondrial respiratory activity. Biochem Biophys Res Commun 407:333–338

Chapter 8
Mitochondrial DNA Multiple Deletion Syndromes, Autosomal Dominant and Recessive (POLG, POLG2, TWINKLE and ANT1)

Margherita Milone

Introduction

Mammalian cells contain hundreds of copies of mitochondrial DNA (mtDNA) depending on their developmental stage and energy demand [1, 2]. In human mitotic and nonmitotic cells, the mtDNA is continuously replicated to maintain mtDNA copy number. The minimal mitochondrial replisome consists of the DNA polymerase gamma (pol γ), the mtDNA helicase Twinkle, and the mitochondrial single-stranded binding protein (mtSSB) (Fig. 8.1), all encoded by nuclear genes [3]. Pol γ is the sole polymerase responsible for mtDNA replication and repair. Twinkle catalyzes the ATP-dependent unwinding of the double-stranded mtDNA with distinct polarity (5′–3′ direction) and creates the single-stranded DNA, which is used by pol γ to synthesize the complementary DNA strand. The mtSSB has a stimulatory effect on the DNA-unwinding rate. A pool of nucleotides, strictly regulated by nuclear-encoded factors, is also required for the correct function of the mitochondrial replisome. Mutations in the pol γ and Twinkle impair mtDNA maintenance and give rise to cellular dysfunction and diseases that are associated with multiple mtDNA deletions or mtDNA depletion. In addition to the proteins composing the basic mitochondrial replisome, additional proteins, such as adenine nucleotide translocator (ANT1), also play a role in mtDNA maintenance, and as such, if mutated, can lead to mitochondrial disorders with phenotype similar to that caused by mutated pol γ or Twinkle [4].

The Polymerase Gamma, POLG and POLG2

The human holoenzyme polymerase gamma (pol γ) is a heterotrimer which consists of one p140 catalytic subunit encoded by the gene *POLG* (OMIM 174763) and two identical p55 accessory subunits encoded by the gene *POLG2* (OMIM

M. Milone (✉)
Department of Neurology, Neuromuscular Division, Mayo Clinic,
200 First St., SW, Rochester, MN 55905, USA
e-mail: Milone.Margherita@mayo.edu

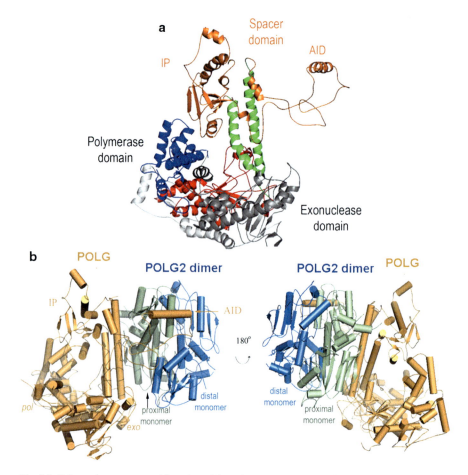

Fig. 8.1 Schematic structure and function of the polymerase gamma (pol γ). **a** The catalytic subunit POLG has a polymerase domain and an exonuclease domain linked by the spacer domain that contains the intrinsic processivity (IP) subdomain and the accessory-interacting determinant (AID) subdomain. **b** POLG and the two identical accessory subunits POLG2 form pol γ holoenzyme. (From [5]. Reprinted with kind permission from Elsevier)

604983; see Fig. 8.1) [5]. The catalytic subunit has two functionally different domains, a polymerase domain located on the C-terminus and an exonuclease domain with proofreading activity located on the N-terminus. These two domains are connected by a spacer domain, which possesses two subdomains, an intrinsic processivity (IP) subdomain and an accessory-interacting determinant (AID) subdomain. The IP subdomain is important for the intrinsic processivity of the catalytic subunit while the AID subdomain creates an interface with the accessory subunit POLG2 that is essential to increase processivity of the holoenzyme. The catalytic subunit interacts primarily with one monomer of the dimeric accessory protein and makes limited contact with the other monomer. Each monomer of the p55 dimer has a distinct role in conferring processivity to the holoenzyme. The proximal monomer strengthens the interaction with the DNA, while the distal monomer is essential for the polymerization rate enhancement [6]. Compared with other processivity factors that increase processivity by increasing the enzyme affinity for the DNA template, the accessory protein enhances processivity not only by increasing affinity and accelerating polymerization, but also by suppressing exonuclease activity [7]. The high processivity of the holoenzyme is indicated by its ability to synthesize DNA up to about 100 nucleotides per enzyme-binding event [8], compared with the 1–25 nucleotides of other polymerases [6]. Disruption of the dimeric structure of the accessory protein results in decreased stability of the holoenzyme and loss of acceleration of polymerization rate [6]. The accessory subunit shows also a catalytic subunit- and substrate DNA-dependent dimerization. In addition, the accessory subunit is a key element in the organization of mtDNA in nucleoprotein complexes or nucleoids, determining the mtDNA copy number per nucleoid [9]. Mutations in either the catalytic or accessory subunits of the pol γ result in mitochondrial dysfunction and diseases.

POLG-Related Disorders

Clinical Manifestations

POLG mutations are a major cause of mitochondrial diseases [10]. *POLG* mutations lead to a continuum broad spectrum of clinical phenotypes ranging from Alpers syndrome to ataxia-neuropathy syndrome with or without epilepsy to progressive external ophthalmoplegia (PEO). Distinct phenotypes can be associated with identical mutations and these can be inherited with either an autosomal-dominant or autosomal-recessive pattern. In addition, phenotypic variability within the same family can occur and the disease onset may range from neonatal period to adult life [11–14].

Alpers Syndrome

Alpers syndrome is one of the most severe clinical manifestations of *POLG*-recessive mutations. It presents in infancy with psychomotor regression, intractable epilepsy, and liver failure. As Alpers syndrome is often associated with mtDNA depletion and as a chapter is specifically dedicated to Alpers syndrome, this syndrome will not be discussed here.

Spectrum of Ataxia Neuropathy Syndrome and Epilepsy

Recessive *POLG* mutations can give rise to an ataxia neuropathy syndrome which may or may not be accompanied by epilepsy or may have epilepsy as the dominating feature. Additional clinical findings, such as PEO or myopathy, can accompany the ataxia. The various combination of clinical manifestations has led to the description of distinct syndromes, such as sensory ataxia neuropathy dysarthria and ophthalmoplegia (SANDO) (OMIM 607459) [15, 16], spinocerebellar ataxia with epilepsy (SCAE) [17], myoclonic epilepsy myopathy sensory ataxia (MEMSA) [16], mitochondrial-recessive ataxic syndrome (MIRAS) [11, 17]. However, phenotypes often overlap and it may be difficult to fit patients into a specific syndrome. MEMSA still distinguishes itself from the other ataxic syndromes because of the lack of PEO and the presence of predominant myoclonus that makes it clinically similar to MERRF (myoclonic epilepsy and ragged-red fibers). However, contrary to MERRF, patients with MEMSA show no ragged-red fibers in muscle.

Ataxia is the result of a sensory and/or cerebellar dysfunction. The sensory ataxia is due to a predominantly sensory axonal peripheral neuropathy or pure sensory neuronopathy or ganglionopathy, as suggested by the reduced amplitudes of the sensory nerve action potentials and preserved nerve conduction velocities. The electrophysiological findings are corroborated by the pathological findings of neuronal cell loss and reduction in cell body size in the dorsal root ganglia. The affected neurons have severe reduction in mtDNA copy number resulting in complexes I and IV deficiency that in turn may be responsible for the histological changes [18]. The involvement of the central proprioceptive pathways, as electrophysiologically documented by somatosensory evoked potentials that may demonstrate slowing along the central proprioceptive pathway, may contribute to the ataxia [12]. The abnormal somatosensory evoked potentials find a pathological correlate in the axonal and myelin loss detected in the posterior spinal cord, in particular in the posterior spinal funiculus [11, 18]. Tremor and pseudo-athetoid finger and toe movements can be observed as the consequence of the proprioception loss [17].

Cerebellar dysfunction can contribute to the ataxia or be the dominating cause of the ataxia. *POLG* has emerged as a leading gene for cerebellar ataxia in Central Europe when the ataxia manifests under the age of 30 years and when associated with PEO and/or sensory neuropathy [19]. Cerebellar atrophy can be detected by MRI (see Fig. 8.2a). Pathological studies have shown marked global neuronal cell loss in the olivary-cerebellum areas of subjects with recessive *POLG* mutations. These pathological findings distinguish POLG-related cerebellar atrophy from that observed in patients harboring m.3243A>G mutation who instead tend to develop focal Purkinje cell loss secondary to microinfarcts and have relatively preserved neuronal cells in the olivary and dentate nuclei [20]. The severity of neuronal cell loss correlates with the clinical severity of disease. Sensory and cerebellar ataxia may coexist. The peripheral neuropathy can precede the onset of the cerebellar ataxia by decades and patients may manifest with a Charcot-Marie-Tooth phenotype in childhood and develop cerebellar ataxia in their adult life [21].

POLG mutations have been identified also in a single patient with multiple system atrophy of cerebellar subtype [22], but additional studies are needed to establish an eventual role of *POLG* in the etiology of multiple system atrophy.

PEO, limb myopathy, bulbar weakness, optic atrophy, limb or palatal myoclonus, levodopa-responsive parkinsonism, dystonia, chorea, stroke-like events, migraine, dementia, psychosis, sensorineural deafness, gastrointestinal dysmotility, and diabetes may accompany the ataxia [12, 23–27].

Fig. 8.2 Brain MRI from patients with POLG-related disease. **a** Left cerebellar infarct in a 19-year-old woman with refractory epilepsy (Axial FLAIR). **b** Cerebellar atrophy in a 51-year-old man with ataxia, PEO, and sensory neuropathy (T1 sagittal FLAIR)

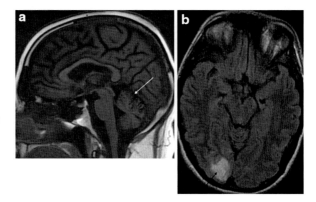

Epilepsy may manifest in childhood or adult life. The mean age of seizure presentation was 18 years (range 6–58 years) in a series of 19 patients [28]. The seizure may have a different semeiology including simple, complex partial, clonic, and myoclonic seizures. Epilepsia partialis continua and convulsive status epilepticus are common and the latter usually carries a very poor prognosis with high risk of death [28, 29]. The epileptic syndrome often manifests with occipital seizures resulting in positive or negative visual phenomena, including flickering lights, visual loss, dysmorphopsia, and palinopsia. Focal occipital predilection is a feature of POLG-related epilepsy and the occipital predilection often persists also in the setting of multifocal epilepsy [28]. The seizures may result in MRI abnormalities with stroke-like appearance mimicking MELAS (see Fig. 8.2b).

There is no standardized antiepileptic regimen to treat the seizures that are often refractory. Valproate can trigger liver failure and therefore it is contraindicated [24, 29–32]. Clonazepam and topiramate may successfully control the myoclonic activity. Sodium channel blockers, such as carbamazepine, oxcarbamazepine, phenytoin, and lamotrigine, in combination with a benzodiazepine, may temporarily prevent the secondary generalization of the seizures. A combination of several antiepileptic drugs, including intravenous magnesium, barbiturate-induced coma, and anesthesia, may be necessary to abort the refractory status epilepticus [28, 33, 34].

PEO, Autosomal Dominant and Recessive

PEO is genetically heterogeneous and can be the result of a single mtDNA mutation or the phenotypic manifestation of nuclear gene mutations inherited with autosomal-dominant (ad) or autosomal-recessive (ar) trait. So far, six nuclear genes have been associated with adPEO and only two nuclear genes are known to underlie arPEO (see Table 8.1) [4, 35–40]. *POLG* mutations can result in adPEO or arPEO and account for up to 45 % of familial cases in some cohorts [41].

PEO is characterized by weakness of the extraocular muscles that can occur as the sole clinical manifestation of the disease or in association with other clinical features, in particular myopathy and exercise intolerance [35]. Additional clinical features may

Table 8.1 Nuclear genes causing PEO

Gene	Protein	Major function
adPEO		
POLG	Pol γ, catalytic subunit	mtDNA Replication
POLG2	Pol γ, accessory subunit	mtDNA Replication
C10ORF2	Twinkle	mtDNA Helicase
SLC25A	ANT1	ADP–ATP exchange across mitochondrial membrane
OPA1	Optic atrophy 1	Mitochondrial fusion and mt cristae maintenance
RRM2B	Ribonucleotide reductase p53R2, small subunit	Conversion of ribonucleoside diphosphates into deoxyribonucleoside diphosphates (essential for DNA synthesis)
arPEO		
POLG	Pol γ, catalytic subunit	mtDNA Replication
TK2	Thymidine kinase 2	mt Nucleotide pool maintenance

In addition to *POLG, POLG2, C10ORF2,* and *SLC25A*, also mutations in *OPA1, RRM2B* and TK2 can result in PEO and multiple mtDNA deletions in muscle

include peripheral neuropathy, levodopa-responsive parkinsonism, optic atrophy, hearing loss, psychiatric symptoms, stroke-like MR abnormalities, CSF oligoclonal bands, diabetes, hypogonadism, gastrointestinal dysmotility, cardiac arrhythmia, and cardiomyopathy, but PEO remains the core clinical feature of the disease [13, 23, 42–45]. Gastrointestinal dysmotility can be so severe that the phenotype may clinically resemble mitochondrial neurogastrointestinal encephalomyopathy (MNGIE), but the normal plasma thymidine level and the lack of extensive leukoencephalopathy distinguish the POLG-related disease from MNGIE [27, 46].

Rare Phenotypes

Recessive *POLG* mutations have been reported in association with isolated distal asymmetric myopathy of the upper extremities without ocular muscle involvement and in association with fatal congenital myopathy leading to death by respiratory failure [47, 48].

Laboratory Findings

While subjects with severe early-onset POLG-related disease, such as Alpers syndrome, often have mtDNA depletion, patients with predominantly milder muscle phenotype show multiple mtDNA deletions in the affected tissues [13, 14]. Occasionally, mtDNA depletion with or without associated multiple mtDNA deletions has been identified in patients with milder phenotypes, such as PEO or distal myopathy [47, 49]. The search of multiple mtDNA deletions is usually poorly informative in blood samples and the analysis of the affected tissues, such as muscle or liver specimen, is often essential to demonstrate the multiple mtDNA deletions [13].

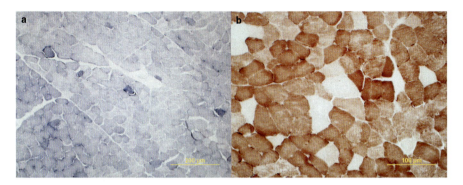

Fig. 8.3 Muscle biopsy from a patient with arPEO and peripheral neuropathy. Few ragged-blue fibers and numerous COX negative fibers in SDH (**a**) and cytochrome *c* oxidase-stained section (**b**), respectively

Histological muscle biopsy studies of patients with POLG-related disease may show normal findings or signs of mitochondrial dysfunction, as indicated by the presence of ragged-red, ragged-blue, or cytochrome *c* oxidase (COX)-negative fibers (Fig. 8.3). COX-negative fibers seem more frequently observed than ragged-red and ragged-blue fibers. The histological findings do not correlate with the severity of the phenotype and the lack of histological signs of mitochondrial dysfunction can be observed in children with Alpers syndrome or in adults with various phenotypes [12, 50, 51].

Muscle biochemical assay of respiratory chain complexes may be normal or may show single or multiple complexes deficiency independently from the age [12, 50, 52].

More than 180 mutations have been identified across the *POLG* gene with no evidence for hot spots. These include missense, nonsense, splice-site mutations, deletions, and insertions (http://tools.niehs.nih.gov/polg/). Missense mutations account for more than 90 % of all *POLG* mutations [13, 14]. The p.A467T is the most common recessive pathogenic mutation, accounting for ∼31 % of all mutant alleles and is associated with a broad constellation of clinical features [13]. From the functional standpoint, *POLG* disease-associated mutations have been classified into three groups [5]. Class I mutations are located in the active-site of the polymerase domain and result in reduced catalytic enzyme activity; class II mutations occur in the putative DNA-binding channel decreasing the DNA-binding affinity; class III mutations cluster at the interface between POLG and POLG2 and disrupt their interaction, thus reducing pol γ processivity. There is no consistent correlation between genotype and phenotype; homozygosity for the most common mutation p.A467T has been observed in infants with Alpers syndrome and in late-onset PEO.

Further insight into the disease mechanism has been provided by animal model of the POLG human disease. Homozygous knock-in mice expressing a proof-reading-deficient POLG show premature aging, increased levels of mtDNA point mutations and deletions [53]. The mutator mice has elevated mtDNA replication pausing possibly due to replisome instability and may have mitochondrial dysfunction in somatic stem cells secondary to sensitivity of the somatic stem cells to mtDNA mutagenesis [54, 55].

POLG2-Related Disorders

Clinical Manifestations and Laboratory Findings

Mutations in *POLG2*, the gene encoding for the accessory subunit of pol γ, are rare causes of adPEO or multisystem disease resembling POLG-related disorders. In 2006, Longley et al. [37] identified a *POLG2* heterozygous missense mutation in a patient with late-onset PEO, exercise intolerance, mild facial and limb weakness, cardiac conduction defect, and impaired glucose tolerance. The patient's muscle biopsy revealed a small percent of COX-negative fibers and multiple mtDNA deletions concentrated in the COX-negative fibers. Biochemical studies showed that the mutant accessory subunit retained the ability to bind DNA but weakened its interaction with the catalytic subunit failing to enhance strong binding of pol γ to the DNA. Haploinsufficiency of POLG2 resulting in decreased availability of functional pol γ was suggested as possible disease-mechanism. Following this report, a patient with late-onset ptosis and myopathy, but without PEO, was found to carry a heterozygous *POLG2* 24 bp in frame insertion in exon 7 leading to skipping of exon 7 [56]. The patient's muscle biopsy showed no histological or biochemical signs of mitochondria dysfunction; Southern blot and long-range PCR analysis detected faint multiple mtDNA deletions in muscle and not in blood. Although it was suggested that skipping of exon 7 might impair the C-terminal domain of the accessory subunit and its interaction with the catalytic subunit, the mutation was detected also in the patient's asymptomatic brother and functional studies of the mutated protein were not performed. Recently, additional mutations in *POLG2* were detected and biochemically characterized [57]. An infant with phenotype resembling Alpers syndrome carried a missense mutation p.P205R, which demonstrated a fourfold weaker ability to bind DNA but preserved ability to assemble into a homodimer and to associate with the catalytic subunit. A child with developmental delay, hypotonia, seizures, cortical blindness, cerebellar atrophy, and elevated liver enzymes was found to have a frameshift mutation and truncation, p.L475DfsX2, which results in inability to bind the catalytic subunit of pol γ and the double-stranded DNA, and in the formation of aberrant oligomeric complexes. Another missense mutation, p.R369G, was identified in a 19-year-old woman with ptosis, PEO, muscle weakness, respiratory failure, delayed gastric emptying, and lactic acidosis. The p.R369G POLG2 variant, which is close to the AID subdomain of the catalytic subunit POLG, has decreased primer-extension activity and reduced affinity for the catalytic subunit. The Arg369 is also in close proximity to the Ala467 of the catalytic subunit, site of the most common mutation p.A467T detected in *POLG*-related disease. This mutation was previously proven to compromise the interaction with the accessory subunit [58]. Therefore, there is convincing evidence that p.P205R, p.L475DfsX2, and p.R369G disrupt mtDNA synthesis and are likely disease-causing mutations. It remains unknown if these mutations result in histological signs of mitochondrial dysfunction and/or multiple mtDNA deletions or depletion, as no muscle tissue was available from these patients. Additional heterozygous *POLG2* mutations were detected in subjects with multisystem disease, but there was no biochemical evidence for their pathogenicity [57].

In addition to disease-causing mutations, *POLG2* polymorphisms have been linked to cancer. Specifically, *POLG2* polymorphisms have been associated with the risk of head and neck cancer in the French population [59] and with the invasiveness of urinary bladder cancer in Japanese males [60].

The Mitochondrial DNA Helicase Twinkle

Twinkle (*C10ORF2*, OMIM 606075) is the replicative helicase and colocalizes with mtDNA in mitochondrial nucleoids with a pattern reminiscent of the twinkling stars to which it owns the name [36]. Elegant experiments have supported the function of Twinkle as the replicative helicase and its essential role in mtDNA maintenance. Indeed, the reduced expression of Twinkle in cultured human cells by RNA interference led to decreased mtDNA copy number. Conversely, overexpression of Twinkle in transgenic mice resulted in increased mtDNA copy number, suggesting the additional role of Twinkle as independent mtDNA copy number regulator [61].

Twinkle oligomerizes in a stable hexameric ring-shaped structure. Contrary to other replicative helicases, which require a cofactor, such as magnesium or NTP, to oligomerize, Twinkle forms stable hexamers in solution even in the absence of cofactors [62, 63]. Twinkle can also load onto closed circular single-stranded mtDNA without the help of a specialized helicase loader, binding stably to it and allowing initiation of DNA replication on the closed circular double-stranded DNA in combination with pol γ [63]. Twinkle has three functional domains, a C-terminal helicase domain, a short linker region, and an N-terminal primase domain [64, 65]. The helicase domain is highly conserved and faces the double-stranded DNA during unwinding [66, 67]. The linker region is conserved across eukaryotes and is involved in protein oligomerization, which in turn is needed for the helicase activity [68]. The N-terminal domain is required for efficient binding of Twinkle to the single-stranded DNA and unwinding double-stranded DNA. Truncation of the N-terminal region decreases the helicase activity and mtDNA replisome processivity [62].

Clinical Manifestations of Twinkle-Related Disorders

PEO, Autosomal Dominant

Heterozygous mutations in Twinkle were first discovered in 2001 in individuals of various ethnic origins with adPEO and multiple mtDNA deletions linked to chromosome 10q24 [36]. Although ptosis and ophthalmoplegia were the major clinical findings, some affected subjects had limb and respiratory muscle weakness and psychiatric disease consisting of severe depression and avoidant personality, which in some cases preceded the onset of PEO [69]. Homozygous individuals from a consanguineous family manifested at a younger age and had a more severe phenotype than affected individuals with a single heterozygous mutation. Although the mean age of onset of the Twinkle-linked adPEO is classically in the adulthood, the disease can manifest as early as in childhood [70]. *C10ORF2* has emerged as the most commonly mutated gene in familial adPEO [71]. It became soon evident that heterozygous-dominant mutations in Twinkle give rise to a spectrum of clinical presentations which may include one or more of the following features: PEO, ptosis, predominant proximal myopathy, bulbar weakness, myalgia, axonal peripheral neuropathy, ataxia, epilepsy, dementia, migraine, sensorineural hearing loss, cataract, parkinsonism, diabetes, hepatopathy, obesity, cardiomyopathy, and cardiac conduction defect [65, 72–76]. Mild proximal myopathy and cardiac abnormalities, often consisting of ventricular enlargement and nonfatal arrhythmias, are the most common extraocular manifestations, occurring in up to 33 and 24 % of patients, respectively [70]. Occasionally, muscle weakness can result in respiratory failure requiring intubation [75, 77].

Infantile-Onset Spinocerebellar Ataxia (IOSCA), Autosomal Recessive

Recessive Twinkle mutations result in a more severe clinical phenotype with earlier onset, such as infantile-onset spinocerebellar ataxia (IOSCA). IOSCA manifests in infancy with ataxia, athetosis, sensory axonal neuropathy, hypotonia, followed by ophthalmoplegia, migraine, episodic psychosis, epileptic encephalopathy, hearing loss, and female hypogonadism [78, 79]. Epilepsia partialis continua is common and can lead to generalized status epilepticus and death. Seizures can be accompanied by stroke-like lesions on brain MRI, but contrary to POLG-related disorders, they lack the occipital lobe predilection observed in the latter [28]. The brain MRI lesions can be small or very large extending from cortex to basal ganglia and thalamus and often result in brain tissue necrosis and atrophy, confirmed histologically [79]. The seizures are often intractable and valproate-induced elevation of the liver enzymes can occur, as observed in subjects with *POLG* mutations. Compound heterozygotes tend to have earlier disease onset and faster progression than homozygotes and can also manifest with early-onset epileptic encephalopathy, hepatopathy, and mtDNA depletion which closely resembles Alpers syndrome caused by recessive *POLG* mutations [80]. Essentially, *Twinkle* and *POLG* mutations share the same complexity of clinical phenotypes and of autosomal-dominant and autosomal-recessive inheritance.

Laboratory Findings

Multiple mtDNA deletions are detected by Southern blotting and/or PCR-based assay in muscle of subjects with autosomal-dominant *Twinkle* mutations. There is not an obvious correlation between mtDNA deletion load and clinical severity. Muscle biopsy of these patients show histological signs of mitochondrial dysfunction of different severity, most often consisting of cytochrome c oxidase negative fibers [70].

Subjects with recessive *Twinkle* mutations, manifesting with IOSCA or early-onset hepatocerebral disease, have no or minimal histological signs of mitochondrial dysfunction and no mtDNA deletions or depletion in muscle, but they have mtDNA depletion in brain (IOSCA) or liver (early-onset hepatocerebral disease) [81, 82].

The pathogenic autosomal-dominant *C10ORF2* mutations reported so far are missense or in-frame duplication and occur in any of the three domains of Twinkle, but more commonly within or near the linker region. No clear genotype–phenotype correlation has emerged [70]. Nonsense, frame-shift, and splice site mutations have not been detected up-to-date, raising the question of their incompatibility with life. Mutational studies have shown that some point mutations in the linker region have deleterious effects by impairing protein hexamerization and DNA helicase activity, while other mutations compromise DNA loading, a key step in the initiation of DNA replication [67, 83]. Pathogenic mutations in the N-terminal domain of Twinkle also have been shown to result in defective DNA helicase activity and may impair the interplay between single-stranded DNA binding and ATP hydrolysis, which is essential in the catalytic cycle of the helicase [84]. It remains unclear why most mutations cause dominant milder disease manifesting with PEO as major clinical feature while recessive mutations result in a much more severe hepatocerebral phenotype.

Transgenic mice expressing Twinkle with autosomal-dominant PEO patient mutations replicated the histological and molecular findings of the human disease, but these changes had little functional effect in mice. Indeed, despite the accumulation of multiple mtDNA deletions during life, the mutant mice seemed to have good exercise capacity and strength [85]. Additional studies in transgenic animals and cultured cells expressing mutant Twinkle associated with adPEO have shed light on the mechanism by which *Twinkle* mutations cause mtDNA deletions, showing accumulation of mtDNA replication intermediates. This suggested that autosomal-dominant *Twinkle* mutations lead to mtDNA replication stalling which predisposes to multiple mtDA deletions [39, 86].

The Adenine Nucleotide Translocator 1 (ANT1)

The human adenine nucleotide translocator (ANT) is a protein located in the inner mitochondrial membrane. Various ANT isoforms are expressed in mammalian cells and ANT1 (*SLC25A*, OMIM 103220) is the heart- and muscle-specific isoform, although expressed to a lesser extent in brain and kidney [87]. In its functional status, ANT1 forms a homodimer [88]. However, various oligomeric and conformational states (cytosolic or matricial) of ANT1, varying with the cell energy state, have been also reported [89, 90]. ANT1 creates a channel by which the ATP synthesized in the mitochondrial matrix is exchanged with ADP generated in the cytoplasm by the hydrolysis of ATP; the stoichiometry of ADP/ATP exchange occurs one-to-one [91]. Therefore, ANT1 has a crucial role in the cell energy supply. From the structural standpoint, ANT1 was predicted to consist of six transmembrane helices with both the amino and carboxy termini oriented toward the intermembrane space [92, 93]. More recent crystallography studies have shown that ANT1 consists of six alpha-helices forming a compact transmembrane domain [91]. This has a deep depression at the surface toward the space between the inner and outer mitochondrial membranes and at its bottom contains the hexapeptide carrying the signature of the nucleotide carriers. The transport substrates would bind to the bottom of the cavity and the translocation would occur with a transition form a "pit" to a "channel" conformation. ANT1 has also a key role in apoptosis as core component of the permeability transition pore complex. Indeed, under various apoptotic stimuli, such as calcium upload, ANT1 structural conformational change can lead to collapse of the mitochondrial membrane potential [90]. ANT1 exerts also a fundamental role in mtDNA maintenance, as suggested by the multiple mtDNA deletions detected in association with *ANT1* mutation [4].

Clinical Manifestations of ANT1-Related Disorders

PEO, Autosomal Dominant

The first human *ANT1* mutation was reported in 2000 in subjects with adPEO manifesting before the age of 45 years and with multiple mtDNA deletions in muscle [4]. The disease was confined to eye and facial muscles in some families or associated with generalized muscle weakness in other families, but no evidence for other organ involvement. Since the discovery of the first disease-causing *ANT1* mutation, several cases of adPEO segregating with *ANT1* mutations have been reported. Up to about 15 % of patients with familial adPEO carry an *ANT1* mutation making it the second most common gene responsible for adPEO [71]. The onset of the ptosis and/or PEO varies from the second decade to midadulthood [94, 95]. Accompanying clinical features may include pigmentary retinopathy, generalized weakness, exercise intolerance, ataxia, bipolar disorder, and schizoaffective disorder [94–97].

Exercise Intolerance and Cardiomyopathy, Autosomal Recessive

Although *ANT1* mutations are in general inherited with autosomal-dominant trait, a recessive homozygous mutation in *ANT1* has been reported in a sporadic patient with hypertrophic cardiomyopathy, exercise intolerance, and lactic acidosis whose muscle biopsy contained ragged-red fibers and multiple mtDNA deletions [98]. The detected mutation (p.A123D) results in loss of function as suggested by the loss of ADP–ATP carrier activity of the mutant ANT1.

Laboratory Findings

Multiple muscle mtDNA deletions are often detected by Southern blot analysis or long-range PCR amplification in subjects with *ANT1* mutations [94, 95, 97]. Most subjects have histological signs of mitochondrial dysfunction with ragged-red fibers and/or COX-negative fibers although some individuals may show only minimal histological signs of mitochondrial dysfunction [94, 95, 97].

Mitochondrial respiratory chain complexes may be normal in subjects harboring ANT1 mutations [97].

Mice with inactivated *ANT1* show a phenotype similar to humans with *ANT1* mutations, exhibiting exercise intolerance, multiple muscle mtDNA deletions, and also hypertrophic cardiomyopathy [99]. Functional studies of mutated ANT1 in mammalian cells have been limited by the resulting apoptotic cell death. Recent studies of PEO-associated *ANT1* mutation expressed in mouse myotubes showed that the mutant ANT1 causes decreased ADP–ATP exchange rate and abnormal translocator reversal potential. So, a chronic exchange defects in ADP–ATP occurring at higher than normal membrane potential could result in energy depletion and nucleotide imbalance which in turn may give rise to the mtDNA abnormalities [100].

Combined Genetic Defects

Coexisting heterozygous mutations in genes responsible for adPEO have been reported, such as the combination of *ANT1* and *C10ORF2* mutations, but the clinical phenotype has not been more severe that that associated with a single mutation [101].

Acknowledgments Margherita Milone thanks Lee-Jun Wong for her constructive comments on the manuscript. Margherita Milone receives research support by the Mayo Clinic CTSA through NIH/NCRR grant number UL1 RR024150.

References

1. Masuyama M, Iida R, Takatsuka H, Yasuda T, Matsuki T (2005) Quantitative change in mitochondrial DNA content in various mouse tissues during aging. Biochimica Et Biophysica Acta 1723(1–3):302–308
2. Shoubridge EA, Wai T (2007) Mitochondrial DNA and the mammalian oocyte. Curr Top Dev Biol 77:87–111
3. Korhonen JA, Pham XH, Pellegrini M, Falkenberg M (2004) Reconstitution of a minimal mtDNA replisome in vitro. EMBO J 23(12):2423–2429
4. Kaukonen J, Juselius JK, Tiranti V et al (2000) Role of adenine nucleotide translocator 1 in mtDNA maintenance. Science 289(5480):782–785
5. Lee YS, Kennedy WD, Yin YW (2009) Structural insight into processive human mitochondrial DNA synthesis and disease-related polymerase mutations. Cell 139(2):312–324

6. Lee YS, Lee S, Demeler B, Molineux IJ, Johnson KA, Yin YW (2010) Each monomer of the dimeric accessory protein for human mitochondrial DNA polymerase has a distinct role in conferring processivity. J Biol Chem 285(2):1490–1499
7. Johnson AA, Johnson KA (2001) Exonuclease proofreading by human mitochondrial DNA polymerase. J Biol Chem 276(41):38097–38107
8. Johnson AA, Tsai Y, Graves SW, Johnson KA (2000) Human mitochondrial DNA polymerase holoenzyme: reconstitution and characterization. Biochemistry 39(7):1702–1708
9. Di Re M, Sembongi H, He J et al (2009) The accessory subunit of mitochondrial DNA polymerase gamma determines the DNA content of mitochondrial nucleoids in human cultured cells. Nucleic Acids Res 37(17):5701–5713
10. Hudson G, Chinnery PF (2006) Mitochondrial DNA polymerase-gamma and human disease. Hum Mol Genet 15(Spec No 2):R244–R252
11. Van Goethem G, Luoma P, Rantamaki M et al (2004) POLG mutations in neurodegenerative disorders with ataxia but no muscle involvement. Neurology 63(7):1251–1257
12. Milone M, Brunetti-Pierri N, Tang LY et al (2008) Sensory ataxic neuropathy with ophthalmoparesis caused by POLG mutations. Neuromuscul Disord 18(8):626–632
13. Tang S, Wang J, Lee NC et al (2011) Mitochondrial DNA polymerase gamma mutations: an ever expanding molecular and clinical spectrum. J Med Genet 48(10):669–681
14. Wong LJ, Naviaux RK, Brunetti-Pierri N et al (2008) Molecular and clinical genetics of mitochondrial diseases due to POLG mutations. Hum Mutat 29(9):E150–E172
15. Fadic R, Russell JA, Vedanarayanan VV, Lehar M, Kuncl RW, Johns DR (1997) Sensory ataxic neuropathy as the presenting feature of a novel mitochondrial disease. Neurology 49(1):239–245
16. Van Goethem G, Martin JJ, Dermaut B et al (2003) Recessive POLG mutations presenting with sensory and ataxic neuropathy in compound heterozygote patients with progressive external ophthalmoplegia. Neuromuscul Disord 13(2):133–142
17. Hakonen AH, Heiskanen S, Juvonen V et al (2005) Mitochondrial DNA polymerase W748S mutation: a common cause of autosomal recessive ataxia with ancient European origin. Am J Hum Genet 77(3):430–441
18. Lax NZ, Whittaker RG, Hepplewhite PD et al (2012) Sensory neuronopathy in patients harbouring recessive polymerase gamma mutations. Brain: J Neurol 135(Pt 1):62–71
19. Schicks J, Synofzik M, Schulte C, Schols L (2010) POLG, but not PEO1, is a frequent cause of cerebellar ataxia in Central Europe. Movement Disord: Official J Movement Disord Soc 25(15):2678–2682
20. Lax NZ, Hepplewhite PD, Reeve AK et al (2012) Cerebellar ataxia in patients with mitochondrial DNA disease: a molecular clinicopathological study. J Neuropathol Exp Neurol 71(2):148–161
21. Harrower T, Stewart JD, Hudson G et al (2008) POLG1 mutations manifesting as autosomal recessive axonal Charcot-Marie-Tooth disease. Arch Neurology 65(1):133–136
22. Mehta AR, Fox SH, Tarnopolsky M, Yoon G (2011) Mitochondrial mimicry of multiple system atrophy of the cerebellar subtype. Movement Disord: Official J Movement Disorder Soc 26(4):753–755
23. Milone M, Wang J, Liewluck T, Chen LC, Leavitt JA, Wong LJ (2011) Novel POLG splice site mutation and optic atrophy. Arch Neurol 68(6):806–811
24. Horvath R, Hudson G, Ferrari G et al (2006) Phenotypic spectrum associated with mutations of the mitochondrial polymerase gamma gene. Brain 129(Pt 7):1674–1684
25. Hinnell C, Haider S, Delamont S, Clough C, Hadzic N, Samuel M (2012) Dystonia in mitochondrial spinocerebellar ataxia and epilepsy syndrome associated with novel recessive POLG mutations. Movement Disord: Official J Movement Disord Soc 27(1):162–163
26. Deschauer M, Tennant S, Rokicka A et al (2007) MELAS associated with mutations in the POLG1 gene. Neurology 68(20):1741–1742
27. Tang S, Dimberg EL, Milone M, Wong LJ (2011) Mitochondrial neurogastrointestinal encephalomyopathy (MNGIE)-like phenotype: an expanded clinical spectrum of POLG1 mutations. J Neurol

28. Engelsen BA, Tzoulis C, Karlsen B et al (2008) POLG1 mutations cause a syndromic epilepsy with occipital lobe predilection. Brain 131(Pt 3):818–828
29. Tzoulis C, Engelsen BA, Telstad W et al (2006) The spectrum of clinical disease caused by the A467T and W748S POLG mutations: a study of 26 cases. Brain 129(Pt 7):1685–1692
30. Kollberg G, Moslemi AR, Darin N et al (2006) POLG1 mutations associated with progressive encephalopathy in childhood. J Neuropathol Exp Neurol 65(8):758–768
31. Uusimaa J, Hinttala R, Rantala H et al (2008) Homozygous W748S mutation in the POLG1 gene in patients with juvenile-onset Alpers syndrome and status epilepticus. Epilepsia 49(6):1038–1045
32. Saneto RP, Lee IC, Koenig MK et al (2010) POLG DNA testing as an emerging standard of care before instituting valproic acid therapy for pediatric seizure disorders. Seizure 19(3):140–146
33. Chinnery PF, Zeviani M (2008) 155th ENMC workshop: polymerase gamma and disorders of mitochondrial DNA synthesis, 21–23 September 2007, Naarden, The Netherlands. Neuromuscul Disord. 18(3):259–267
34. Visser NA, Braun KP, Leijten FS, van Nieuwenhuizen O, Wokke JH, van den Bergh WM (2011) Magnesium treatment for patients with refractory status epilepticus due to POLG1-mutations. J Neurol 258(2):218–222
35. Van Goethem G, Dermaut B, Lofgren A, Martin JJ, Van Broeckhoven C (2001) Mutation of POLG is associated with progressive external ophthalmoplegia characterized by mtDNA deletions. Nat Genet 28(3):211–212
36. Spelbrink JN, Li FY, Tiranti V et al (2001) Human mitochondrial DNA deletions associated with mutations in the gene encoding Twinkle, a phage T7 gene 4-like protein localized in mitochondria. Nat Genet 28(3):223–231
37. Longley MJ, Clark S, Yu Wai Man C et al (2006) Mutant POLG2 disrupts DNA polymerase gamma subunits and causes progressive external ophthalmoplegia. Am J Human Genet 78(6):1026–1034
38. Hudson G, Amati-Bonneau P, Blakely EL et al (2008) Mutation of OPA1 causes dominant optic atrophy with external ophthalmoplegia, ataxia, deafness and multiple mitochondrial DNA deletions: a novel disorder of mtDNA maintenance. Brain 131(Pt 2):329–337
39. Goffart S, Cooper HM, Tyynismaa H, Wanrooij S, Suomalainen A, Spelbrink JN (2009) Twinkle mutations associated with autosomal dominant progressive external ophthalmoplegia lead to impaired helicase function and in vivo mtDNA replication stalling. Hum Mol Genet 18(2):328–340
40. Tyynismaa H, Sun R, Ahola-Erkkila S et al (2012) Thymidine kinase 2 mutations in autosomal recessive progressive external ophthalmoplegia with multiple mitochondrial DNA deletions. Hum Mol Genet 21(1):66–75
41. Lamantea E, Tiranti V, Bordoni A et al (2002) Mutations of mitochondrial DNA polymerase gammaA are a frequent cause of autosomal dominant or recessive progressive external ophthalmoplegia. Annals Neurol 52(2):211–219
42. Luoma P, Melberg A, Rinne JO et al (2004) Parkinsonism, premature menopause, and mitochondrial DNA polymerase gamma mutations: clinical and molecular genetic study. Lancet 364(9437):875–882
43. Horvath R, Hudson G, Ferrari G et al (2006) Phenotypic spectrum associated with mutations of the mitochondrial polymerase gamma gene. Brain: J Neurol 129(Pt 7):1674–1684
44. Pagnamenta AT, Hargreaves IP, Duncan AJ et al (2006) Phenotypic variability of mitochondrial disease caused by a nuclear mutation in complex II. Mol Genet Metab 89(3):214–221
45. Echaniz-Laguna A, Chassagne M, de Seze J et al (2010) POLG1 variations presenting as multiple sclerosis. Arch Neurol 67(9):1140–1143
46. Van Goethem G, Schwartz M, Lofgren A, Dermaut B, Van Broeckhoven C, Vissing J (2003) Novel POLG mutations in progressive external ophthalmoplegia mimicking mitochondrial neurogastrointestinal encephalomyopathy. Eur J Hum Genet 11(7):547–549
47. Giordano C, Pichiorri F, Blakely EL et al (2010) Isolated distal myopathy of the upper limbs associated with mitochondrial DNA depletion and polymerase gamma mutations. Arch Neurol 67(9):1144–1146

48. Giordano C, Powell H, Leopizzi M et al (2009) Fatal congenital myopathy and gastrointestinal pseudo-obstruction due to POLG1 mutations. Neurol 72(12):1103–1105
49. Tzoulis C, Papingji M, Fiskestrand T, Roste LS, Bindoff LA (2009) Mitochondrial DNA depletion in progressive external ophthalmoplegia caused by POLG1 mutations. Acta Neurol Scand Suppl 2009(189):38–41
50. Isohanni P, Hakonen AH, Euro L et al (2011) POLG1 manifestations in childhood. Neurology 76(9):811–815
51. Winterthun S, Ferrari G, He L et al (2005) Autosomal recessive mitochondrial ataxic syndrome due to mitochondrial polymerase gamma mutations. Neurol 64(7):1204–1208
52. de Vries MC, Rodenburg RJ, Morava E et al (2007) Multiple oxidative phosphorylation deficiencies in severe childhood multi-system disorders due to polymerase gamma (POLG1) mutations. Eur J Pediat 166(3):229–234
53. Trifunovic A, Wredenberg A, Falkenberg M et al (2004) Premature ageing in mice expressing defective mitochondrial DNA polymerase. Nat 429(6990):417–423
54. Bailey LJ, Cluett TJ, Reyes A et al (2009) Mice expressing an error-prone DNA polymerase in mitochondria display elevated replication pausing and chromosomal breakage at fragile sites of mitochondrial DNA. Nucleic Acids Res 37(7):2327–2335
55. Ahlqvist KJ, Hamalainen RH, Yatsuga S et al (2012) Somatic progenitor cell vulnerability to mitochondrial DNA mutagenesis underlies progeroid phenotypes in Polg mutator mice. Cell Metab 15(1):100–109
56. Walter MC, Czermin B, Muller-Ziermann S et al (2010) Late-onset ptosis and myopathy in a patient with a heterozygous insertion in POLG2. J Neurol 257(9):1517–1523
57. Young MJ, Longley MJ, Li FY, Kasiviswanathan R, Wong LJ, Copeland WC (2011) Biochemical analysis of human POLG2 variants associated with mitochondrial disease. Hum Mol Genet 20(15):3052–3066
58. Chan SS, Longley MJ, Copeland WC (2005) The common A467T mutation in the human mitochondrial DNA polymerase (POLG) compromises catalytic efficiency and interaction with the accessory subunit. J Biol Chem 280(36):31341–31346
59. Michiels S, Danoy P, Dessen P et al (2007) Polymorphism discovery in 62 DNA repair genes and haplotype associations with risks for lung and head and neck cancers. Carcinogenesis 28(8):1731–1739
60. Nakano M, Tashiro K (2011) Association studies getting broader: a commentary on 'A polymorphism of the POLG2 gene is genetically associated with the invasiveness of urinary bladder cancer in Japanese males'. J Hum Genet 56(8):550–551
61. Tyynismaa H, Sembongi H, Bokori-Brown M et al (2004) Twinkle helicase is essential for mtDNA maintenance and regulates mtDNA copy number. Hum Mol Genet 13(24):3219–3227
62. Farge G, Holmlund T, Khvorostova J, Rofougaran R, Hofer A, Falkenberg M (2008) The N-terminal domain of TWINKLE contributes to single-stranded DNA binding and DNA helicase activities. Nucleic Acids Res 36(2):393–403
63. Jemt E, Farge G, Backstrom S, Holmlund T, Gustafsson CM, Falkenberg M (2011) The mitochondrial DNA helicase TWINKLE can assemble on a closed circular template and support initiation of DNA synthesis. Nucleic Acids Res 39(21):9238–9249
64. Guo S, Tabor S, Richardson CC (1999) The linker region between the helicase and primase domains of the bacteriophage T7 gene 4 protein is critical for hexamer formation. J Biol Chem 274(42):30303–30309
65. Van Hove JL, Cunningham V, Rice C et al (2009) Finding twinkle in the eyes of a 71-year-old lady: a case report and review of the genotypic and phenotypic spectrum of TWINKLE-related dominant disease. Am J Med Genet Part A 149A(5):861–867
66. Patel SS, Picha KM (2000) Structure and function of hexameric helicases. Annu Rev Biochem 69:651–697
67. Korhonen JA, Pande V, Holmlund T et al (2008) Structure-function defects of the TWINKLE linker region in progressive external ophthalmoplegia. J Mol Biol 377(3):691–705
68. Shutt TE, Gray MW (2006) Twinkle, the mitochondrial replicative DNA helicase, is widespread in the eukaryotic radiation and may also be the mitochondrial DNA primase in most eukaryotes. J Mol Evol 62(5):588–599

69. Suomalainen A, Majander A, Wallin M et al (1997) Autosomal dominant progressive external ophthalmoplegia with multiple deletions of mtDNA: clinical, biochemical, and molecular genetic features of the 10q-linked disease. Neurology 48(5):1244–1253
70. Fratter C, Gorman GS, Stewart JD et al (2010) The clinical, histochemical, and molecular spectrum of PEO1 (Twinkle)-linked adPEO. Neurology 74(20):1619–1626
71. Virgilio R, Ronchi D, Hadjigeorgiou GM et al (2008) Novel Twinkle (PEO1) gene mutations in mendelian progressive external ophthalmoplegia. J Neurol 255(9):1384–1391
72. Kiechl S, Horvath R, Luoma P et al (2004) Two families with autosomal dominant progressive external ophthalmoplegia. J Neurol Neurosurg Psychiatry 75(8):1125–1128
73. Hudson G, Deschauer M, Busse K, Zierz S, Chinnery PF (2005) Sensory ataxic neuropathy due to a novel C10Orf2 mutation with probable germline mosaicism. Neurology 64(2):371–373
74. Baloh RH, Salavaggione E, Milbrandt J, Pestronk A (2007) Familial parkinsonism and ophthalmoplegia from a mutation in the mitochondrial DNA helicase twinkle. Arch Neurol 64(7):998–1000
75. Bohlega S, Van Goethem G, Al Semari A et al (2009) Novel Twinkle gene mutation in autosomal dominant progressive external ophthalmoplegia and multisystem failure. Neuromuscul Disord 19(12):845–848
76. Echaniz-Laguna A, Chanson JB, Wilhelm JM et al (2010) A novel variation in the Twinkle linker region causing late-onset dementia. Neurogenetics 11(1):21–25
77. Lewis S, Hutchison W, Thyagarajan D, Dahl HH (2002) Clinical and molecular features of adPEO due to mutations in the Twinkle gene. J Neurol Sci 201(1–2):39–44
78. Nikali K, Suomalainen A, Saharinen J et al (2005) Infantile onset spinocerebellar ataxia is caused by recessive mutations in mitochondrial proteins Twinkle and Twinky. Hum Mol Genet 14(20):2981–2990
79. Lonnqvist T, Paetau A, Valanne L, Pihko H (2009) Recessive twinkle mutations cause severe epileptic encephalopathy. Brain: J Neurol 132(Pt 6):1553–1562
80. Hakonen AH, Davidzon G, Salemi R et al (2007) Abundance of the POLG disease mutations in Europe, Australia, New Zealand, and the United States explained by single ancient European founders. Eur J Hum Genet 15(7):779–783
81. Sarzi E, Goffart S, Serre V et al (2007) Twinkle helicase (PEO1) gene mutation causes mitochondrial DNA depletion. Annals Neurol 62(6):579–587
82. Hakonen AH, Isohanni P, Paetau A, Herva R, Suomalainen A, Lonnqvist T (2007) Recessive Twinkle mutations in early onset encephalopathy with mtDNA depletion. Brain: J Neurol 130(Pt 11):3032–3040
83. Patel G, Johnson DS, Sun B et al (2011) A257T linker region mutant of T7 helicase-primase protein is defective in DNA loading and rescued by T7 DNA polymerase. J Biol Chem 286(23):20490–20499
84. Holmlund T, Farge G, Pande V, Korhonen J, Nilsson L, Falkenberg M (2009) Structure-function defects of the twinkle amino-terminal region in progressive external ophthalmoplegia. Biochimica Et Biophysica Acta 1792(2):132–139
85. Tyynismaa H, Mjosund KP, Wanrooij S et al (2005) Mutant mitochondrial helicase Twinkle causes multiple mtDNA deletions and a late-onset mitochondrial disease in mice. Proc Natl Acad Sci U S A 102(49):17687–17692
86. Pohjoismaki JL, Goffart S, Spelbrink JN (2011) Replication stalling by catalytically impaired Twinkle induces mitochondrial DNA rearrangements in cultured cells. Mitochondrion 11(4):630–634
87. Levy SE, Chen YS, Graham BH, Wallace DC (2000) Expression and sequence analysis of the mouse adenine nucleotide translocase 1 and 2 genes. Gene 254(1–2):57–66
88. Huang SG, Odoy S, Klingenberg M (2001) Chimers of two fused ADP/ATP carrier monomers indicate a single channel for ADP/ATP transport. Arch Biochem Biophys 394(1):67–75
89. Faustin B, Rossignol R, Rocher C, Benard G, Malgat M, Letellier T (2004) Mobilization of adenine nucleotide translocators as molecular bases of the biochemical threshold effect observed in mitochondrial diseases. J Biol Chem 279(19):20411–20421

90. Benjamin F, Rodrigue R, Aurelien D et al (2011) The respiratory-dependent assembly of ANT1 differentially regulates Bax and Ca2+ mediated cytochrome c release. Front Biosci (Elite Ed) 3:395–409
91. Pebay-Peyroula E, Dahout-Gonzalez C, Kahn R, Trezeguet V, Lauquin GJ, Brandolin G (2003) Structure of mitochondrial ADP/ATP carrier in complex with carboxyatractyloside. Nature 426(6962):39–44
92. Brandolin G, Boulay F, Dalbon P, Vignais PV (1989) Orientation of the N-terminal region of the membrane-bound ADP/ATP carrier protein explored by antipeptide antibodies and an arginine-specific endoprotease. Evidence that the accessibility of the N-terminal residues depends on the conformational state of the carrier. Biochemistry 28(3):1093–1100
93. Walker JE, Runswick MJ (1993) The mitochondrial transport protein superfamily. J Bioenerg Biomembr 25(5):435–446
94. Napoli L, Bordoni A, Zeviani M et al (2001) A novel missense adenine nucleotide translocator-1 gene mutation in a Greek adPEO family. Neurology 57(12):2295–2298
95. Komaki H, Fukazawa T, Houzen H, Yoshida K, Nonaka I, Goto Y (2002) A novel D104G mutation in the adenine nucleotide translocator 1 gene in autosomal dominant progressive external ophthalmoplegia patients with mitochondrial DNA with multiple deletions. Annals Neurol 51(5):645–648
96. Siciliano G, Tessa A, Petrini S et al (2003) Autosomal dominant external ophthalmoplegia and bipolar affective disorder associated with a mutation in the ANT1 gene. Neuromuscul Disord 13(2):162–165
97. Deschauer M, Hudson G, Muller T, Taylor RW, Chinnery PF, Zierz S (2005) A novel ANT1 gene mutation with probable germline mosaicism in autosomal dominant progressive external ophthalmoplegia. Neuromuscul Disord 15(4):311–315
98. Palmieri L, Alberio S, Pisano I et al (2005) Complete loss-of-function of the heart/muscle-specific adenine nucleotide translocator is associated with mitochondrial myopathy and cardiomyopathy. Hum Mol Genet 14(20):3079–3088
99. Graham BH, Waymire KG, Cottrell B, Trounce IA, MacGregor GR, Wallace DC (1997) A mouse model for mitochondrial myopathy and cardiomyopathy resulting from a deficiency in the heart/muscle isoform of the adenine nucleotide translocator. Nat Genet 16(3):226–234
100. Kawamata H, Manfredi G (2011) Introduction to neurodegenerative diseases and related techniques. Methods Mol Biol 793:3–8
101. Park KP, Kim HS, Kim ES, Park YE, Lee CH, Kim DS (2011) SLC25A4 and C10ORF2 mutations in autosomal dominant progressive external ophthalmoplegia. J Clin Neurol 7(1):25–30

Chapter 9
Defects in Mitochondrial Dynamics and Mitochondrial DNA Instability

Patrick Yu-Wai-Man, Guy Lenaers and Patrick F. Chinnery

Abbreviations

DRP1   Dynamin-related protein 1
GDAP1   Ganglioside-induced differentiation-associated protein 1
hFIS1   Fission 1 homolog
MFN1   Mitofusin 1
MFN2   Mitofusin 2
OPA1   Optic atrophy 1
OPA3   Optic atrophy 3

Introduction

The first mitochondrial DNA (mtDNA) mutations linked with human disease were identified in 1988, with the seminal finding of mtDNA deletions in muscle biopsies obtained from patients with myopathies [1], followed shortly afterwards by the report of the m.11778G>A mutation in families with Leber hereditary optic neuropathy (LHON) [2]. The next decade saw an exponential increase in the

---

P. Yu-Wai-Man (✉) · P. F. Chinnery
Wellcome Trust Centre for Mitochondrial Research, Institute of Genetic Medicine,
International Centre for Life, Newcastle University, Newcastle Upon Tyne, NE1 3BZ, UK
e-mail: Patrick.Yu-Wai-Man@ncl.ac.uk

Department of Ophthalmology, Neuro-Ophthalmology Division, Royal Victoria Infirmary,
Newcastle Upon Tyne, NE1 4LP, UK

G. Lenaers
Neuropathies Optiques Héréditaires, Institut des Neurosciences de Montpellier,
Université Montpellier I et II, INSERM U1051, Montpellier, France

P. F. Chinnery
Department of Neurology, Royal Victoria Infirmary,
Newcastle Upon Tyne, NE1 4LP, UK

number of pathogenic mtDNA mutations and an increasing realisation among clinicians that mitochondrial disease has a much broader phenotypic spectrum than initially considered, ranging from the more classical mitochondrial syndromes, such as mitochondrial encephalomyopathy with lactic acidosis and stroke-like episodes (MELAS) and chronic progressive external ophthalmoplegia (CPEO), to more non-specific clinical manifestations limited to fatigue, gastrointestinal disturbance, infertility, and psychiatric disturbances [3–5]. Mitochondria have a rather limited genetic autonomy and in 2001, nuclear genetic defects involving *POLG* and *PEO1* were conclusively linked to autosomal-dominant CPEO in several independent families [6, 7]. This led to a paradigm shift in our understanding of fundamental aspects of mitochondrial biology, in particular, the pathological consequences resulting from disturbed mtDNA replication and the subsequent accumulation of high levels of somatic mtDNA deletions within at-risk tissues [8]. *POLG1* is the stereotypical mtDNA maintenance disorder, but for over a decade, the mechanisms leading to the formation and clonal expansion of these somatic mtDNA abnormalities have remained largely unresolved [9, 10]. In this chapter, the focus is on a number of related nuclear mitochondrial disorders where the accumulation of mtDNA deletions is not related to a dysfunctional replication machinery *per se*, but to a primary defect in mitochondrial network dynamics, revealing fascinating insights on disease mechanisms intrinsically linked with both mitochondrial structure and function.

Mitochondrial Fusion and Fission

Mitochondria are not discrete organelles located at static positions within the cell. Instead, mitochondrial distribution and concentration are intimately linked to the local energetic demands and sophisticated systems are required, especially in more complex cellular environments such as neurones, which can extend over relatively long anatomical distances [11–13]. High-resolution *in vitro* and time-lapsed imaging has revealed in exquisite details the highly-interconnected and dynamic nature of the mitochondrial network. This syncitium extends throughout the cell and it is in a constant state of flux with the balance between fusional and fissional forces regulating the overall morphology—a fluid state dictated by physical as well as physiological constraints [14, 15]. The major players co-ordinating mitochondrial fusion and fission were first identified in elegant experiments performed in yeast models (Fig. 9.1). Subsequently, it became clear that these mediators have been highly conserved throughout evolution, a clear indication of the critical role played by these proteins in both simple and complex organisms [16]. It is therefore not surprising that mutations in the main pro-fusion (*OPA1* and *MFN2*) and pro-fission nuclear genes (*OPA3*, *DRP1* and *GDAP1*) have been identified as causing human disease, with phenotypes ranging from severe, early-onset and invariably fatal encephalomyopathies, to isolated optic atrophy and peripheral neuropathy, and to more complex late-onset multi-system neurological disorders (Table 9.1) [17–20].

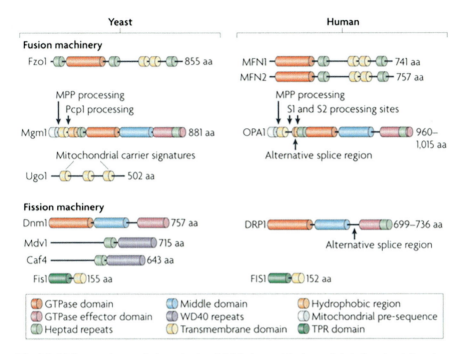

Fig. 9.1 Major proteins regulating mitochondrial fusion and fission and their functional domains. The major mediators of mitochondrial fusion in yeast are Fzo1 and Mgm1, which are located in the outer and inner membrane, respectively. Fzo1 and Mgm1 are connected by the yeast-specific outer membrane protein Ugo1. Dnm1 is a soluble cytosolic protein that is recruited to mitochondria by the outer membrane pro-fission protein Fis1. Two yeast-specific proteins, Mdv1 and Caf4, serve as molecular adaptors for Dnm1 and Fis1. Matrix-processing peptidase (MPP) removes the matrix-targeting sequences following import and Pcp1 is a membrane protease involved in the alternative processing of Mgm1. The mammalian orthologues of the yeast proteins are shown on the right. aa: amino acids; TPR: tetratricopeptide repeat. (From [12]. Reprinted with kind permission from Nature Publishing Group)

OPA1-Related Disorders

Epidemiology

A recent population-based epidemiological survey of patients with autosomal-dominant optic atrophy (DOA) has estimated the prevalence of *OPA1* mutations at ~1 in 50,000 in the North of England [21]. This figure is likely to be an underestimate as 10–20 % of *OPA1* mutation carriers are non-penetrant and a proportion of patients remain mildly visually affected throughout their life [21, 22]. These two specific categories are therefore much less likely to have been fully ascertained.

Table 9.1 Primary mitochondrial dynamic disorders

Inheritance	Locus	Gene	OMIM	Phenotypes	Reference
AD	1p36.2	MFN2	609260, 601152	Charcot-Marie-Tooth disease—axonal form (CMT2A), hereditary motor and sensory neuropathy 6 (HMSN6)	[84, 85, 96]
	3q28-q29	OPA1	165500	Isolated optic atrophy and syndromic forms of dominant optic atrophy (DOA+)	[31, 32, 45]
	8q13-q21.1	GDAP1	607831	Charcot-Marie-Tooth disease—axonal form (CMT2K)	[100]
	12p11.21	DRP1	603850	Severe infantile neurodegenerative disease	[79]
	19q13.2-q13.3	OPA3	165300	Optic atrophy and premature cataracts (ADOAC)	[72]
AR	8q13-q21.1	GDAP1	214400	Charcot-Marie-Tooth— demyelinating form (CMT4A)	[99]
	19q13.2-q13.3	OPA3	258501	3-methylglutaconic aciduria type III, (Costeff syndrome)	[71]

Visual Failure

The cardinal clinical feature of DOA is one of progressive, insidious, bilateral visual loss starting in early childhood [21, 23]. There is a wide intra- and interfamilial variability in disease severity with visual acuities ranging from 20/20 to the detection of hand movement only. Visual function worsens in 50–75 % of patients with DOA and most of these patients are eventually registered blind [24, 25]. However, the rate of visual deterioration is highly variable and it is difficult to predict the disease course for an individual patient [21, 23]. Colour vision is significantly impaired and most patients will exhibit a central or cecocentral scotoma on visual field testing, reflecting the selective involvement of retinal ganglion cells (RGCs) within the papillomacular bundle (Fig. 9.2). Interestingly, patients with DOA usually do not demonstrate an afferent pupillary defect and this clinical observation is likely related to the relative immunity of melanopsin-containing RGCs to the deleterious consequences of pathogenic *OPA1* mutations [26]. The optic nerve head can look diffusely pale or it can have a characteristic temporal wedge of pallor, especially in early disease, when RGC loss remains mostly limited to the papillomacular bundle (Fig. 9.3) [21, 24].

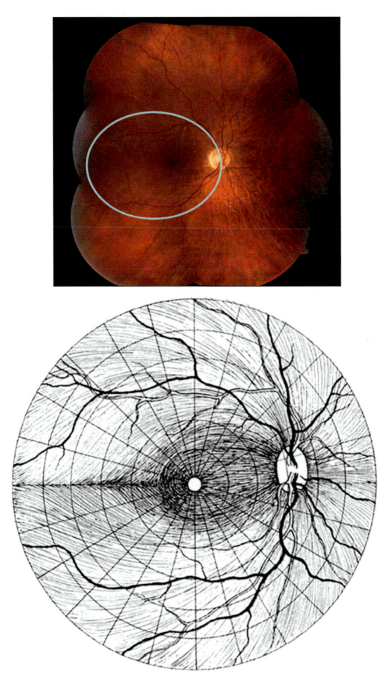

Fig. 9.2 The anatomical location and organisation of the papillomacular bundle within the macula. The papillomacular bundle contains the highest density of RGC axons. These RGCs subserve the central visual field and if compromised, visual acuity and colour discrimination are both severely affected. The *blue oval boundary* encompasses the macula

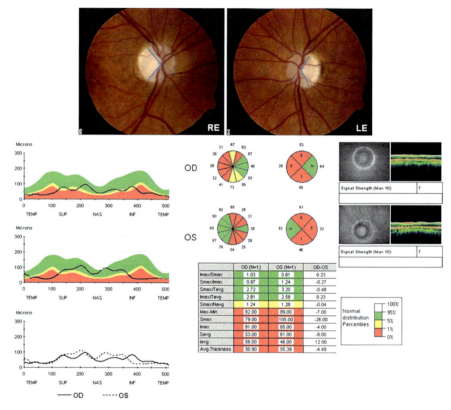

Fig. 9.3 Typical fundal appearance and pattern of retinal ganglion cell loss in autosomal-dominant optic atrophy. The optic disc appearance of a patient with a confirmed pathogenic *OPA1* mutation showing pallor of the neuroretinal rim, which is more marked temporally. The *bottom panels* illustrate the pattern of retinal nerve fibre layer (RNFL) thinning seen in patients with *OPA1* mutations, with relative sparing of the nasal peripapillary quadrant. The RNFL profile for each eye is superimposed on the normal distribution percentiles and compared with each other (*Bottom left panel*). Various measurement parameters are automatically generated by the analysis software including sectoral RNFL thickness for each individual quadrant and clock hour, and an overall value for the average RNFL thickness (*Bottom middle panel*). The normal distribution indices are colour-coded: (i) *red* < 1 %, (ii) *yellow* 1–5 %, (iii) *green* 5–95 %, and (iv) *white* > 95 % (*Bottom right panel*). (From [19]. Reproduced with kind permission from Elsevier)

Mutational Spectrum

The *OPA1* gene (3q28-q29) consists of 30 exons spanning ~ 100 kb of genomic DNA and it codes for a 960 amino acid, dynamin-related GTPase protein-located within the inner mitochondrial membrane (see Table 9.1) [27]. Alternative splicing of exons 4, 4b, and 5b result in eight different mRNA isoforms [28, 29]. There are tissue-specific variations in their relative concentrations and the functional relevance and subcellular localisation of these different mRNA isoforms are currently being investigated [30].

Between 50 and 60 % of families with DOA harbour pathogenic *OPA1* mutations and more than 200 variants have been reported so far in this highly polymorphic gene [31–33]. These mutations cluster in three specific functional regions: the GTPase domain (Exons 8–15), the dynamin central domain (Exons 16–23), and the putative GTPase effector domain at the carboxy-terminal end [33, 34]. The most common mutational subtypes are single base-pair substitutions (69 %), followed by deletions (26 %) and insertions (5 %) [33, 34]. Using more sophisticated multiplex ligation probe amplification (MLPA) assays, large-scale *OPA1* rearrangements have been detected in 10–20 % of patients who were initially thought to be *OPA1* negative by standard polymerase chain reaction (PCR) sequencing methods [35, 36]. The majority of these *OPA1* mutations result in premature termination codons and the resulting truncated mRNA species are highly unstable, being rapidly degraded by protective surveillance mechanisms operating via non-sense-mediated mRNA decay [37].

Expanding Phenotypes

With greater access to *OPA1* genetic testing, a specific *OPA1* mutation in exon 14 (c.1334G>A, p.R445H) was found to be particularly prevalent among patients with DOA and sensorineural deafness [38–41]. More recently, the phenotypes associated with *OPA1* disease have expanded further to encompass a wider range of neurological features including ataxia, myopathy, peripheral neuropathy, and classical CPEO (Fig. 9.4) [42–45]. These so-called DOA+ syndromes are not rare, affecting up to 20 % of *OPA1*-mutation carriers [45], and with *OPA1* screening now increasingly performed as part of diagnostic panels for patients with unexplained neurodegenerative disorders, other novel pathological manifestations are likely to emerge.

Disease Mechanisms

Haploinsufficiency is a major disease mechanism in DOA and the pathological consequences of a dramatic reduction in OPA1 protein level are highlighted by those rare families who are heterozygous for microdeletions spanning the entire *OPA1* coding region [36, 46]. Interestingly, there is a threefold increased risk of developing a more severe DOA+ phenotype with missense *OPA1* mutations involving the catalytic GTPase domain [45]. Although this genotype–phenotype association supports a dominant-negative effect exerted by an aberrant mutant OPA1 protein on mitochondrial function, other mutational subtypes that result in haploinsufficiency can also lead to DOA+, clearly indicating that the mechanisms leading to multi-system neurological involvement, as opposed to pure optic atrophy, are likely to be multifactorial.

OPA1 is highly expressed within the RGC layer, but it is a ubiquitous protein, and abundant levels have also been identified in photoreceptors, and other non-ocular tissues such as the inner ear and the brain [47–49]. The native OPA1 protein has

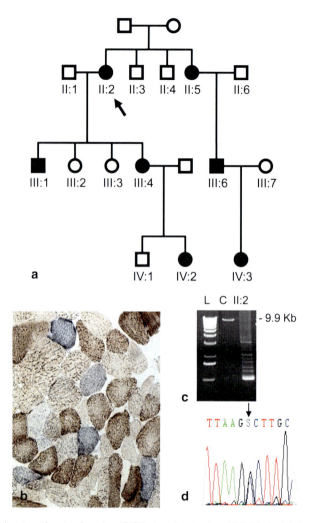

Fig. 9.4 Dominant optic atrophy plus (DOA+) and mitochondrial DNA deletions. **a** A large three-generation British family manifesting early-onset optic atrophy and progressive external ophthalmoplegia with variable neurological features including sensorineural deafness, ataxia, and peripheral neuropathy. The two affected children (*IV:2* and *IV:3*) are currently less than 10-years-old, and as yet, they do not demonstrate any DOA+ features. However, they are both already severely visually impaired with best-corrected visual acuities reduced to less than 20/200. **b** Dual COX-SDH histochemistry performed on 10-micron-thick cryostat sections showing the presence of COX-negative muscle fibres (∼10 %) in the proband's (*II:2*) biopsy specimen. **c** A 9.9-kb long-range PCR assay, encompassing the major arc of the mitochondrial genome, was performed on homogenate skeletal muscle DNA (*II:2*). Characteristic multiple mtDNA deletion bands were identified. *L* DNA ladder; *C* normal control. **d** The initial screen for *POLG1*, *POLG2*, *PEO1*, and *SLC25A4* was negative, and further molecular investigation eventually revealed a c.1635C>A (p.S545R) *OPA1* GTPase mutation segregating in this family. Sequence chromatogram with the *arrow* indicating the location of the pathogenic heterozygous *OPA1* variant. (Adapted with permission from [42])

9 Defects in Mitochondrial Dynamics and Mitochondrial DNA Instability

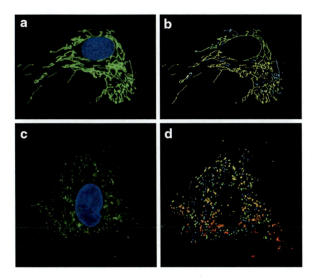

Fig. 9.5 Mitochondrial network fragmentation in *OPA1*-mutant fibroblasts. Staining was performed with MitoTracker™ (Invitrogen, UK), a mitochondrion-specific fluorescent dye, and Hoechst 33342 (Invitrogen, UK) for the nucleus. Images were captured as a series of Z-stacks with 16 slices, each 0.2 µm apart, using an inverted epifluorescence microscope (Leica, UK). Huygens™ generates a data file, which includes information on the total number of mitochondrial fragments per cell, their individual lengths and volumes. A false colour scheme aids visualisation of the overall network morphology. **a** and **b** Control fibroblast, pre- and post-image deconvolution and reconstruction, respectively, showing a highly interconnected mitochondrial network, **c** and **d** Fragmented mitochondrial network in an *OPA1*-mutant fibroblast

an amino-terminal domain with a mitochondrial targeting pre-sequence, which is cleaved by mitochondrial proteases following import into the mitochondrial intermembrane space [30]. Additional post-translational maturational steps generate long (L) and short (S) forms of the protein, which on their own have little functional activity, but when co-expressed, they complement each other, triggering mitochondrial network fusion [50, 51]. A key mitochondrial protease is paraplegin, encoded by *SPG7* (16q24.3), and interestingly, mutations in this gene have been identified in an autosomal recessive form of hereditary spastic paraplegia associated with bilateral optic atrophy (HSP7, OMIM 607259) [52, 53]. OPA1 is firmly anchored to the inner mitochondrial membrane, via its transmembrane domain, in the narrow junctional regions enclosing the mitochondrial cristae. The transmembrane domain is followed by a series of coiled-coil segments that allow OPA1 to homo-polymerise into a cylindrical tubular structure [12, 17, 30].

OPA1 is a multifunctional protein and the highly conserved dynamin GTPase domain is a central regulatory component [12, 17, 30]. One important attribute relates to its pro-fusion properties and unsurprisingly, *OPA1* mutations result in mitochondrial network fragmentation typified by dysmorphic mitochondria with balloon-like enlargements, disrupted cristae, and paracrystalline inclusion bodies (Fig. 9.5) [54–56].

Pro-apoptotic cytochrome c molecules are tightly sequestered within the cristae junctions and their unregulated release into the cytosol potentiates the apoptotic cascade, ultimately contributing to cell death and tissue dysfunction [57–59].

OPA1 is also thought to regulate oxidative phosphorylation by stabilising the mitochondrial respiratory chain complexes and by facilitating the effective coupling of electron transport with adenosine triphosphate (ATP) synthesis [54, 55]. In one study, fibroblasts harbouring *OPA1* mutations showed reduced mitochondrial membrane potential and ATP synthesis was significantly impaired secondary to a predominantly complex I defect [55]. However, another study failed to confirm a complex I defect reporting instead a 25 % reduction in complex IV activity [54]. Although these contrasting *in vitro* results could simply reflect the different biochemical assays used, another attractive explanation is a mutation-specific effect on the various mitochondrial respiratory chain complexes. With the use of *in vivo* phosphorus magnetic resonance spectroscopy (^{31}P-MRS), impaired mitochondrial oxidative phosphorylation has been conclusively demonstrated in the calf muscles of patients harbouring a broad range of *OPA1* mutations [60–62].

In addition to these essential cellular functions, there is now growing evidence that OPA1 plays a critical role in maintaining the integrity of the mitochondrial genome. Central to this argument is the characteristic histochemical finding of cytochrome c oxidase (COX)-negative fibres in skeletal muscle biopsies from patients harbouring *OPA1* mutations (Fig. 9.4b) [42, 43]. At the single-fibre level, the biochemical defect is triggered by the accumulation of high levels of clonally-expanded mitochondrial DNA (mtDNA) deletions [63]. Marked mitochondrial proliferation was also observed probably as a compensatory bioenergetic mechanism to maintain adequate levels of ATP production—in keeping with the *sick mitochondrion* hypothesis [64–66]. These somatic mtDNA abnormalities are visible on long-range PCR analysis of homogenate skeletal muscle DNA as multiple mtDNA deletion bands (Fig. 9.4c), clearly classifying *OPA1* as a novel disorder of mtDNA maintenance.

Animal Models

Three *Opa1* mouse models have been developed harbouring truncative mutations in exon 8 (c.1051C>T) [67], intron 10 (c.1065+5 g>a) [68], and exon 27 (c.2708–2711delTTAG) (Dr Guy Lenaers, unpublished data). In all these models, homozygous mutant mice (*Opa1-/-*) die *in utero* during early embryogenesis, highlighting the central role played by Opa1 in early development. Heterozygous mutant mice (*Opa1±*) exhibit a 50 % reduction in overall protein expression and these mice faithfully replicate the human phenotype, with a slowly progressive bilateral optic neuropathy and reduced visual parameters. Optic nerve degeneration was documented as early as 6 months, but it was much more striking by 2 years of age. Histological and retrograde labelling experiments confirmed a gradual loss of RGCs and the surviving axons had abnormal morphologies with segmental areas of demyelination and myelin aggregation. Interestingly, loss of dendritic arborisation was observed for on-centre but not for off-centre RGCs and these distinct features, indicative of early neuronal stress, preceded the onset of axonal loss [69]. An increased number of autophagosomes was also noted in the RGC layer as a relatively late neurodegenerative feature linked to the accumulation of dysfunctional mitochondria. COX deficiency and mtDNA deletions were not identified in the RGC layer of heterozygous mutant mice harbouring the intron 10 c.1065+5 g>a *Opa1* mutation, with the caveat, however, that this mouse model only manifested pure optic atrophy [70]. The characterisation of DOA+ mouse models will therefore prove essential to further dissect the mechanisms contributing to the development of multi-system neurological involvement in a subgroup of *OPA1* mutation carriers [45].

OPA3-Related Disorders

Mutations in *OPA3* were first identified in Iraqi-Jewish families with autosomal-recessive type III 3-methylglutaconic aciduria (Costeff syndrome), a progressive neurodegenerative disorder characterised by early-onset optic atrophy, hypotonia, ataxia, extrapyramidal dysfunction and cognitive decline (see Table 9.1) [71]. Pathogenic *OPA3* mutations were subsequently described in two French families with a dominantly inherited form of optic atrophy associated with premature cataracts (ADOAC) [72]. Similar to OPA1, the OPA3 protein has a mitochondrial-targeting domain and it was initially thought to localise to the mitochondrial inner membrane [73, 74]. However, more recent evidence suggests instead that OPA3 is an integral mitochondrial outer membrane protein with its carboxyl-terminus projecting into the cytosol [75]. Mitochondrial fragmentation was induced with both mutant forms of the protein (p.G93S) and following overexpression of the wild-type protein, indicating an important profission role.

A zebrafish model of type III 3-methylglutaconic aciduria has been created by disrupting the *opa3* reading frame and engineering a premature termination codon [76]. One-year-old *opa3* null allele zebrafish had significant thinning of the RGC layer and reduced optic nerve diameters. These mutants also developed prominent extraocular features with a marked impairment in swimming behaviour secondary to ataxia, loss of buoyancy control, and hypokinesia. The *opa3* null allele zebrafish had normal mitochondrial oxidative phosphorylation but there was an increased sensitivity to various pro-apoptotic stimuli, consistent with the results obtained in fibroblasts from an affected French patient with ADOAC [72]. An *Opa3* mouse model carrying a c.365T>C (p.L122P) mutation has also been described [77]. Heterozygous mutant mice were phenotypically normal whereas homozygous mutant mice developed multi-system organ failure with a reduced lifespan of less than 4 months. These mice had severely impaired visual function with marked thinning of the RGC layer, again highlighting the selective vulnerability of this specific cell type to disturbed mitochondrial dynamics [19].

DRP1-Related Disorders

Mitochondrial fission is dependent upon two major proteins, DRP1 from the cytosol and hFIS1, which is located within the mitochondrial outer membrane [12, 17]. DRP1 and hFIS1 are the mammalian orthologues of the yeast proteins Dnm1 and Fis1, respectively (see Fig. 9.1). Mitochondrial fission involves hFIS1 recruiting DRP1 to the mitochondrial outer membrane, with both proteins then assembling into rings and spirals to encircle and constrict the mitochondrial tubule [12, 17]. Following GTPase hydrolysis and energy release, these helical structures twist and the conformational changes result in mitochondrial fission. The actual sequence of events is likely to be more complex and additional studies are needed to explore the intermediate steps involved, and crucially, the roles played by the cytoskeleton and endoplasmic reticulum in facilitating this process [78]. The central role of mitochondrial dynamics in cellular homeostatis was irrefutably confirmed when a heterozygous missense *DRP1* mutation was identified in a female infant with a complex, and ultimately fatal, neurodegenerative disorder characterised by optic atrophy, abnormal brain development and disturbed metabolic function (see Table 9.1) [79]. The mitochondrial network in the patient's cultured fibroblasts was elongated with tangled tubular structures clumped around the nucleus. A likely explanation is that this specific *DRP1* mutation leads to a dominant-negative loss of function, resulting in unopposed mitochondrial fusion and altered mitochondria cellular distribution [80]. No other cases have since been reported, which points towards the lethal nature of disturbing DRP1-mediated mitochondrial fission.

MFN2-Related Disorders

Clinical Manifestations

Charcot-Marie-Tooth (CMT) disease is a commonly encountered group of inherited peripheral neuropathies in neurological practice and the prevalence in the general population has been estimated at ~ 1 in 2,500 [81, 82]. The pathological hallmark is irreversible degeneration of the peripheral nerves and the major clinical classification is based on whether this is predominantly axonal or demyelinating [81, 82]. Affected patients develop distal muscle weakness and sensory loss but the age of onset and rate of progression can be highly variable, depending on the actual mutational status [83]. *MFN2* mutations typically result in the most common axonal form of autosomal-dominant Charcot-Marie-Tooth disease (CMT2A, OMIM 609260) [84]. However, there is a degree of phenotypic variability and in some families the peripheral neuropathy is complicated by subacute visual failure and optic atrophy (see Table 9.1) [85]. Although relatively rare compared with pure CMT2A, this specific disease variant, hereditary sensory motor neuropathy type 6 (HSMN6, OMIM 601152), is mechanistically important because of the causal link it establishes with

RGC dysfunction and optic nerve degeneration [19]. Visual acuity is usually reduced to 20/200 or worse in HSMN6, but despite the severity of the initial visual loss, some patients can show a dramatic visual recovery in later life [85].

Pathophysiology

MFN2 and its companion protein MFN1 are mitochondrial outer membrane proteins with dynamin GTPase domains. They are structurally and functionally complementary to OPA1, and these three proteins carefully choreograph the sequence of events leading to mitochondrial outer and then inner membrane fusion [18, 19, 86, 87]. The carboxy-terminal region of both mitofusins faces the cytosol and it is folded into a coiled-coil tertiary structure. An important initial step in the fusional process is the mechanical tethering of neighbouring mitochondria and this is achieved by the formation of homo- or hetero-typic complexes between MFN1 and MFN2 coiled-coil segments [18, 19, 86, 87].

Knockout *Mfn1* and *Mfn2* mice have led to important breakthroughs in our understanding of how mitofusins operate *in vivo* [88–91]. Similar to *Opa1* mouse models, homozygous *Mfn1* and *Mfn2* mutants were embryonically lethal secondary to placental insufficiency, further reinforcing the indispensable nature of these pro-fusion proteins in early development [87]. The mitochondrial network in cultured mouse embryonic fibroblasts was severely fragmented and a dramatic loss of membrane potential was observed in a subpopulation of these fragmented mitochondria. In one transgenic *Mfn2* mouse model, oxygraphic and enzymatic measurements revealed a combined complex II and V defect and reduced ATP synthesis in brain tissues [90]. By using an elegant cre-*loxP* recombinase strategy, a conditional *Mfn2* knockout mouse model was created that spared placental tissues, thereby permitting embryonic development to proceed to term [89]. The most striking histopathological feature of these *Mfn2^{loxP}* mutant mice was seen in the cerebellum where Purkinje cell neurodegeneration was associated with increased levels of apoptosis throughout the cerebellar layers [89]. These aberrant Purkinje cells manifested a marked respiratory chain defect and their mitochondrial networks were grossly fragmented. In another independent study, fibroblasts from patients harbouring *MFN2* mutations were also found to exhibit a mitochondrial coupling defect with impaired membrane potential and depressed oxidative capacity [92]. A significant biochemical defect in the visual cortex has also been reported with ^{31}P-MRS in patients with HSMN6 and impaired vision [93]. A plausible hypothesis is that MFN2 can exert a direct influence on mitochondrial biogenesis by regulating the expression of nuclear-encoded respiratory chain subunits [94]. An even more fundamental observation in these conditional *Mfn2* knockout mice was the direct correlation made between the lack of mitochondrial fusion and the loss of mtDNA nucleoids within affected cells [89, 95].

Somatic mtDNA Deletions

A novel heterozygous *MFN2* GTPase mutation (c.629A>T, p.D210V) was recently identified in a large Tunisian family segregating a variable neurological phenotype characterised by optic atrophy, deafness, cerebellar ataxia, and proximal myopathy, in addition to the more classical CMT2A features of an axonal sensorimotor neuropathy [96]. COX-negative fibres and multiple mitochondrial mtDNA deletions were detected in skeletal muscle biopsies from several affected family members—findings strikingly reminiscent of the pathological features first reported in patients with *OPA1* mutations and DOA+ phenotypes [97]. The obvious corollary is that MFN2 must be involved in mtDNA replication with the mutant protein promoting, directly or indirectly, the formation and clonal expansion of secondary mtDNA deletions. *In vitro* fibroblast studies confirmed the pathogenic nature of this novel *MFN2* mutation by revealing an overt respiratory chain defect and marked fragmentation of the mitochondrial network [97]. Significant mitochondrial proliferation has also been found in blood leukocytes from patients harbouring a broad range of pathogenic *MFN2* mutations [98]. CMT2A should therefore be considered as the newest member of an expanding group of nuclear mitochondrial disorders linked to disturbed mtDNA maintenance.

GDAP1-Related Disorders

GDAP1 mutations were originally described in families with a recessively inherited form of severe peripheral neuropathy associated with demyelinating features (CMT4A) [99]. Although relatively rare, dominant mutations in *GDAP1* have also been linked to a milder axonal form of CMT (CMT2K) with a variable age of onset (see Table 9.1) [100, 101]. Recent work has shed some important light on the rather intriguing fact that *GDAP1* mutations can lead to both dominant and recessive forms of CMT, with contrasting disease severity and pathology [102]. GDAP1 is located in the mitochondrial outer membrane and it is an important pro-fission protein, with recessive *GDAP1* mutations resulting in a reduction in mitochondrial fissional activity [102–104]. GDAP1-induced fission is dependent on hFIS1 and DRP1 and interestingly, the fragmentation induced by specific GDAP1 overexpression does not increase the cell's susceptibility to undergo apoptosis. The underlying disease mechanism in autosomal-dominant CMT2K is more complex. Dominant *GDAP1* mutations have a multi-modal deleterious impact on cell survival resulting in impaired fusion, elevated levels of reactive oxygen species, and increased apoptosis [102]. Fibroblasts from patients with CMT2K had impaired complex I activity and the mitochondrial network was significantly more tubular with larger diameters [105]. GDAP1 therefore seems to be involved in bioenergetic regulation in addition to its role in the regulation of mitochondrial network morphology.

Linking Mitochondrial Dynamics with mtDNA Instability

The emerging pathophysiological links between OPA1, MFN2, and disturbed mtDNA maintenance point strongly towards a common mechanism. The mitochondrial genome is a high copy number genome and for a mutation to cause a biochemical defect, it must exceed a critical threshold level, which will be both mutation- and cell-specific [66]. This process of gradual accumulation of the mutant mtDNA species is known as clonal expansion and it is time dependent, partly accounting for the late onset of the neurological features described in mtDNA maintenance disorders [65, 97]. There are currently two attractive models that are partly supported by mathematical modelling and experimental data [97, 106]. Based on *in silico* modelling, the degree of mitochondrial network fragmentation induced by different *OPA1* mutations is expected to influence the rate at which somatic mtDNA deletions become fixed at the single cell level [65]. Mitochondrial fragmentation effectively partitions the total cellular mtDNA content into smaller discrete units, decreasing the likelihood that the mutant species will be lost by random genetic drift, and reducing the time required for the mtDNA deletion to reach supra-threshold levels [65]. Although not mutually exclusive, a second mechanism that could result in the accumulation of somatic mtDNA abnormalities is an increased mutational rate. MtDNA molecules do not exist in isolation but they are packaged within intricate structures known as nucleoids [107–109]. These mtDNA replicative units are located within the mitochondrial matrix space and recent work suggests that the OPA1 peptide segment encoded by exon 4b plays a crucial role in anchoring nucleoids to the mitochondrial inner membrane [110]. Although speculative, *OPA1* mutations could therefore indirectly disrupt nucleoid morphology leading to an increased rate of mtDNA mutagenesis and the eventual emergence of COX-negative muscle fibres [110].

Multiple mtDNA deletions have so far only been identified in patients harbouring pathogenic *OPA1* and *MFN2* mutations. A possible explanation for their absence in other primary disorders of mitochondrial dynamics, for example *OPA3* and *DRP1*, relates to their early age of onset and their aggressive natural history, with most patients not surviving long enough to allow the accumulation of these somatic mtDNA abnormalities. It also highlights the multiple pathways that can trigger cell loss in this group of disorders, such as impaired mitochondrial oxidative phosphorylation and an increased sensitivity to apoptosis, independent of mtDNA instability and mitochondrial network disruption.

Conclusion

Over a relatively short period of time, the contribution of mitochondrial network plasticity to cellular homeostasis has become firmly established. Mutations in nuclear genes encoding key mediators of mitochondrial fusion and fission lead to a heterogeneous group of disorders characterised by both intra- and inter-familial variability, pointing towards a myriad of secondary factors modulating the final phenotype at the individual level. Although the elucidation of these complex disease mechanisms represents a major challenge, the availability of animal models and recent technological advances provide powerful tools to tackle these fundamental questions. Hopefully, a greater understanding of specific pathological pathways and, more importantly, the ability to manipulate them will lead to effective treatments for these multi-system neurodegenerative disorders.

Conflicts of Interest

None

Acknowledgments PYWM is a Medical Research Council (MRC, UK) Clinician Scientist. GL is supported by INSERM, Université de Montpellier I et II, and the patient associations Retina France and Ouvrir Les Yeux. PFC is a Wellcome Trust Senior Fellow in Clinical Science and a UK National Institute of Health Senior Investigator who also receives funding from the MRC (UK), Parkinson's UK, the Association Francaise contre les Myopathies, and the UK NIHR Biomedical Research Centre for Ageing and Age-related disease award to the Newcastle upon Tyne Hospitals NHS Foundation Trust.

References

1. Holt IJ, Harding AE, Morganhughes JA (1988) Deletions of muscle mitochondrial-DNA in patients with mitochondrial myopathies. Nature 331:717–719
2. Wallace DC, Singh G, Lott MT et al (1988) Mitochondrial DNA mutation associated with Leber's hereditary optic neuropathy. Science 242:1427–1430
3. DiMauro S, Schon EA (2003) Mechanisms of disease: mitochondrial respiratory-chain diseases. N Engl J Med 348:2656–2668
4. Taylor RW, Turnbull DM (2005) Mitochondrial DNA mutations in human disease. Nat Rev Genet 6:389–402
5. McFarland R, Taylor RW, Turnbull DM (2010) A neurological perspective on mitochondrial disease. Lancet Neurol 9:829–840
6. Van Goethem G, Dermaut B, Lofgren A, Martin JJ, Van Broeckhoven C (2001) Mutation of POLG is associated with progressive external ophthalmoplegia characterized by mtDNA deletions. Nat Genet 28:211–212
7. Spelbrink JN, Li FY, Tiranti V et al (2001) Human mitochondrial DNA deletions associated with mutations in the gene encoding Twinkle, a phage T7 gene LF-like protein localized in mitochondria. Nat Genet 28:223–231
8. Chinnery PF, Zeviani M (2008) 155th ENMC workshop: polymerase gamma and disorders of mitochondrial DNA synthesis, 21–23 September 2007, Naarden, The Netherlands. Neuromuscul Disord 18:259–267
9. Krishnan KJ, Reeve AK, Samuels DC et al (2008) What causes mitochondrial DNA deletions in human cells? Nat Genet 40:275–279
10. Khrapko K (2011) The timing of mitochondrial DNA mutations in aging. Nat Genet 43:726–727
11. Bristow EA, Griffiths PG, Andrews RM, Johnson MA, Turnbull DM (2002) The distribution of mitochondrial activity in relation to optic nerve structure. Arch Ophthalmol 120:791–796

12. Westermann B (2010) Mitochondrial fusion and fission in cell life and death. Nat Rev Mol Cell Biol 11:872–884
13. Barron MJ, Griffiths P, Turnbull DM, Bates D, Nichols P (2004) The distributions of mitochondria and sodium channels reflect the specific energy requirements and conduction properties of the human optic nerve head. Br J Ophthalmol 88:286–290
14. Chan DC (2006) Mitochondrial fusion and fission in mammals. Annu Rev Cell Dev Biol 22:79–99
15. Campello S, Scorrano L (2010) Mitochondrial shape changes: orchestrating cell pathophysiology. EMBO Rep 11:678–684
16. Praefcke GJK, McMahon HT (2004) The dynamin superfamily: universal membrane tubulation and fission molecules? Nat Rev Mol Cell Biol 5:133–147
17. Knott AB, Perkins G, Schwarzenbacher R, Bossy-Wetzel E (2008) Mitochondrial fragmentation in neurodegeneration. Nat Rev Neurosci 9:505–518
18. Chen H, Chan DC (2009) Mitochondrial dynamics-fusion, fission, movement, and mitophagy-in neurodegenerative diseases. Hum Mol Genet 18:R169–R176
19. Yu-Wai-Man P, Griffiths PG, Chinnery PF (2011) Mitochondrial optic neuropathies—disease mechanisms and therapeutic strategies. Prog Retin Eye Res 30(2):81–114
20. Huizing M, Brooks BP, Anikster Y (2005) Optic atrophies in metabolic disorders. Mol Genet Metab 86:51–60
21. Yu-Wai-Man P, Griffiths PG, Burke A et al (2010) The prevalence and natural history of dominant optic atrophy due to OPA1 mutations. Ophthalmol 117:1538–1546
22. Cohn AC, Toomes C, Potter C et al (2007) Autosomal dominant optic atrophy: penetrance and expressivity in patients with OPA1 mutations. Amer J Ophthalmol 143:656–662
23. Cohn AC, Toomes C, Hewitt AW et al (2008) The natural history of OPA1-related autosomal dominant optic atrophy. Br J Ophthalmol 24:24
24. Votruba M, Fitzke FW, Holder GE, Carter A, Bhattacharya SS, Moore AT (1998) Clinical features in affected individuals from 21 pedigrees with dominant optic atrophy. Arch Ophthalmol 116:351–358
25. Yu-Wai-Man P, Griffiths PG, Hudson G, Chinnery PF (2009) Inherited mitochondrial optic neuropathies. J Med Genet 46:145–158
26. La Morgia C, Ross-Cisneros FN, Sadun AA et al (2010) Melanopsin retinal ganglion cells are resistant to neurodegeneration in mitochondrial optic neuropathies. Brain 133:2426–2438
27. Davies V, Votruba M (2006) Focus on molecules: the OPA1 protein. Exp Eye Res 83:1003–1004
28. Olichon A, Elachouri G, Baricault L, Delettre C, Belenguer P, Lenaers G (2007) OPA1 alternate splicing uncouples an evolutionary conserved function in mitochondrial fusion from a vertebrate restricted function in apoptosis. Cell Death Diff 14:682–692
29. Akepati VR (2008) Characterization of OPA1 isoforms isolated from mouse tissues. J Neurochem 106:372–383
30. Lenaers G, Reynier P, Elachouri G et al (2009) OPA1 functions in mitochondria and dysfunctions in optic nerve. Int J Biochem Cell Biol 41:1866–1874
31. Alexander C, Votruba M, Pesch UEA et al (2000) OPA1, encoding a dynamin-related GTPase, is mutated in autosomal dominant optic atrophy linked to chromosome 3q28. Nat Genet 26:211–215
32. Delettre C, Lenaers G, Griffoin JM et al (2000) Nuclear gene OPA1, encoding a mitochondrial dynamin-related protein, is mutated in dominant optic atrophy. Nat Genet 26:207–210
33. Ferre M, Amati-Bonneau P, Tourmen Y, Malthiery Y, Reynier P (2005) eOPA1: an online database for OPA1 mutations. Hum Mutat 25:423–428
34. Ferre M, Bonneau D, Milea D et al (2009) Molecular screening of 980 cases of suspected hereditary optic neuropathy with a report on 77 novel OPA1 mutations. Hum Mutat 30:E692–705
35. Fuhrmann N, Alavi MV, Bitoun P et al (2009) Genomic rearrangements in OPA1 are frequent in patients with autosomal dominant optic atrophy. J Med Genet 46:136–144

36. Almind GJ, Gronskov K, Milea D, Larsen M, Brondum-Nielsen K, Ek J (2011) Genomic deletions in OPA1 in Danish patients with autosomal dominant optic atrophy. BMC Med Genet 12:49
37. Schimpf S, Fuhrmann N, Schaich S, Wissinger B (2008) Comprehensive cDNA study and quantitative transcript analysis of mutant OPA1 transcripts containing premature termination codons. Hum Mutat 29:106–112
38. Amati-Bonneau P, Odent S, Derrien C et al (2003) The association of autosomal dominant optic atrophy and moderate deafness may be due to the R445H mutation in the OPA1 gene. Am J Ophthalmol 136:1170–1171
39. Amati-Bonneau P, Guichet A, Olichon A et al (2005) OPA1 R445H mutation in optic atrophy associated with sensorineural deafness. Ann Neurol 58:958–963
40. Li CM, Kosmorsky G, Zhang K, Katz BJ, Ge J, Traboulsi EI (2005) Optic atrophy and sensorineural hearing loss in a family caused by an R445H OPA1 mutation. Am J Med Genet Part A 138A:208–211
41. Payne M, Yang ZL, Katz BJ et al (2004) Dominant optic atrophy, sensorineural hearing loss, ptosis, and ophthalmoplegia: a syndrome caused by a missense mutation in OPA1. Am J Ophthalmol 138:749–755
42. Hudson G, Amati-Bonneau P, Blakely EL et al (2008) Mutation of OPA1 causes dominant optic atrophy with external ophthalmoplegia, ataxia, deafness and multiple mitochondrial DNA deletions: a novel disorder of mtDNA maintenance. Brain 131:329–337
43. Amati-Bonneau P, Valentino ML, Reynier P et al (2008) OPA1 mutations induce mitochondrial DNA instability and optic atrophy plus phenotypes. Brain 131:338–351
44. Spinazzi M, Cazzola S, Bortolozzi M et al (2008) A novel deletion in the GTPase domain of OPA1 causes defects in mitochondrial morphology and distribution, but not in function. Hum Mol Genet 17:3291–3302
45. Yu-Wai-Man P, Griffiths PG, Gorman GS et al (2010) Multi-system neurological disease is common in patients with OPA1 mutations. Brain 133:771–786
46. Marchbank NJ, Craig JE, Leek JP et al (2002) Deletion of the OPA1 gene in a dominant optic atrophy family: evidence that haploinsufficiency is the cause of disease. J Med Genet 39:e47
47. Aijaz S, Erskine L, Jeffery G, Bhattacharya SS, Votruba M (2004) Developmental expression profile of the optic atrophy gene product: OPA1 is not localized exclusively in the mammalian retinal ganglion cell layer. Invest Ophthalmol Visual Sci 45:1667–1673
48. Bette S, Schlaszus H, Wissinger B, Meyermann R, Mittelbronn M (2005) OPA1, associated with autosomal dominant optic atrophy, is widely expressed in the human brain. Acta Neuropathol 109:393–399
49. Wang AG, Fann MJ, Yu HY, Yen MY (2006) OPA1 expression in the human retina and optic nerve. Exp Eye Res 83:1171–1178
50. Pellegrini L, Scorrano L (2007) A cut short to death: Parl and Opa1 in the regulation of mitochondrial morphology and apoptosis. Cell Death Diff 14:1275–1284
51. Martinelli P, Rugarli EI (2010) Emerging roles of mitochondrial proteases in neurodegeneration. Biochim Biophys Acta 1797
52. Rugarli EI, Langer T (2006) Translating m-AAA protease function in mitochondria to hereditary spastic paraplegia. Trends Mol Med 12:262–269
53. Casari G, De Fusco M, Ciarmatori S et al (1998) Spastic paraplegia and OXPHOS impairment caused by mutations in paraplegin, a nuclear-encoded mitochondrial metalloprotease. Cell 93:973–983
54. Chevrollier A (2008) Hereditary optic neuropathies share a common mitochondrial coupling defect. Ann Neurol 63:794–798
55. Zanna C, Ghelli A, Porcelli AM et al (2008) OPA1 mutations associated with dominant optic atrophy impair oxidative phosphorylation and mitochondrial fusion. Brain 131:352–367
56. Song ZY, Chen H, Fiket M, Alexander C, Chan DC (2007) OPA1 processing controls mitochondrial fusion and is regulated by mRNA splicing, membrane potential, and Yme1L. J Cell Biol 178:749–755

57. Cipolat S, Rudka T, Hartmann D et al (2006) Mitochondrial rhomboid PARL regulates cytochrome c release during apoptosis via OPA1-dependent cristae remodeling. Cell 126:163–175
58. Frezza C, Cipolat S, de Brito OM et al (2006) OPA1 controls apoptotic cristae remodeling independently from mitochondrial fusion. Cell 126:177–189
59. Suen DF, Norris KL, Youle RJ (2008) Mitochondrial dynamics and apoptosis. Genes Dev 22:1577–1590
60. Lodi R, Tonon C, Valentino ML et al (2004) Deficit of in vivo mitochondrial ATP production in OPA1-related dominant optic atrophy. Ann Neurol 56:719–723
61. Yu-Wai-Man P, Trenell MI, Hollingsworth KG, Griffiths PG, Chinnery PF (2011) OPA1 mutations impair mitochondrial function in both pure and complicated dominant optic atrophy. Brain 134:e164
62. Lodi R, Tonon C, Valentino ML et al (2011) Defective mitochondrial adenosine triphosphate production in skeletal muscle from patients with dominant optic atrophy due to OPA1 mutations. Arch Neurol 68:67–73
63. Yu-Wai-Man P, Sitarz KS, Samuels DC et al (2010) OPA1 mutations cause cytochrome c oxidase deficiency due to loss of wild-type mtDNA molecules. Hum Mol Genet 19:3043–3052
64. Capps GJ, Samuels DC, Chinnery PF (2003) A model of the nuclear control of mitochondrial DNA replication. J Theor Biol 221:565–583
65. Chinnery PF, Samuels DC (1999) Relaxed replication of mtDNA: a model with implications for the expression of disease. Am J Hum Genet 64:1158–1165
66. Durham SE, Samuels DC, Cree LM, Chinnery PF (2007) Normal levels of wild-type mitochondrial DNA maintain cytochrome c oxidase activity for two pathogenic mitochondrial DNA mutations but not for m.3243A→G. Am J Hum Genet 81:189–195
67. Davies VJ, Hollins AJ, Piechota MJ et al (2007) Opa1 deficiency in a mouse model of autosomal dominant optic atrophy impairs mitochondrial morphology, optic nerve structure and visual function. Hum Mol Genet 16:1307–1318
68. Alavi MV, Bette S, Schimpf S et al (2007) A splice site mutation in the murine Opa1 gene features pathology of autosomal dominant optic atrophy. Brain 130:1029–1042
69. Williams PA, Morgan JE, Votruba M (2011) Opa1 deficiency in a mouse model of dominant optic atrophy leads to retinal ganglion cell dendropathy. Brain 133:2942–2951
70. Yu-Wai-Man P, Davies VJ, Piechota MJ, Cree LM, Votruba M, Chinnery PF (2009) Secondary mtDNA defects do not cause optic nerve dysfunction in a mouse model of dominant optic atrophy. Invest Ophthalmol Vis Sci 50:4561–4566
71. Anikster Y, Kleta R, Shaag A, Gahl WA, Elpeleg O (2001) Type III 3-methylglutaconic aciduria (optic atrophy plus syndrome, or Costeff optic atrophy syndrome): identification of the OPA3 gene and its founder mutation in Iraqi Jews. Am J Hum Genet 69:1218–1224
72. Reynier P, Amati-Bonneau P, Verny C et al (2004) OPA3 gene mutations responsible for autosomal dominant optic atrophy and cataract. J Med Genet 41(9):e110
73. Da Cruz S, Xenarios I, Langridge J, Vilbois F, Parone PA, Martinou JC (2003) Proteomic analysis of the mouse liver mitochondrial inner membrane. J Biol Chem 278:41566–41571
74. Huizing M, Dorward H, Ly L et al (2010) OPA3, mutated in 3-methylglutaconic aciduria type III, encodes two transcripts targeted primarily to mitochondria. Mol Genet Metab 100:149–154
75. Ryu SW, Jeong HJ, Choi M, Karbowski M, Choi C (2010) Optic atrophy 3 as a protein of the mitochondrial outer membrane induces mitochondrial fragmentation. Cell Mol Life Sci 67:2839–2850
76. Pei WH, Kratz LE, Bernardini I et al (2010) A model of Costeff syndrome reveals metabolic and protective functions of mitochondrial OPA3. Dev 137:2587–2596
77. Davies VJ, Powell KA, White KE et al (2008) A missense mutation in the murine Opa3 gene models human Costeff syndrome. Brain 131:368–380
78. Friedman JR, Lackner LL, West M, DiBenedetto JR, Nunnari J, Voeltz GK (2011) ER tubules mark sites of mitochondrial division. Science 334:358–362

79. Waterham HR, Koster J, van Roermund CWT, Mooyer PAW, Wanders RJA, Leonard JV (2007) A lethal defect of mitochondrial and peroxisomal fission. N Engl J Med 356:1736–1741
80. Chang CR, Manlandro CM, Arnoult D et al (2010) A lethal de novo mutation in the middle domain of the dynamin-related GTPase Drp1 impairs higher order assembly and mitochondrial division. J Biol Chem 285:32494–32503
81. Pareyson D, Marchesi C (2009) Diagnosis, natural history, and management of Charcot-Marie-Tooth disease. Lancet Neurol 8:654–667
82. Zuchner S, Vance JM (2006) Mechanisms of disease: a molecular genetic update on hereditary axonal neuropathies. Nat Clin Pract Neurol 2:45–53
83. Amati-Bonneau P, Milea D, Bonneau D et al (2009) OPA1-associated disorders: phenotypes and pathophysiology. Int J Biochem Cell Biol 41:1855–1865
84. Zuchner S, Mersiyanova IV, Muglia M et al (2004) Mutations in the mitochondrial GTPase mitofusin 2 cause Charcot-Marie-Tooth neuropathy type 2A. Nat Genet 36:449–451
85. Zuchner S, De Jonghe P, Jordanova A et al (2006) Axonal neuropathy with optic atrophy is caused by mutations in mitofusin 2. Ann Neurol 59:276–281
86. Cartoni R, Martinou JC (2009) Role of mitofusin 2 mutations in the physiopathology of Charcot-Marie-Tooth disease type 2A. Exp Neurol 218:268–273
87. Chen HC, Detmer SA, Ewald AJ, Griffin EE, Fraser SE, Chan DC (2003) Mitofusins Mfn1 and Mfn2 coordinately regulate mitochondrial fusion and are essential for embryonic development. J Cell Biol 160:189–200
88. Detmer SA, Velde CV, Cleveland DW, Chan DC (2008) Hindlimb gait defects due to motor axon loss and reduced distal muscles in a transgenic mouse model of Charcot-Marie-Tooth type 2A. Hum Mol Genet 17:367–375
89. Chen H, McCaffery JM, Chan DC (2007) Mitochondrial fusion protects against neurodegeneration in the cerebellum. Cell 130:548–562
90. Guillet V, Gueguen N, Cartoni R et al (2011) Bioenergetic defect associated with mKATP channel opening in a mouse model carrying a mitofusin 2 mutation. Faseb J 25:1618–1627
91. Detmer SA, Chan DC (2007) Complementation between mouse Mfn1 and Mfn2 protects mitochondrial fusion defects caused by CMT2A disease mutations. J Cell Biol 176:405–414
92. Loiseau D, Chevrollier A, Verny C et al (2007) Mitochondrial coupling defect in Charcot-Marie-Tooth type 2A disease. Ann Neurol 61:315–323
93. Del Bo R, Moggio M, Rango M et al (2008) Mutated mitofusin 2 presents with intrafamilial variability and brain mitochondrial dysfunction. Neurology 71:1959–1966
94. Pich S, Bach D, Briones P et al (2005) The Charcot-Marie-Tooth type 2A gene product, Mfn2, up-regulates fuel oxidation through expression of OXPHOS system. Hum Mol Genet 14:1405–1415
95. Chen HC, Vermulst M, Wang YE et al (2010) Mitochondrial fusion is Rrquired for mtDNA stability in skeletal muscle and tolerance of mtDNA mutations. Cell 141:280–289
96. Rouzier C, Bannwarth S, Chaussenot A et al (2012) The MFN2 gene is responsible for mitochondrial DNA instability and optic atrophy 'plus' phenotype. Brain 135:23–34
97. Yu-Wai-Man P, Chinnery PF (2012) Dysfunctional mitochondrial maintenance: what breaks the circle of life? Brain 135:9–11
98. Sitarz KS, Yu-Wai-Man P, Pyle A, Stewart JD, Rautenstrauss B, Seeman P, Reilly MM, Horvath R, Chinnery PF (2012) MFN2 mutations cause compensatory mitochondrial DNA proliferation. Brain. Epub ahead of print. PMID:22492563
99. Baxter RV, Ben Othmane K, Rochelle JM et al (2002) Ganglioside-induced differentiation-associated protein-1 is mutant in Charcot-Marie-Tooth disease type 4A/8q21. Nat Genet 30:21–22
100. Claramunt R, Pedrola L, Sevilla T et al (2005) Genetics of Charcot-Marie-Tooth disease type 4A: mutations, inheritance, phenotypic variability, and founder effect. J Med Genet 42:358–365
101. Zimon M, Baets J, Fabrizi GM et al (2011) Dominant GDAP1 mutations cause predominantly mild CMT phenotypes. Neurology 77:540–548

102. Niemann A, Wagner KM, Ruegg M, Suter U (2009) GDAP1 mutations differ in their effects on mitochondrial dynamics and apoptosis depending on the mode of inheritance. Neurobiol Dis 36:509–520
103. Pedrola L, Espert A, Wu X, Claramunt R, Shy ME, Palau F (2005) GDAP1, the protein causing Charcot-Marie-Tooth disease type 4A, is expressed in neurons and is associated with mitochondria. Hum Mol Genet 14:1087–1094
104. Niemann A, Ruegg M, La Padula V, Schenone A, Suter U (2005) Ganglioside-induced differentiation associated protein 1 is a regulator of the mitochondrial network: new implications for Charcot-Marie-Tooth disease. J Cell Biol 170:1067–1078
105. Cassereau J, Chevrollier A, Gueguen N et al (2009) Mitochondrial complex I deficiency in GDAP1-related autosomal dominant Charcot-Marie-Tooth disease (CMT2K). Neurogenetics 10:145–150
106. Payne BA, Wilson IJ, Hateley CA et al (2011) Mitochondrial aging is accelerated by antiretroviral therapy through the clonal expansion of mtDNA mutations. Nat Genet 43:806–810
107. Cao LQ, Shitara H, Horii T et al (2007) The mitochondrial bottleneck occurs without reduction of mtDNA content in female mouse germ cells. Nat Genet 39:386–390
108. Cree LM, Samuels DC, Lopes S et al (2008) A reduction of mitochondrial DNA molecules during embryogenesis explains the rapid segregation of genotypes. Nat Genet 40:249–254
109. Khrapko K (2008) Two ways to make an mtDNA bottleneck. Nat Genet 40:134–135
110. Elachouri G, Vidoni S, Zanna C et al (2011) OPA1 links human mitochondrial genome maintenance to mtDNA replication and distribution. Genome Res 21:12–20

Chapter 10
Depletion of mtDNA with MMA: *SUCLA2* and *SUCLG1*

Nelson Hawkins Jr and Brett H. Graham

Introduction

Mitochondria are dynamic organelles involved in a broad variety of cellular processes within the eukaryotic cell. They are constantly undergoing fission, fusion, and orient themselves in a variety of different morphologies and numbers to best fit the needs of the cellular niche in which they reside [1]. While mitochondria play a fundamental role in energy metabolism and oxidative phosphorylation, they are also involved in many other cellular functions such as the intrinsic pathway of apoptosis, intracellular Ca^{2+} buffering, intermediate metabolism, and cellular homeostasis [2–4]. As an organelle, mitochondria have a characteristic structure and topology, consisting of double lipid bilayers that define an intermembrane space (the space between the mitochondrial outer and inner membranes) and matrix (the space within the mitochondrial inner membrane). The components of the TCA cycle are located in the matrix, while the electron transport chain (ETC) is embedded in the mitochondrial inner membrane with oxidative phosphorylation occurring at the interface of the inner membrane and the matrix [5].

The biology of mitochondria is notable for the organelle having its own circular genome, the mitochondrial DNA (mtDNA), which encodes 13 polypeptides (a subset of ETC complex subunits), 22 transfer RNAs, and 2 ribosomal RNAs [6, 7]. While mtDNA encodes 13 proteins, the mitochondrion contains at least 1300 proteins that contribute to all of its functions, 99% of which are nuclear encoded and imported into the mitochondrion [8]. These nuclear encoded proteins include components of the ETC as well as the protein machinery—distinct from the nucleus and cytoplasm—for

---

B. H. Graham (✉)
Department of Molecular and Human Genetics, Baylor College of Medicine,
One Baylor Plaza, MS:BCM225, Houston, TX 77030, USA
e-mail: bgraham@bcm.edu

Medical Genetics, Texas Children's Hospital, Houston, TX, USA

N. Hawkins Jr
Department of Molecular and Human Genetics, Baylor College of Medicine,
One Baylor Plaza, MS:BCM225, Houston, TX 77030, USA

replication and maintenance of the mtDNA, and transcription and translation of the mtDNA [8].

Mitochondrial disease can be caused by mutations in mtDNA or in nuclear encoded mitochondrial genes. As described elsewhere in this volume, mitochondrial disease can be caused by mutations in nuclear encoded mitochondrial genes that are involved in a variety of aspects of energy metabolism and oxidative phosphorylation.

One subset of genes are genes that, when mutated, can cause mtDNA depletion syndromes (MDS). MDS are characterized by depletion of mtDNA content in skeletal muscle and/or liver and can present clinically as a mitochondrial myopathy, encephalomyopathy, or encephalohepatopathy [9]. This chapter is devoted to MDS caused by mutations in *SUCLA2* or *SUCLG1*, which encode β and α subunits, respectively, of the TCA cycle enzyme Succinyl-CoA Synthetase (SCS, also known as Succinyl-CoA Ligase or Succinate Thiokinase) and cause mtDNA depletion and mild methylmalonic aciduria (MMA). The following sections will review the current state of knowledge regarding SCS biology and MDS caused by mutations in *SUCLA2* or *SUCLG1*.

Succinyl-CoA Synthetase Biology

SCS in Prokaryotes

The molecular, biochemical, and structural characteristics of SCS in the prokaryote *Escherichia coli* (*E. coli*) have been studied for several decades. SCS is a component of the citric acid (TCA) cycle and catalyzes the reversible conversion of succinyl-CoA to succinate, which is coupled to the phosphorylation of GDP to GTP or ADP to ATP. The enzyme is composed of two subunits, α and β. In *E. coli*, the genes that encode the α and β subunits of SCS are encoded by *sucD* and *sucC*, respectively, and are part of a larger TCA cycle operon [10]. The quaternary enzyme structure in gram negative bacteria is a $(\alpha\beta)_2$ heterotetramer, while in gram positive bacteria and eukaryotes it is a heterodimer [11]; and examination of SCS structure by x-ray crystallography has elucidated the enzyme active site and nucleotide diphosphate binding site [12, 13]. Biochemical and structural functional studies in *E. coli* have demonstrated that the SCS active site is formed at the interface of the α and β subunits [13]. In addition, two glutamate residues (Glu 208α and Glu 197β) are critical for SCS enzyme activity and are required for the phosphorylation and dephosphorylation of His 246α, which is located within the active site and is transiently phosphorylated during catalysis [14].

SCS in Eukaryotes

SCS is conserved throughout eukaryoytes [15]. The enzyme activity in SCS deficient *E. coli* can be rescued by ectopic coexpression of pig SCS α and β subunits, demonstrating functional conservation [16]. After the initial purification of eukaryotic SCS in the 1950s [17, 18], it was generally thought there was only a GDP-binding β isoform in animals until an ADP-specific SCS was purified from multiple sources, including insects [19] and pigeon [20, 21] in the 1970s and 1980s. In the 1990s, with the application of PCR based technologies, the identification and cloning of distinct GDP and ADP β gene isoforms was demonstrated across a wide variety of multicellular eukaryotes [15]. While the ADP and GDP forms of SCS contain distinct β subunits (encoded by *SUCLA2* and *SUCLG2*, respectively, in humans), both forms contain the same α subunit (encoded by *SUCLG1* in humans) [22]. Additional studies in mammals have shown that *SUCLA2* and *SUCLG2* are widely expressed in tissues, but exhibit distinct tissue specific expression patterns, with *SUCLA2* strongly expressed in brain, skeletal muscle, heart and testes and *SUCLG2* strongly expressed in liver and kidney [23].

SCS and MDS

SUCLA2 and MDS

Mitochondrial DNA (mtDNA) depletion syndrome or MDS resulting from mutations in subunits of succinyl-CoA synthetase can manifest in a range of disease phenotypes including infantile lethal lactic acidosis and encephalomyopathy with sensorineural hearing loss [9]. Mutations in *SUCLA2* were first identified in a consanguineous Middle Eastern family by homozygosity mapping which identified a complex indel mutation spanning the exon 6-intron region [24]. This mutation causes aberrant mRNA splicing affecting exon 6 and/or exon 7 resulting in either a frameshift and truncated transcript or a transcript with an intragenic deletion of 71 residues [24]. Affected members of this family exhibited an encephalomyopathic phenotype including Leigh disease features on MRI of the brain, multiple respiratory chain deficiencies in muscle sparing complex II, and depletion of mtDNA in muscle [24]. Subsequently, a founder splice site mutation in *SUCLA2* (IVS4 + 1G > A) was identified by homozygosity mapping in multiple individuals from the Faroe Islands with autosomal recessive mitochondrial encephalomyopathy with elevated methylmalonic acid [25, 26]. The elevated methylmalonic acid (typically a more modest elevation than seen with methylmalonic aciduria caused by deficiency of methylmalonyl-CoA mutase) presumably results from the secondary inhibition of methylmalonyl-CoA mutase by elevated succinyl-CoA resulting from SCS deficiency (Fig. 10.1) [26]. Other biochemical abnormalities described in individuals with *SUCLA2* mutations include elevations in methylmalonyl carnitine ester (C4DC), methylcitrate, 3-hydroxypropionate, propionylcarnitine (C3), 3-hydroxyisovaleric acid, and 3-methylglutaconic acid [27]. Over 30 patients with *SUCLA2* mutations have been reported to date and the mutations range from missense mutations to frameshift mutations resulting from splice site mutations and indels [28].

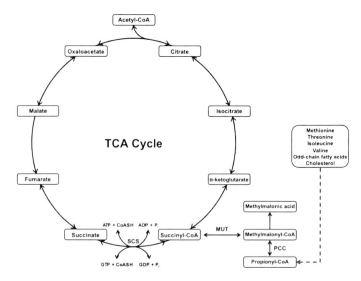

Fig. 10.1 Succinyl-CoA, Methylmalonic Acid, and the TCA Cycle. Propionyl-CoA generated from the catabolism of methionine, threonine, isoleucine, valine, odd-chain fatty acids, and cholesterol is metabolized to methylmalonyl-CoA by propionyl-CoA carboxylase (PCC). Methylmalonyl-CoA mutase (MUT) catalyzes the isomerization of methylmalonyl-CoA to succinyl-CoA for entry into the TCA cycle as a substrate for succinyl-CoA synthetase (SCS). When SCS is deficient, succinyl-CoA and methylmalonyl-CoA accumulates, thereby resulting in an increased level of methylmalonic acid

SUCLG1 and MDS

The first cases of MDS resulting from a mutation in *SUCLG1* were described in 2007 from a consanguineous Pakistani family presenting with fatal infantile lactic acidosis [29]. The affected individuals from the family exhibited lactic acidosis, encephalopathy, liver dysfunction, hypotonia, and methylmalonic aciduria with respiratory chain deficiencies and mtDNA depletion in liver and muscles. Homozygosity mapping identified a 2-bp deletion in *SUCLG1* [29]. Subsequently, multiple individuals with hepatoencephalomyopathy and *SUCLG1* mutations ranging from missense to frameshift mutations have been reported [30–36]. A subset of these patients with *SUCLG1* mutations has been reported to have congenital anomalies, including brain, cardiovascular, and genitourinary malformations (Anderson H. and Wong L.J., personal communication) [31, 34].

The Role of SCS and mtDNA Depletion

Since the first report of a *SUCLA2* mutation associated with mtDNA depletion, it has been hypothesized that SCS deficiency may affect mitochondrial deoxynucleotide pools through a disruption of mitochondrial nucleotide diphosphate kinase (NDPK) activity [24]. This hypothesis derives from an observation that SCS directly interacts with NDPK as demonstrated by immunoprecipitation studies [37]. More recently, knockdown of *SUCLG2* in *SUCLA2* deficient fibroblasts was shown to be associated with reduced NDPK activity, further supporting this hypothesis [38]. Additional studies are necessary to fully understand the pathogenic role of SCS in MDS.

Conclusion

Succinyl-CoA synthetase is a component of the Krebs cycle and is essential for aerobic respiration and mtDNA maintenance. Mutations in the α (*SUCLG1*) and ADP-specific β (*SUCLA2*) subunits of SCS cause MDS associated with modest elevations of methylmalonic acid. Efficacious treatments for mtDNA depletion syndromes including those that result from SCS deficiency are lacking. New and existing models must be generated and explored in order to expand the understanding of the role of SCS in normal mtDNA biology, in the pathogenesis of MDS, and, ultimately, for the development of novel therapeutic options.

References

1. Bereiter-Hahn J, Voth M (1994) Dynamics of mitochondria in living cells: Shape changes, dislocations, fusion, and fission of mitochondria. Microsc Res Tech 27(3):198–219
2. Rambold AS, Lippincott-Schwartz J (2011) Mechanisms of mitochondria and autophagy crosstalk. Cell Cycle 10(23):4032–4038
3. Green DR, Galluzzi L, Kroemer G (2011) Mitochondria and the autophagy-inflammation-cell death axis in organismal aging. Sci 333(6046):1109–1112
4. Parone P, Priault M, James D, Nothwehr SF, Martinou JC (2003) Apoptosis: Bombarding the mitochondria. Essays Biochem 39:41–51
5. Bereiter-Hahn J, Jendrach M (2010) Mitochondrial dynamics. Int Rev Cell Mol Biol 284:1–65
6. Suzuki T, Nagao A (2011) Human mitochondrial tRNAs: Biogenesis, function, structural aspects, and diseases. Annu Rev Genet 45:299–329
7. O'Brien TW (2003) Properties of human mitochondrial ribosomes. IUBMB Life 55(9):505–513
8. Pagliarini DJ, Calvo SE, Chang B et al (2008) A mitochondrial protein compendium elucidates complex I disease biology. Cell 134(1):112–123
9. Spinazzola A, Invernizzi F, Carrara F et al (2009) Clinical and molecular features of mitochondrial DNA depletion syndromes. J Inherit Metab Dis 32(2):143–158
10. Buck D, Spencer ME, Guest JR (1985) Primary structure of the succinyl-CoA synthetase of Escherichia coli. Biochem 24(22):6245–6252
11. Weitzman PD, Kinghorn HA (1978) Occurrence of 'large' or 'small' forms of succinate thiokinase in diverse organisms. FEBS Lett 88(2):255–258
12. Joyce MA, Fraser ME, James MN, Bridger WA, Wolodko WT (2000) ADP-binding site of Escherichia coli succinyl-CoA synthetase revealed by x-ray crystallography. Biochem 39(1):17–25
13. Fraser ME, James MN, Bridger WA, Wolodko WT (1999) A detailed structural description of Escherichia coli succinyl-CoA synthetase. J Mol Biol 285(4):1633–1653

14. Fraser ME, Joyce MA, Ryan DG, Wolodko WT (2002) Two glutamate residues, Glu 208 alpha and Glu 197 beta, are crucial for phosphorylation and dephosphorylation of the active-site histidine residue in succinyl-CoA synthetase. Biochem 41(2):537–546
15. Johnson JD, Mehus JG, Tews K, Milavetz BI, Lambeth DO (1998) Genetic evidence for the expression of ATP- and GTP-specific succinyl-CoA synthetases in multicellular eucaryotes. J Biol Chem 273(42):27580–27586
16. Bailey DL, Wolodko WT, Bridger WA (1993) Cloning, characterization, and expression of the beta subunit of pig heart succinyl-CoA synthetase. Protein Sci 2(8):1255–1262
17. Sanadi DR, Gibson M, Ayengar P (1954) Guanosine triphosphate, the primary product of phosphorylation coupled to the breakdown of succinyl coenzyme A. Biochim Biophys Acta 14(3):434–436
18. Ayengar P, Gibson DM, Sanadi DR (1954) A new coenzyme for phosphorylation. Biochim Biophys Acta 13(2):309–310
19. Hansford RG (1973) An adenine nucleotide-linked succinic thiokinase of animal origin. FEBS Lett 31(3):317–320
20. Allen DA, Ottaway JH (1986) Succinate thiokinase in pigeon breast muscle mitochondria. FEBS Lett 194(1):171–175
21. Severin SE, Feigina MM (1976) alpha-keto acid dehydrogenases and acyl-CoA synthetases from pigeon breast muscle. Adv Enzyme Regul 15:1–21
22. Johnson JD, Muhonen WW, Lambeth DO (1998) Characterization of the ATP- and GTP-specific succinyl-CoA synthetases in pigeon. The enzymes incorporate the same alpha-subunit. J Biol Chem 273(42):27573–27579
23. Lambeth DO, Tews KN, Adkins S, Frohlich D, Milavetz BI (2004) Expression of two succinyl-CoA synthetases with different nucleotide specificities in mammalian tissues. J Biol Chem 279(35):36621–36624
24. Elpeleg O, Miller C, Hershkovitz E et al (2005) Deficiency of the ADP-forming succinyl-CoA synthase activity is associated with encephalomyopathy and mitochondrial DNA depletion. Am J Hum Genet 76(6):1081–1086
25. Carrozzo R, Dionisi-Vici C, Steuerwald U et al (2007) *SUCLA2* mutations are associated with mild methylmalonic aciduria, Leigh-like encephalomyopathy, dystonia and deafness. Brain 130(Pt 3):862–874
26. Ostergaard E, Hansen FJ, Sorensen N et al (2007) Mitochondrial encephalomyopathy with elevated methylmalonic acid is caused by *SUCLA2* mutations. Brain 130(Pt 3):853–861
27. Morava E, Steuerwald U, Carrozzo R et al (2009) Dystonia and deafness due to *SUCLA2* defect; Clinical course and biochemical markers in 16 children. Mitochondrion 9(6):438–442
28. Poulton J, Hirano M, Spinazzola A et al (2009) Collated mutations in mitochondrial DNA (mtDNA) depletion syndrome (excluding the mitochondrial gamma polymerase, POLG1). Biochim Biophys Acta 1792(12):1109–1112
29. Ostergaard E, Christensen E, Kristensen E et al (2007) Deficiency of the alpha subunit of succinate-coenzyme A ligase causes fatal infantile lactic acidosis with mitochondrial DNA depletion. Am J Hum Genet 81(2):383–387
30. Sakamoto O, Ohura T, Murayama K et al (2011) Neonatal lactic acidosis with methylmalonic aciduria due to novel mutations in the *SUCLG1* gene. Pediatr Int 53(6):921–925
31. Randolph LM, Jackson HA, Wang J et al (2011) Fatal infantile lactic acidosis and a novel homozygous mutation in the *SUCLG1* gene: A mitochondrial DNA depletion disorder. Mol Genet Metab 102(2):149–152
32. Rouzier C, Le Guedard-Mereuze S, Fragaki K et al (2010) The severity of phenotype linked to *SUCLG1* mutations could be correlated with residual amount of *SUCLG1* protein. J Med Genet 47(10):670–676
33. Van Hove JL, Saenz MS, Thomas JA et al (2010) Succinyl-CoA ligase deficiency: a mitochondrial hepatoencephalomyopathy. Pediatr Res 68(2):159–164
34. Rivera H, Merinero B, Martinez-Pardo M et al (2010) Marked mitochondrial DNA depletion associated with a novel *SUCLG1* gene mutation resulting in lethal neonatal acidosis, multi-organ failure, and interrupted aortic arch. Mitochondrion 10(4):362–368

35. Valayannopoulos V, Haudry C, Serre V et al (2010) New *SUCLG1* patients expanding the phenotypic spectrum of this rare cause of mild methylmalonic aciduria. Mitochondrion 10(4):335–341
36. Ostergaard E, Schwartz M, Batbayli M et al (2010) A novel missense mutation in *SUCLG1* associated with mitochondrial DNA depletion, encephalomyopathic form, with methylmalonic aciduria. Eur J Pediatr 169(2):201–205
37. Kowluru A, Tannous M, Chen HQ (2002) Localization and characterization of the mitochondrial isoform of the nucleoside diphosphate kinase in the pancreatic beta cell: evidence for its complexation with mitochondrial succinyl-CoA synthetase. Arch Biochem Biophys 398(2):160–169
38. Miller C, Wang L, Ostergaard E, Dan P, Saada A (2011) The interplay between *SUCLA2*, *SUCLG2*, and mitochondrial DNA depletion. Biochim Biophys Acta 1812(5):625–629

Chapter 11
RRM2B-Related Mitochondrial Disease

Gráinne S. Gorman, Robert D. S. Pitceathly, Douglass M. Turnbull and Robert W. Taylor

Abbreviations

RRM2B	Ribonucleoside-diphosphate reductase subunit M2 B
POLG	Polymerase (DNA directed), gamma
POLG2	Polymerase (DNA directed), gamma 2, accessory subunit
PEO1	Progressive external ophthalmoplegia 1 protein
C10orf2	Chromosome 10 open reading frame 2
TK2	Thymidine kinase 2
DGUOK	Deoxyguanosine kinase
SUCLA2	Succinate-CoA ligase, ADP-forming, beta subunit
SUCLG1	Succinyl-CoA ligase, GDP-forming, alpha subunit
TYMP	Thymidine phosphorylase
SLC25A4(ANT1)	Solute carrier family 25 (mitochondrial carrier; adenine nucleotide translocator), member 4 (Adenine Nucleotide Translocase1

Introduction

To date, nuclear-encoded mitochondrial genes implicated in mitochondrial DNA (mtDNA) replication disorders appear to fall into two distinct genetic categories: first, those which affect mtDNA replication proteins, such as *POLG, POLG2*, and PEO1 (also called *C10orf2*, encoding the Twinkle helicase) or second, genes which encode

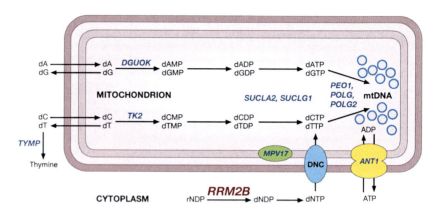

Fig. 11.1 A general overview of mitochondrial nucleotide metabolism for mtDNA synthesis, replication and repair identifying the major genes (*italicized in blue*) which have been associated with either mtDNA depletion or multiple mtDNA deletion syndromes [based on Figure from reference 18]. The *RRM2B* gene (*highlighted in red*) encodes a cytoplasmic protein, the small subunit of the p53-inducible ribonucleotide reductase (p53R2) and is responsible for the *de novo* conversion of ribonucleoside 5′-diphosphates (rNDP) to deoxyribonucleoside 5′-diphosphates (dNDP), resulting in the synthesis of dNTPs

proteins involved in nucleotide metabolism by supplying mitochondria with deoxyribonucleotide triphosphate (dNTP) pools needed for mtDNA replication. These include genes such as *TK2, DGUOK, SUCLA2, SUCLG1, TYMP* and *RRM2B* [1] (see Fig. 11.1).

Ribonucleotide reductase (RNR) M2 B, (MIM 604712, *RRM2B*), is a nuclear-encoded mitochondrial gene that encodes the small subunit of p53-inducible ribonucleotide reductase (p53R2), so called because transcription of RNR is tightly regulated by the tumor suppressor protein p53 [2, 3]. p53R2 is a component of the heterotetrameric catalytic enzyme R1/p53R2 that is responsible for the de novo synthesis of dNTPs by direct reduction of cytoplasmic ribonucleoside 5′-diphosphates to deoxyribonucleoside 5′-diphosphates. This reaction occurs in response to DNA damage [2, 4], and has been confirmed in *RRM2B*-null mice, where impairment of the pathway results in mtDNA depletion associated with renal failure, growth retardation, and early mortality [5]. Furthermore, p53R2 supplements the mitochondrion's own deoxynucleotide salvage pathway with the provision of constant dNTP pools for mtDNA synthesis as evidenced by the constant levels of both R1 and p53R2 in post-mitotic cells [6, 7] and the clear reciprocity between *RRM2B* mutations and the quantitative loss of mtDNA in specific tissues in mtDNA depletion syndromes [8–12]. In this chapter, we briefly review the secondary mtDNA defects associated with *RRM2B* mutation, highlighting the expanding clinical spectrum associated with this specific mtDNA maintenance gene disorder.

Clinical Spectrum of *RRM2B*-Associated Disease

The phenotypic expression of each of the mitochondrial maintenance gene defects including *POLG*, *POLG2*, *SLC25A4* (ANT1), and *PEO1* (Fig. 11.1) can, for the most part, be divided into mutations that cause mtDNA depletion, which usually present with severe multisystem disease and are often fatal in early life and mutations that predispose to accumulation of clonally-expanded (multiple) mtDNA deletions causing focal cytochrome *c* oxidase (COX) deficiency in affected tissues which may manifest as either recessive or dominant disease. Recessive disease, associated with multiple mtDNA deletions, generally presents during early childhood/adulthood with marked multisystem involvement, whilst dominant disease, typically develops in later adulthood, is usually less severe, and is often tissue-specific. However, we recognize that these are broad generalizations that the most common mitochondrial maintenance gene defects, including *POLG*, *POLG2*, *SLC25A4*, *PEO1* and *RRM2B* do not necessarily conform to. Specifically, in relation to *RRM2B* mutations, there are some notable exceptions. First, identical *RRM2B* mutations are associated with varied phenotypic severity depending on whether they exist in homozygous or heterozygous states. Second, mtDNA depletion, which is commonly associated with clinically severe, recessively-inherited mtDNA maintenance gene mutations in children has now been reported in one adult with a compound heterozygous *RRM2B* mutation [13], and third, *RRM2B*-related mtDNA depletion can potentially cause a relatively mild clinical phenotype.

RRM2B mutations are an important cause of both pediatric and adult-onset mitochondrial disease and are emerging as the third most common cause of multiple mtDNA deletions, following *POLG* and *PEO1* mutations [14, 15]. Through 2011, detailed clinical information on 48 individuals from 29 families has been published in the literature and is shown in Table 11.1. The *RRM2B* gene mutations have been associated with both sporadic and familial mitochondrial disease characterized by either autosomal recessive mtDNA depletion syndrome or recessive and dominant mutations which cause the accumulation of multiple mtDNA deletions. Given that the clonal expansion of mtDNA deletions requires time, these are only generally observed in the postmitotic tissues of adult patients.

Autosomal Recessive mtDNA Depletion Syndrome

The first reported human diseases attributed to *RRM2B* mutations were associated with a quantitative loss of mtDNA copies, the so-called mtDNA depletion syndromes. The mtDNA depletion syndrome due to *recessive* mutations in the *RRM2B* gene are characterised by myopathy, gastrointestinal dysmotility, respiratory insufficiency, lactic acidosis and renal proximal tubulopathy with nephrocalcinosis [8, 9]. Central nervous system (CNS) features can include hearing loss, seizures, microcephaly, and global developmental delay. These syndromes are early-onset and invariably fatal [8–12]. The first report of adult-onset, autosomal recessive mtDNA depletion

Table 11.1 A summary table of the molecular and clinical features of 48 individuals from 29 families with RRM2B-related mitochondrial disease, published in the literature to date, associated with both mtDNA depletion syndromes and multiple mtDNA deletion syndromes

Amino acid change	Nucleotide change	Genotype	Effect on mtDNA	Affected individuals/ families	Clinical features (number of affected individuals in brackets)	Age of onset	Outcome	Reference
p.Q284X	c.850C>T	Homozygous	Depletion	3/1	Gastrointestinal dysmotility (1), High lactate (7), Renal proximal tubulopathy (6), Respiratory insufficiency (5), Seizures (2), Truncal hypotonia (7)	<2 months	Died <4 months	[8]
p.E194K	c.580G>A c.IVS3-2A>G	Compound heterozygous	Depletion	2/1				
p.W64R	c.190T>C	Compound heterozygous	Depletion	1/1				
p.E194G	c.581A>G							
p.E85del	c.253_255delGAG	Compound heterozygous	Depletion	1/1				
p.C236F	c.707G>T							
p.I224S	c.671T>G	Homozygous	Depletion	1/1	Congenital deafness (1), Failure to thrive (3), Gastrointestinal dysmotility (2), High lactate (3), Microcephaly (2), Respiratory failure (2), Truncal hypotonia (3)	Birth	Died 8 months	[9]
p.M282I	c.846G>C	Compound heterozygous	Depletion	1/1		1–4 months	Alive 4 years	
p.N307IfsX11	c.920delA							
p.L317V	c.949T>G	Compound heterozygous	Depletion	1/1		6 months	Alive 27 months	
p.G195EfxX14	c.584delG							
p.R41P	c.122G>C	Homozygous	Depletion	1/1	Congenital deafness (1), Glycosuria (1), Hypotonia (2), Renal proximal tubulopathy (1)	Birth	Died 4–5 months	[12]
p.R110C	c.328C>T IVS3-2A>C	Compound heterozygous	Depletion	1/1		2 months	Died 1 year	

Table 11.1 (continued)

Amino acid change	Nucleotide change	Genotype	Effect on mtDNA	Affected individuals/ families	Clinical features (number of affected individuals in brackets)	Age of onset	Outcome	Reference
p.G229V	c.686G>T	Homozygous	Depletion	2/1	Cerebral atrophy (1), Failure to thrive (2), Hypocalcaemia (1), Hypotonia (2), Lactic acidosis (2), Muscle atrophy (1), Raised CK (1), Renal proximal tubulopathy (1), Respiratory insufficiency (2), Seizures (2)	2 months (subject 1 and 2)	Died 3 months (subject 1) and 5 months (subject 2)	[11]
p.P123S	c.386T>C	Probable compound heterozygous	Depletion	1/1	Congenital deafness, Dystonia, Focal seizures, Generalised central hypomyelination, Nephrocalcinosis, Nystagmus, PEO, Poor weight gain, Proximal renal tubulopathy, Ptosis, Raised plasma lactate, Respiratory distress, Truncal hypotonia, UMN signs, Vomiting	Birth	Died 4 months	[10]

Table 11.1 (continued)

Amino acid change	Nucleotide change	Genotype	Effect on mtDNA	Affected individuals/ families	Clinical features (number of affected individuals in brackets)	Age of onset	Outcome	Reference
p.R110H p.R121H	c.329G>A c.362G>A		Depletion	1/1	Ataxia, Dysarthria, Gastrointestinal dysmotility, Cachexia, Hearing loss, PEO, Peripheral neuropathy, Ptosis	30 years	Alive 42 years	[13]
p.R327X	c.979 C>T		Multiple deletions	17/2	Ataxia, Cognitive dysfunction, Exercise intolerance, Hypoacussis, Mood disturbance, PEO, Ptosis, Reduced reflexes	3rd decade (subject 3, family 2)	Not available	[16]
p.T144I	c.431C>T c.632G>A	Compound heterozygous	Multiple deletions	1/1	Areflexia (1), Ataxia (1), Cataracts (1), Cognitive impairment (1),	Birth	Died 25 years	[14]
p.R211K	c.606T>A c.817G>A	Compound heterozygous	Multiple deletions	1/1	Developmental delay (1),	4 years	Alive 14 years	
p.F202L	c.950delT	Heterozygous	Multiple deletions	3/3	Diabetes Mellitus (1), Dysarthria (2), Dysphagia (5), Dysphonia (2),	30–53 years	Alive 52–73 years	

11 *RRM2B*-Related Mitochondrial Disease

Table 11.1 (continued)

Amino acid change	Nucleotide change	Genotype	Effect on mtDNA	Affected individuals/ families	Clinical features (number of affected individuals in brackets)	Age of onset	Outcome	Reference
p.G273S	c.952G>T	Heterozygous	Multiple deletions	1/1	Encephalopathy (1), Facial weakness (1),	53 years	Alive 64 years	
p.L317X	c.965dupA	Heterozygous	Multiple deletions	3/3	Fatigue (4), Gastrointestinal disturbance (2),	26–54 years	Alive 61–64 years	
p.E318X	c.122G>A	Heterozygous	Multiple deletions	1/1	Glaucoma (1), Hearing loss (5),	59 years	Alive 71 years	
	c.583G>A	Heterozygous	Multiple deletions	1/1	Hypogonadism (1)	50 years	Alive 75 years	
p.N322KfsX4 p.R41Q p.G195R p.G229V	c.686G>T	Heterozygous	Multiple deletions	1/1	Hypoparathyroidism (1), Myopathy (5), PEO (8) Ptosis (7), Renal failure (1), Stroke-like episodes (1),	15 years	Alive 25 years	
p.R41Q p.E131Lys	c.122G>A c.391G>A	Compound heterozygous	Multiple deletions	1/1	Cachexia (1), Delayed puberty (1), Fatigue (2)	4 years	Died 22 years	[15]

Table 11.1 (continued)

Amino acid change	Nucleotide change	Genotype	Effect on mtDNA	Affected individuals/families	Clinical features (number of affected individuals in brackets)	Age of onset	Outcome	Reference
p.E85del	c.253_255delGAG	Heterozygous	Multiple deletions	1/1	Hearing loss (1), Hydronephrosis (1) PEO (2), Pigmentary retinopathy (1), Proximal muscle weakness (1), Ptosis (2) Raised CK (1), Raised CSF lactate (1), Raised CSF protein (1), Short stature (1)	58 years	Died 66 years	
p.P33S	c.341G>A	Homozygous	Multiple deletions	1/1	PEO Ptosis Deafness Muscle weakness Pigmented retinopathy Gonadal atrophy Depression	16 years	Alive at 43 years	[17]

PEO progressive external ophthalmoplegia, *CK* creatine kinase, *CSF* cerebrospinal fluid, *MRI* magnetic resonance imaging

syndrome was in a 42-year-old woman with mitochondrial neurogastrointestinal encephalopathy (MNGIE) who had compound heterozygous missense mutations in the *RRM2B* gene and documented evidence of mtDNA depletion in clinically-relevant tissues [13].

RRM2B Mutation and Multiple mtDNA Deletion Presentations

Mutations within *RRM2B* associated with multiple mtDNA deletions are now known to be associated with either autosomal recessive disease, which usually presents during childhood and is characterized by severe multisystem involvement, or autosomal dominant disease, which is usually clinically milder, later in onset—typically developing in adulthood—and is often tissue-specific. The first cases to be described were from two large, unrelated families with autosomal dominant PEO where affected individuals were shown to harbor a heterozygous nonsense mutation in exon 9 (c.979 C>T predicting p.Arg327Ter) causing truncation of the translated p53R2 protein [16]. In these families, in which the disease locus was first mapped to chromosome 8q22.1-q23.3 facilitating the identification of the disease gene, onset of symptoms occurred from the third decade of life, and the cardinal clinical features were PEO and ptosis in addition to ataxia, cognitive dysfunction, exercise intolerance, hypoacussis, mood disturbance, and peripheral neuropathy. More recently, *RRM2B* mutations have been shown to be frequent in familial PEO with muscle-restricted multiple mtDNA deletions [14]. The frequency of *RRM2B* mutations in a large cohort of patients with chronic PEO and multiple mtDNA deletions in muscle in whom mutations in all known candidate genes (i.e. *POLG, POLG2, SLC25A4,* and *PEO1*) had been excluded, identified 10 different *RRM2B* variants in 12 subjects [14]. All 10 *RRM2B* changes were novel. Two patients harbored compound heterozygous, missense variants implying autosomal recessive inheritance, phenotypically characterized by severe, early onset disease including PEO, ptosis, deafness, renal failure, developmental delay, hypoparathyroidism, hypogonadism, areflexia, and dysphagia. Seven patients harbored single heterozygous, truncating mutations within exon 9, all of whom had family histories consistent with autosomal dominant inheritance together with supportive genetic segregation data in one family. These patients presented from the third decade of life with a predominantly myopathic phenotype including severe PEO, ptosis, proximal myopathy, ataxia, dysphagia, and fatigue. Single heterozygous, missense variants were detected in a further 3 patients (MLPA excluded exonic copy number variation in *trans*), however, the pathogenicity of these 3 variants was provisional in the absence of further supporting evidence. Two of these patients presented significantly later in the fifth decade of life, whilst the third patient presented at age 15 years with a MNGIE-like phenotype [14]. Most recently, a homozygous missense variant in *RRM2B*, suggested to cause autosomal recessive PEO, has been described in a 42-year-old man presenting with mood disorder extending the clinical phenotype further [17].

A clinical presentation of Kearns-Sayre Syndrome (KSS) has also been linked to mutation of *RRM2B*. Two novel missense heterozygous *RRM2B* mutations (c.122G>A predicting p.Arg41Gln and c.391G>A predicting p.Glu131Lys) were identified in an adult patient by Pitceathly and colleagues in which clinical and biochemical markers analogous to KSS were reported, including PEO and pigmentary retinopathy with onset before the age of 20 years in association with sensorineural deafness and raised CSF protein [15]. The authors also demonstrated impaired assembly of R1/p53R2 as a direct consequence of *RRM2B* dysfunction using blue-native gel electrophoresis, thus providing mechanistic insight into the deleterious effects *RRM2B* mutations have on the heterotetrameric structure of the enzyme.

When to Screen for *RRM2B* Mutations?

Deleterious mutations in *RRM2B* are associated with a phenotype that includes hypotonia, lactic acidosis, and renal tubulopathy in pediatric-related mitochondrial disease. The presence of these clinical features in association with mitochondrial respiratory chain defects and severe mtDNA depletion in skeletal muscle should prompt screening of the *RRM2B* gene, over other nuclear-encoded genes, in which muscle weakness is the dominant clinical feature.

PEO, ptosis, and proximal muscle weakness are recognized as the predominant clinical characteristics seen in adult patients with *POLG* and *PEO1* (Twinkle) mutations. Bulbar dysfunction, hearing loss, and gastrointestinal problems, including Irritable Bowel Syndrome-like symptoms and low BMI, are also additional key features seen in *RRM2B*-related mitochondrial disease, and appear to occur more often than with *POLG* and *PEO1* mutations. Interestingly, gastrointestinal disturbance is often a severe finding in children with *RRM2B* mutations. CNS features characteristic of other syndromic presentations of mitochondrial disease, however, are present less frequently [14]. Thus, the prominence of bulbar dysfunction, hearing loss, and gastrointestinal problems in the absence of overt CNS features, would support early prioritization for screening of *RRM2B* over *POLG* and *PEO1* in adult patients with PEO and multiple mtDNA deletions in skeletal muscle. Furthermore, *RRM2B* gene analysis should be considered early in KSS when there appears to be a Mendelian pattern of inheritance or evidence of multiple mtDNA deletions in a diagnostic biopsy. Finally, *RRM2B* screening should be considered in patients with MNGIE if deoxyuridine and thymidine levels in both blood and urine are negative, and thymidine phosphorylase activity is normal in white cells and platelets and analysis of the *TYMP* gene does not identify causative mutations.

Conclusion

Defects in *RRM2B* are now recognized as an important cause of nucleotide metabolism dysfunction that causes mtDNA instability resulting in mitochondrial disease. Although originally associated with depletion disorders, it is now increasingly recognized as a cause of adult onset mitochondrial disease. *RRM2B* gene screening should be considered early in the differential diagnosis of children with muscle weakness, lactic acidosis, and renal tubulopathy with severe mtDNA depletion in skeletal muscle and in adults with multiple mtDNA mutations once *POLG* and *PEO1* mutations have been excluded.

Acknowledgments DMT and RWT are supported by a Wellcome Trust Strategic Award (906919) and the UK NHS Specialised Services "Rare Mitochondrial Disorders of Adults and Children" Diagnostic Service (http://www.mitochondrialncg.nhs.uk). GSG, DMT and RWT are also supported by the Medical Research Council (UK, G1000848) as part of the MRC Centre for Neuromuscular Diseases. RDSP is funded by MRC grant number G0800674.

References

1. Copeland WC (2012) Defects in mitochondrial DNA replication and human disease. Crit Rev Biochem Mol Biol 47:64–74
2. Tanaka H, Arakawa H, Yamaguchi T et al (2000) A ribonucleotide reductase gene involved in a p53-dependent cell-cycle checkpoint for DNA damage. Nat 404:2–49
3. Nakano K, Bálint E, Ashcroft M et al (2000) A ribonucleotide reductase gene is a transcriptional target of p53 and p73. Oncogene 19:4283–9
4. Yamaguchi T, Matsuda K, Sagiya Y et al (2001) p53R2-dependent pathway for DNA synthesis in a p53-regulated cell cycle checkpoint. Cancer Res 61:8256–62
5. Kimura T, Takeda S, Sagiya Y et al (2003) Impaired function of p53R2 in Rrm2b-null mice causes severe renal failure through attenuation of dNTP pools. Nat Genet 34:440–5
6. Hákansson P, Hofer A, Thelander L (2006) Regulation of mammalian ribonucleotide reduction and dNTP pools after DNA damage and in resting cells. J Biol Chem 281:7834–41
7. Chabes A, Georgieva B, Domkin V et al (2003) Survival of DNA damage in yeast directly depends on increased dNTP levels allowed by relaxed feedback inhibition of ribonucleotide reductase. Cell 112:391–401
8. Bourdon A, Minai L, Serre V et al (2007) Mutation of *RRM2B*, encoding p53-controlled ribonucleotide reductase (p53R2), causes severe mitochondrial DNA depletion. Nat Genet 39:776–780
9. Bornstein B, Area E, Flanigan KM et al (2008) Mitochondrial DNA depletion syndrome due to mutations in the *RRM2B* gene. Neuromuscul Disord 18:453–459
10. Acham-Roschitz B, Plecko B, Lindbichler F et al (2009) A novel mutation of the RRM2B gene in an infant with early fatal encephalomyopathy, central hypomyelination and tubulopathy. Mol Genet Metab 98:300–4
11. Kollberg G, Darin N, Benan K et al (2009) A novel homozygous *RRM2B* missense mutation in association with severe mtDNA depletion. Neuromuscul Disord 19:147–150
12. Spinazzola A, Invernizzi F, Carrara F et al (2009) Clinical and molecular features of mitochondrial DNA depletion syndromes. J Inherit Metab Dis 32:143–58
13. Shaibani, A, Shchelochkov O, Zhang S et al (2009) Mitochondrial neurogastrointestinal encephalopathy due to mutations in *RRM2B*. Arch Neurol 66:1028–1032
14. Fratter C, Raman P, Alston CL et al (2011) *RRM2B* mutations are frequent in familial PEO with multiple mtDNA deletions. Neurol 76:2032–4
15. Pitceathly RD, Fassone E, Taanman JW et al (2011) Kearns–Sayre syndrome caused by defective R1/p53R2 assembly. J Med Genet 48:610–7
16. Tyynismaa H, Ylikallio E, Patel M et al (2009) A heterozygous truncating mutation in *RRM2B* causes autosomal-dominant progressive external ophthalmoplegia with multiple mtDNA deletions. Am J Hum Genet 85:290–295

17. Takata A, Kato M, Nakamura M et al (2011) Exome sequencing identifies a novel missence variant in *RRM2B* associated with autosomal recessive progressive external ophthalmoplegia. Genome Biol 12:R92
18. Di Mauro S, Schon EA (2008) Mitochondrial disorders in the nervous system. Annu Rev Neurosci 31:91–123

Part III
Complex Subunits and Assembly Genes

Chapter 12
Complex Subunits and Assembly Genes: Complex I

Ann Saada (Reisch)

Introduction

Complex I Structure and Function

Mitochondria are the major providers of ATP in animal cells by oxidative phosphorylation (OXPHOS). OXPHOS utilizes reduced coenzymes, reduced nicotinamide adenine dinucleotide (NADH), and reduced flavin adenine dinucleotide (FADH$_2$), mainly formed in the TCA (tricarboxylic acid) cycle and by fatty acid oxidation. The electrons derived from these reducing equivalents are shuttled through the mitochondrial respiratory chain (MRC) which is comprised of five multiprotein complexes (CI-CV) located in the mitochondrial inner membrane (MIM). MRC complex I (CI, NADH ubiquinone oxidoreductase, NADH Coenzyme Q reductase, EC 1.6.5.3) performs the first electron transfer step. This large multisubunit enzyme catalyzes the transfer of electrons from NADH, via a flavin mononucleotide (FMN) and a series of eight iron-sulfur clusters (Fe-S), to ubiquinone (Q), a small lipid-soluble electron carrier in the MIM. Electron transfer from succinate through FADH$_2$ to Q is carried out by CII (succinate ubiquinone oxidoreductase). Subsequently, electrons are shuttled from Q through complex III (ubiquinol cytochrome c oxidoreductase/cytochrome bc_1 complex) via cytochrome c, a small soluble electron carrier protein, followed by complex IV (cytochrome c oxidase) to molecular oxygen producing water. CI, CIII, and CIV utilize energy derived from the electron transfer to

---

The nomenclatures of genes C20orf7 and C8orf38 has recently been updated by the Human Genome Organization (HUGO) in accordance with the Guidelines for Human Gene Nomenclature as follows: C20orf7 is now designated as NDUFAF5 and C8orf38 is now designated as NDUFAF6.

A. Saada (Reisch) (✉)
Department of Genetic and Metabolic Diseases and Monique and Jacques Roboh Department of Genetic Research, Hadassah Hebrew University Medical Center,
P. O. Box 12000, 91120 Jerusalem, Israel
e-mail: Annsr@hadassah.org.il

pump protons from the matrix across the MIM to the intermembrane space, creating an electrochemical gradient (Δ(delta)ψ(psi)) which induces a conformational change in complex V (F1F$_o$ATP synthase) resulting in the conversion of ADP and inorganic phosphate to ATP [1]. The mechanism of proton pumping by CI is still elusive but experimental data show that the passage of two electrons through CI is coupled to the translocation of four protons from the mitochondrial matrix to the intermembrane space and contributes ~40 % of the proton-motive force needed for ATP production [2]. Mitochondrial CI is the largest and most intricate of the ETC complexes. Its main structure is conserved from bacterial homologues to bovine mitochondria. Electron microscopy has revealed that it is an L-shaped structure, with one hydrophobic arm embedded in the MIM and a hydrophilic peripheral arm containing the FMN moiety, protruding into the mitochondrial matrix [3]. Mitochondrial CI interacts with CIII and CIV and together they combine into a $I_1III_2IV_1$ "supercomplex" or "respirasome" [4, 5]. Studies of the bovine enzyme performed to define the composition of the closely related human enzyme, showed that the mammalian mitochondrial CI is composed of 45 protein subunits with a combined molecular weight approaching ~1 MDa [6]. This complex, can be disassembled into subcomplexes I α-λ, designated into three functional modules: N (harboring the NADH-oxidizing site and the FMN moiety), Q (harboring the ubiquinone reduction site), and P (harboring the proton-pumping machinery) [6, 7]. Fourteen subunits are the "core subunits" essential for catalysis, and are conserved from bacteria to humans. Recent solution of the crystal structure of the *E. Coli* and *Thermus thermophilus* enzyme has enabled the positioning of these subunits within the complex [8]. Over the course of evolution, a number of the so-called accessory or supernumerary subunits have been added and these vary between species. In humans, mitochondria accessory subunits contribute about one-third of the total CI mass providing a scaffold which links the peripheral arm and the distal proton-pumping module in the membrane arm. The exact localization of most of these has yet to be determined (Fig. 12.1) [9, 10]. Many of the accessory subunits are posttranscriptionally modified by myristoylation, acetylation, methylation, and phosphorylation (Fig. 12.1) [9–13]. CI activity can also be regulated by glutathione levels via S-nitrosation and is inhibited by many compounds, among them are the pesticide, rotenone. Even some Q analogs are potential inhibitors rather than substrates [14–18]. Other subunits are less defined, albeit many of them proved to be essential for CI assembly and function, as mutations in them were linked to human diseases (Fig. 12.1) [19, 20]. Complex I subunits are encoded by two genomes, the mitochondrial (mtDNA) and the nuclear (nDNA) [21]. Seven subunits originate from mtDNA, and are designated by an "ND" prefix (ND1, ND2, ND3, ND4, ND4L, ND5, and ND6). All mtDNA-encoded proteins are hydrophobic core subunits, located in the membrane arm. The remaining 38 subunits including 7 core subunits and all accessory subunits are encoded by the nDNA, designated by a "NDU" prefix. They are transcribed in the cytosol and are imported through the outer and inner membranes via the TOM and TIM transporter machineries, although many lack defined import sequences [22, 23].

12 Complex Subunits and Assembly Genes: Complex I

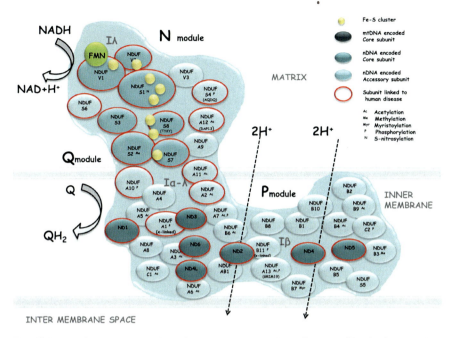

Fig. 12.1 This figure depicts a simplified schematic overview of human CI subunits, based on published structural and genetic data obtained from human—bovine and bacterial CI

Complex I Assembly

The CI assembly was first studied in CI mutants of the fungus *Neurospora crassa* where a separate assembly of the membrane and peripheral arms was observed and then joined together in a stepwise fashion to form the holoenzyme. Two additional proteins CIA30 and CIA84, were found to play a role in this process by transiently associating with the large membrane arm subcomplex proteins. These proteins (CI assembly factors) were not components of the mature holoenzyme; however, mutating them blocked complex assembly [24]. This finding highlighted the existence and importance of CI assembly factors, the human homolog of CIA30 (NDUFAF1) was characterized, and was found to be associated with CI deficiency in humans [25]. Another approach in the search for CI assembly factors was a comparative bioinformatics approach, where genes from a yeast strain with a functional CI were subtracted from fermentative yeast lacking true CI. This study generated a list of candidate assembly factors. Among these, a paralog (B17.2L, NDUFAF2) of the B17.2 structural subunit was found to be mutated in the first patient identified with a CI assembly factor defect [26]. Phylogenetic profiling combined with mitochondrial proteomics and siRNA led to the identification of 24 additional putative assembly factor candidates. Among these was C8orf38, a new CI assembly factor associated with mitochondrial disease. Peculiarly, some of the putative CI assembly factors were already associated with nonmitochondrial diseases. Additional human CI assembly factors were identified by "reverse genetics", i.e., by locating disease-causing mutations in previously uncharacterized nDNA-encoded proteins essential for CI function and assembly but not a part of the final CI complex [27–29]. Linkage analysis combined with Sanger sequencing identified C6orf66 (NDUFAF4), its interacting partner C3orf 60 (NDUFA3), C20orf7 (probable methyltransferase) and ACAD9 (acyl coA dehydrogenase 9) [30–33]. High-throughput sequencing identified mutations in FOXRED1 (FAD-dependent oxidoreductase domain, which may be involved in electron transfer) and NUBLP (IND 1,nucleotide binding protein-like) in CI deficient patients [34, 35].

Till date, nine assembly factors have been identified and some of them were renamed and designated with a "NDUFAF" prefix. In addition, two other proteins interacting with assembly factors and essential for CI biogenesis were identified: ECSIT (a Toll pathway intermediate) which interacts with NDUFAF 1 and MidA (a putative methyltransferase) interacts with NDUFS2 [12, 36]. Detailed analysis of the CI assembly process in humans was obtained by blue native polyacrylamide gel electrophoretic analysis (BN-PAGE) of stalled assembly intermediates in CI-deficient patient cells or analyzing CI formation after inhibition of protein synthesis [37–39]. Assembly complexes were also analyzed by fluorescence recovery after photobleaching (FRAP) in living cells and by mass spectrometry [40, 41]. Based on information available, two models of human CI assembly were proposed. One model suggested two assembly modules corresponding to the membrane and peripheral arms, while the other identified six or seven mixed intermediates. Analysis of several mouse lines proposed that the mammalian assembly pathway consists of five different assembly steps [42]. The current combined model proposes that initially the hydrophilic core subunits NDUFS2, NDUFS3, NDUFS7, and NDUFS8 assemble. They are joined with a hydrophobic subcomplex containing ND1 to form a ~400 kDa assembly intermediate. This intermediate combines with another hydrophobic ~460 kDa subcomplex containing ND2, ND3, ND6, and ND4L to form the ~650 kDa intermediate. Subsequently, ND4, ND5, and the subunits of the distal portion of the P module are added to form the ~830 kDa intermediate. In the last step, the N module is added and the ~980 kDa final CI complex is complete. The role of assembly factors is not fully elucidated but C20orf7 and C8orf38 probably play roles in the translation and/or stabilization of ND1 and the formation of the ~400 kDa intermediate. NDUFAF3 and NDUFAF4 are involved early in the assembly process as well, by forming/stabilizing the ~400 kDa intermediate. NDUFAF1 and ECSIT appear to be involved at an intermediate stage and NDUFAF2 plays a role in the later. NUBLP is essential for the formation of Fe-S clusters (43) and ACAD9 affects the levels of NDUFAF1 and ECSIT (Fig. 12.2) [28, 29, 43]. Of note, most probably additional CI assembly factors will be discovered, bearing in mind that the much smaller complex IV, which has only 13 subunits, requires more than 15 assembly factors for its assembly [44]. As new information becomes available, presumably the current model of CI assembly will be modified in the future. It is also important to recognize that defective CI assembly may also affect the stability of other complexes: a specific decrease of fully assembled CIII was noted in patients with mutations in NDUFS2 and NDUFS4. In addition, mutations in C20orf7 and C8orf38 reported to negatively affect CIV activity. These secondary effects are possibly a result of destabilized $I_1III_2IV_1$ supercomplex [29, 30, 32, 37, 45].

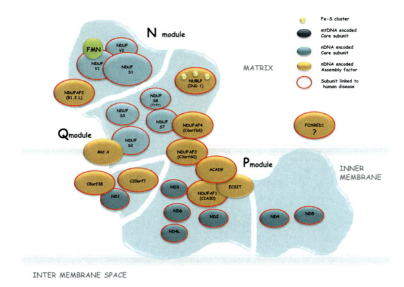

Fig. 12.2 This figure depicts a simplified schematic overview of human CI assembly intermediates and assembly factors according to currently published knowledge

Dual Roles of Complex I Components

Interestingly, some accessory subunits and assembly factors are homologous, or identical, to proteins of known function with distinct roles besides their role in complex I. The accessory subunit NDUFA13 was previously designated GRIM19 (gene associated with retinoic acid and interferon-β-induced mortality-19) which interacts with STAT3 (signal transducer and activator of transcription) [46]. NDUFS3 is selectively degraded by granzymes (proteases excreted by cytotoxic T cells and natural killer cells of the immune system) and is thereby directly linked to caspase independent cell death [47]. Two CI assembly factors, NDUFAF2 and NDUFAF4 have been associated with cancer. NDUFAF2 is identical with mimitin (myc-induced mitochondrial protein) and NDUFAF4 is identical to HRPAP20, a protein playing a part in breast cancer invasiveness [26, 30]. ECSIT is essential for inflammatory response and embryonic development [36]. Moreover, designations and homology can sometimes be misleading as was in the case of ACAD9. This protein was previously identified as a member of the acyl-CoA dehydrogenase family closely resembling very long-chain fatty acid dehydrogenase (VLCAD). While exhibiting acyl-CoA dehydrogenase activity in vitro, apparently ACAD9 does not participate in β-oxidation of fatty acids [33].

Also, proteins which are not a part of, or do not directly interact with CI or CI assembly factors may still be essential for CI function. This became evident when impaired CI biogenesis was detected in the Harlequin mouse. This animal model expresses low levels of apoptosis-inducing factor (AIF), a mitochondrial flavoprotein which translocates to the nucleus upon mitochondrial outer membrane permeabilization. In the nucleus, this protein mediates nuclear features of apoptosis such as chromatin condensation and large-scale DNA degradation [48, 49]. Another example is the testicular nuclear receptor (TR4), a member of the nuclear receptor superfamily of transcription factors which appears to regulate complex I activity by binding to hormone response elements in the NDUFAF1 promoter sequence [50]. The above data suggest that CI is involved not only in OXPHOS per se but is involved in a variety of cellular pathways.

Inherited, Isolated Complex I Deficiency of Nuclear Origin

Isolated Complex I Deficiency and Diagnosis

Congenital disorders of OXPHOS are common inborn errors of metabolism with an incidence of 1:5000–1:8000 in live births [51, 52]. The first cases of isolated CI deficiency (MIM #252010) were reported in 1979 in two sisters with exercise and intermittent lactic acidosis [53]. Since then, the clinical spectrum has expanded to include a wide range of disorders: severe neonatal lactic acidosis (lethal-infantile mitochondrial disease, LIMD); Leigh syndrome or Leigh-like syndrome, (L.S.); mitochondrial encephalopathy, lactic acidosis, and stroke-like episodes, MELAS), encephalopathy, leukodystrophy, cerebellar ataxia, hypotonia, cardiomyopathy, hepatopathy, tubulopathy, diabetes mellitus, optic atrophy (LHON, Leber hereditary optic neuropathy); retinitis pigmentosa (RP), ophthalmoplegia, growth retardation, and other presentations [19, 20, 28, 54, 55]. Given the complexity of CI structure and assembly, it is not surprising that defective CI is the most common diagnosis in patients with inherited mitochondrial disease. Isolated CI deficiency accounts for more than a third of all patients referred for OXPHOS evaluation [19, 56, 57]. The diagnosis is made by measuring CI (NADH CoQ reductase) and CI+CIII (NADH cytochrome c reductase) activities in mitochondrial-enriched fractions obtained from muscle biopsy or other affected tissues by spectrophotometric methods. The CI activity is dependent on the purity of the mitochondria, due to the contaminating microsomal "NADH- dehydrogenase". Therefore, rotenone sensitivity serves as an indicator of correct measurement. This is also why pathology staining is not informative. The CI and CI+III activities are also compared with other ETC enzymes and to normal controls. Diagnosis is frequently complemented by measuring oxygen consumption or ATP production with CI-dependent substrates (pyruvate, malate, glutamate), by BN-PAGE and/or western blot. As the biochemical workup and interpretation are complicated, they are usually performed only in specialized laboratories [37, 58–61].

The clinical phenotype of complex I deficiency is extremely variable and the symptoms may overlap with a variety of other clinical conditions. Thus, frequently a dilemma arises as to whether to proceed to an invasive biopsy procedure for diagnostic purposes. Lactic acidosis is nonspecific as it may be present in many cases but it is also frequently absent. Fibroblast growth factor 21 (FGF21) in plasma has recently been suggested to be a more suitable biomarker, and, in combination with lactate measurement, the correlation with respiratory chain deficiencies in muscle, especially in children, is increased [62]. Brain MRI has also been shown to be a practical, noninvasive diagnostic procedure and may direct molecular investigation, possibly circumventing the need for muscle biopsy [63]. The investigation of CI deficiency is not complete until the molecular basis of the disease is verified. The finding of disease-causing mutations in a complex I subunit or assembly factor concludes the diagnosis.

Mutations in Nuclear-Encoded Structural Genes

Due to its bigenomic origin, CI deficiency may result from mutations in either CI-structural protein subunits (mtDNA or nDNA encoded) or CI assembly factors (nDNA encoded). Disease-causing mutations have been reported in all seven mtDNA genes and are mainly associated with LHON, MELAS, and L.S. These mutations are maternally inherited and accordingly the clinical phenotype and tissue expression is highly dependent on the degree of heteroplasmy. This subject lies outside the scope of this chapter and the reader is kindly referred to a number of excellent reviews (see Fig. 12.1) [19–21, 28, 64, 65].

Mutations in nuclear genes (designated with a "NDU" prefix) encoding CI subunits are inherited in a Mendelian fashion. Accordingly, genetic counseling and molecular prenatal diagnosis is feasible. It is therefore imperative to rule out maternal inheritance and to establish the causative nuclear gene mutations. Finding a new missense mutation in a single patient is challenging due to the high frequency of benign missense mtDNA polymorphisms [65]. Thus, the pathogenicity of a novel missense variant needs to be verified by some form of experimental proof such as complementation with the wild-type gene and mutation screening in an appropriate population.

Although far from all CI deficiencies have been deciphered, significant progress has been made since the first mutation in a nuclear-encoded CI subunit was detected in 1998 [66] (a list of mutated genes, clinical characteristics and references are provided in the table). Mutations have been identified in genes encoding all seven nDNA-encoded core subunits; *NDUFV1*(MIM 161015) [34, 57, 67–69], *NDUFV2* (MIM 1600532) [70, 71], *NDUFS1*(MIM 157665) [68, 71–76], *NDUFS2* (MIM 602985) [73, 77–79], *NDUFS3* (MIM 603 846) [80], *NDUFS7* (MIM 601825) [81, 82], *NDUFS8* (MIM 602141) [34, 66, 71, 83], and in the genes of seven accessory subunits; *NDUFS4* (MIM 602694) [84–89], *NDUFS6* (MIM 602694) [90, 91], *NDUFA1* (MIM 300078) [92–94], *NDUFA2* (MIM 602137) [95], *NDUFA10* (MIM 603835) [96], *NDUFA11* (MIM 612638) [97], and *NDUFA12* [98] (Table 12.1). Some genes, *NDUFV1*, *NDUFS1*, and *NDUFS4*, are mutational hotspots, as reflected by a large number of citations in the literature. Mutations in other genes are less frequently detected. For example, mutation in *NDUFA12* has been detected in only one case of 122 patients with CI deficiency [98]. Most genes are autosomal recessive and the mutations are homozygous. An exception is *NDUFA1* which is X-linked and hemizygous males present with CI deficiency. One report also described a heterozygous female patient with skewed X-inactivation expressing the CI-deficient phenotype [92–94]. Compound heterozygosity is less common but has occurred [34, 68]. The involvement of two different CI genes has also been reported in a case of encephalomyopathy associated with heterozygous mutations in *NDUFA8* and *NDUFS2* [73].

With respect to phenotype, early onset and grave outcome is characteristic of nuclear-encoded mutations in the structural CI subunits. The symptoms are variable but mostly neurological with some patients presenting with cardiomyopathy (Table 12.1). The tissue distribution is peculiar since these genes are nDNA encoded and the mutations are expected to affect all tissues [20, 28]. The scarcity of reports of some and the lack of many other accessory subunits may be due to a number of reasons, among them, atypical clinical presentation leading to misdiagnosis or incompatibility with life. However, most likely, new mutations will be identified in the future with the increasing availability of next-generation sequencing.

Mutations in Nuclear-Encoded Nonstructural Genes

At least half of all patients with isolated complex I deficiency remain without molecular diagnosis despite normal sequence of all CI-structural genes. Failure to assemble a functional CI complex in many of these patients suggested that factors involved with assembly or maintenance of complex I are important causes of disease. Extended genetic and molecular investigation including linkage analysis and sequencing concluded that some of these patients harbor mutations in nDNA-encoded proteins involved in CI assembly and/or stabilization. As most assembly proteins were previously uncharacterized, a new mutation necessitated experimental corroboration by restoring function through insertion of the wild-type gene [28, 99]. Pathogenic mutations have been detected in genes encoding nine assembly factors; *NDUFAF1* (CIA30, MIM 606934) [25, 100], *NDUFAF2* (B17.2 L, MIM 609653 [26, 34, 101, 102], *NDUFAF3* (C3orf60, MIM 612911) [31], *NDUFAF4* (C6orf66, MIM 611776) [30], C20orf7 (MIM 612360) [32, 45, 103], C8orf38 (MIM 612392) [27, 29], *ACAD9* (MIM 611126) [104, 105], *FOXRED1* (MIM 613622) [34, 35], and *NUBPL* (Ind- 1, MIM 613621) [34]. Generally, CI assembly defects present early in life. The prognosis is grave and the clinical spectrum overlaps that of CI deficiency due to mutations in structural genes (Table 12.1).

In addition, to assembly proteins per se, mutations have also been reported in the X-linked *AIFM1* gene-encoding AIF (MIM 30169). Differing from the harlequin mouse model, patients show a combined OXPHOS defect including CI, CIII, and CIV and a normal or decrease in mtDNA content, possibly suggesting that AIF is involved also in mtDNA maintenance [49, 106].

12 Complex Subunits and Assembly Genes: Complex I

Table 12.1 Genes with known mutations associated with isolated CI deficiency in human

Gene/protein	Function/location	Onset/outcome	Clinical symptoms associated with mutations	Reference
NDUFV1/51 kDa	Core, Flavo—Fe–S protein/N module	Infancy/fatal-grave	Failure to thrive, hypotonia, cerebellar ataxia, dystonia, myoclonus, seizures leukoencephalopathy, encephalopathy, L.S, brain-stem lesions, ptosis, ophtalmoplegia, nystagmus, lactic acidosis	[34, 57, 67, 68, 69]
NDUFV2/24 kDa	Core, Fe–S protein/N module	Infancy/fatal	Hypertropic cardiomyopathy, encephalopathy	[70, 71]
NDUFS1/75 kDa	Core, Fe–S protein/N module	Infancy/fatal-severe	Microcephaly, muscle dystrophy, hypotonia, L.S., leukoencephalopathy, wanishing white matter, nystagmus, optic atrophy, lactic acidosis, hepatomegaly, anemia, lactic acidosis	[68, 71, 72, 73, 74, 75, 76]
NDUFS2/49 kDa	Core/Q module	Infancy/fatal-severe	Failure to thrive, hypertrophic cardiomyopathy, L.S., encephalomyopathy, hypotonia, dystonia, nystagmus, lactic acidosis	[73, 77, 78, 79]
NDUFS3/30 kDa	Core/Q module	Childhood/fatal	L.S.	[80]
NDUFS7/20 KDa	Core, Fe–S protein/Q module	Infancy/fatal-severe	L.S., hepatomegaly, lactic acidosis	[81, 82]
NDUFS8/23 kDa (TYKY)	Core, Fe–S protein/Q module	Infancy-childhood/fatal-severe	Cardiomyopathy, mitochondrial encephalopathy, leukodystrophy, L.S. lactic acidosis	[34, 66, 71, 83]
NDUFS4/18K kDa (AQDQ)	Accessory, protein/N module	infancy/fatal	L.S., microcephaly, hypertropic cardiomyopathy, strabismus, lactic acidosis	[84, 85, 86, 87, 88, 89]
NDUFS6/13 kDa	Accessory/N module	Infancy/fatal	Lactic acidosis	[90, 91]
NDUFA1/7.5 kDa	Accessory (Xlinked)	Infancy-childhood/fatal-severe	L.S., hypotonia, ataxia, retinitis pigmentosa, cerebellar atrophy, seizure, lactic acidosis	[92, 93, 94]
NDUFA2/8 kDa	Accessory	Infancy/fatal	L.S., cerebellar atrophy, hypertropic cardiomyopathy	[95]
NDUFA10/42 kDa	Accessory	Infancy/fatal	Hypotonia, L.S., lactic acidosis	[96]
NDUFA11/14.7 kDa	Accessory	Infancy/fatal	Encephalopathy, cardiomyopathy, lactic acidosis	[97]
NDUFA12/(B17.2, DAP13)	Accessory	Childhood/severe	L.S., muscular atrophy, dystonia	[98]

Table 12.1 (continued)

Gene/protein	Function/location	Onset/outcome	Clinical symptoms associated with mutations	Reference
NDUFAF1/(CIA 30)	Assembly factor	Infancy/fatal-severe	Encephalopathy, hypertropic cardiomyopathy, FTT, retinopathy, lactic acidosis	[25, 100]
NDUFAF2/(B17.2.L, Mimitin)	Assembly factor	Childhood/severe	Encephalopathy, L.S., ataxia, hypertropic cardiomyopathy, elevated CSF lactate	[25, 34, 100, 102]
NDUFAF3/ (C3orf60)	Assembly factor	Infancy/fatal	Encephalomypathy, Lactic acidosis	[31]
NDUFAF4 (C6orf66)	Assembly factor	Antenatal-infancy/ fatal-severe	Encephalomypathy, cardiomyopathy, optic atrophy, lactic acidosis	[30]
C20orf7	Assembly factor	Antenatal-infancy/ lethal-severe	Encpahalopathy, L.S., optic atrophy, dysmorphism, lactic acidosis	[32, 45, 103]
C8orf38	Assembly factor	Infancy/fatal-severe	L.S. lactic acidosis	[27, 29]
ACAD9	Assembly factor	Infancy-childhood/ lethal	Muscle weekness, hypertropic cardiomyopathy, hearing loss, liver dysfunction, hypoglycemia, lactic acidosis	[104, 105]
FOXRED1	Assembly factor	Infancy-childhood/ severe	L.S. microcephaly, cerebellar atrophy, hypertropic cardiomyopathy hypotonia, cortical blindness	[34, 35]
NUBPL/(IND 1)	Assembly factor/ Fe–S transfer	Childhood/severe	L.S.	[34]
AIFM1/(AIF)	Apoptosis factor	Prenatal-childhood/ severe	Encephalomyopathy, lactic acidosis	[49, 106]

L.S. Leigh syndrome or Leigh-like syndrome

To summarize; an increasing number of isolated CI deficiencies are now molecularly defined. However, many cases of CI deficiency still remain without a molecular diagnosis. High-throughput genetic sequencing will allow the identification of additional mutations in known CI genes as well as in new assembly factors. Likely, candidates are ECSIT or MidA and other putative assembly factors [27].

Pathomechanism, Treatment, and Model Systems

Pathomechanism

The obvious result of mitochondrial ETC dysfunction in CI deficiency is the disruption of $\Delta\Psi$ leading to energy depletion and impaired calcium homeostasis. However, CI is also a major site of oxygen-free radical (reactive oxygen species, ROS) production which is aggravated in MRC dysfunction. Thus, defective CI activity not only diminishes NADH oxidation and ATP production but also exerts a critical effect on calcium homeostasis, cell growth, death, and degeneration caused by oxidative stress. Moreover, due to its structural complexity CI is particularly vulnerable to oxidative stress which in turn exacerbates CI dysfunction and so forth [20, 107]. As mentioned above, defective CI function and assembly may also affect other respiratory chain complexes, metabolic pathways and other cellular functions by direct interaction with certain subunits [45–50, 105, 108].

Additionally, mitochondrial CI dysfunction and oxidative stress has been linked to other neurodegenerative diseases such as Parkinson disease (PD) and Alzheimer and possibly also to schizophrenia. The complex etiology of these disorders is not in the scope of this chapter and the reader is referred to the existing literature on this subject [109–111].

Treatment Strategies

Despite major advances in the biochemical and molecular diagnostics and the deciphering of the CI structure, function, assembly, and pathomechanism, there is currently no satisfactory cure for patients with mitochondrial complex I defects. Strategies to ameliorate the devastating effects of CI dysfunction include the administration of vitamins (vitamin K, riboflavin, B1, B2), cofactors (CoQ$_{10}$, carnitine, creatine), ROS scavengers (vitamin E, CoQ$_{10}$) and dichloroacetate (to decrease lactate levels) [112]. The effect of administration of single or a mixture of compounds in patients is difficult to evaluate due to the wide variety of CI defects and the lack of appropriate control subjects. However, there have been reports of a measurable positive effect for some strategies. For example, improvement was detected with riboflavin supplementation in ACAD9 deficiency and a ketogenic diet improved some of the parameters in NDUFV1 deficiency [69, 104]. Recently, a controlled trial with idebenone treatment in acute LHON (of mtDNA origin(suggested that early and prolonged idebenone (a Q analogue) treatment may significantly improve the frequency of visual recovery and possibly change the natural history of the disease [113]. Then again, as mentioned above, the administration of a short-chain Q analogue in systematic CI deficiency could be problematic [17]. Another Q analogue, engineered to specifically target and enter mitochondria, MitoQ$_{10}$ is presently undergoing clinical evaluation for PD [114].

Model Systems

On the cellular level, many studies have been conducted on patient's fibroblasts, examining the effect of various compounds. For example, riboflavin, trollox (an antioxidant) and Mito Q10 were shown to be beneficial in fibroblasts from CI-deficient patients [20, 60, 115]. Small molecules aimed to increase mitochondrial biogenesis through the activation of the PGC-1α (peroxisome proliferator-activated receptor gamma-coactivator 1-alpha) pathway bezafibrate and AICAR are interesting new compounds [116, 117]. Another practical cell line is the Chinese hamster complex I-deficient cell line NDUFA1 [94]. In addition, to studies in tissue culture, some animal models proved to be valuable for studying treatments. A *Caenorhabditis elegans* model carrying an *NDUFV1* mutation showed cytochrome *c* oxidase deficiency, oxidative stress and lactic acidosis. This phenotype could be partially corrected by riboflavin, thiamine, or sodium dichloroacetate, resulting in significant increases in animal fitness [118].

Recently, transgenic mouse models have been developed and are promising models for treatment. In one model, the *NDUFS4* gene was disrupted and the mice developed a neurological phenotype similar to Leigh (Leigh-like) syndrome in patients [119]. In another mouse model, with targeted disruption of *NDUFA1*, the optic neuropathy was suppressed by mitochondrial superoxide dismutase gene transfer. This study confirmed the importance of ROS and demonstrated the possibility of gene therapy [120]. As previously mentioned, Harlequin mice may also be a valuable model as it resembles the phenotypical variability of human complex deficiency [121]. Although the direct translation from in vitro systems and animal models to patient treatment is disputable, valuable information can still be obtained with respect to pathomechanism and the results could direct future clinical trials.

Summary and Conclusions

Isolated CI deficiency of nuclear origin is a relatively common mitochondrial disease caused by mutations in structural subunits or assembly factors. However, many still lack molecular diagnosis. The clinical symptoms differ but are mostly neurological, onset is early and the prognosis is grave. Although much is known about CI assembly, function and pathomechanism satisfactory treatments are presently lacking. Therefore, identification of the underlying molecular cause is imperative in order to provide genetic counseling.

Acknowledgments The author is supported by research grants from the Israel Science Foundation and by the Chief Scientist Office Ministry of Health Israel.

References

1. Hatefi Y (1985) The mitochondrial electron transport and oxidative phosphorylation system. Annu Rev Biochem 54:1015–1069
2. Brandt U (2011) A two-state stabilization-change mechanism for proton-pumping complex I. Biochim Biophys Acta 1807:1364–1369
3. Clason T, Ruiz T, Schägger H et al (2010) The structure of eukaryotic and prokaryotic complex I. J Struct Biol 169:81–88
4. Althoff T, Mills DJ, Popot JL, Kühlbrandt W (2011) Arrangement of electron transport chain components in bovine mitochondrial supercomplex I(1)III(2)IV(1). EMBO J 30:4652–4664. doi:10.1038/emboj
5. Schagger H (2001) Respiratory chain supercomplexes. IUBMB Life 52:119–128
6. Carroll J, Fearnley IM, Skehel JM, Shannon RJ, Hirst J, Walker JE (2006) Bovine complex I is a complex of 45 different subunits. J Biol Chem 281:32724–32727
7. Hunte C, Zickermann V, Brandt U (2010) Functional modules and structural basis of conformational coupling in mitochondrial complex I. Science 329:448–551
8. Efremov RG, Sazanov LA (2011) Respiratory complex I: 'steam engine' of the cell? Curr Opin Struct Biol 21:532–540
9. Angerer H, Zwicker K, Wumaier Z et al (2011) A scaffold of accessory subunits links the peripheral arm and the distal proton-pumping module of mitochondrial complex I. Biochem J 437:279–288
10. Szklarczyk R, Wanschers BF, Nabuurs SB, Nouws J, Nijtmans LG, Huynen MA (2011) NDUFB7 and NDUFA8 are located at the intermembrane surface of complex I. FEBS Lett 585:737–743
11. Carroll J, Fearnley IM, Skehel JM et al (2005) The post-translational modifications of the nuclear encoded subunits of complex I from bovine heart mitochondria. Mol Cell Proteomics 4:693–699
12. Carilla-Latorre S, Gallardo ME, Annesley SJ et al (2010) MidA is a putative methyltransferase that is required for mitochondrial complex I function. J Cell Sci 123:1674–1683
13. De Rasmo D, Palmisano G, Scacco S et al (2010) Phosphorylation pattern of the NDUFS4 subunit of complex I of the mammalian respiratory chain. Mitochondrion 10:464–471
14. Burwell LS, Nadtochiy SM, Tompkins AJ, Young S, Brookes PS (2006) Direct evidence for S-nitrosation of mitochondrial complex I. Biochem J 394:627–634
15. Chinta SJ, Andersen JK (2006) Reversible inhibition of mitochondrial complex I activity following chronic dopaminergic glutathione depletion in vitro: implications for Parkinson's disease. Free Radic Biol Med 41:1442–1448
16. Belaiche C, Holt A, Saada A (2009) Nonylphenol ethoxylate plastic additives inhibit mitochondrial respiratory chain complex I. Clin Chem 55:1883–1884
17. Degli Esposti M (1998) Inhibitors of NADH-ubiquinone reductase: an overview. Biochim Biophys Acta 1364:222–235
18. Bénit P, Slama A, Rustin P (2008) Decylubiquinol impedes mitochondrial respiratory chain complex I activity. Mol Cell Biochem 314:45–50

19. Janssen RJ, Nijtmans LG, van den Heuvel LP, Smeitink JA (2006) Mitochondrial complex I: structure, function and pathology. J Inherit Metab Dis 29:499–515
20. Valsecchi F, Koopman WJ, Manjeri GR, Rodenburg RJ, Smeitink JA, Willems PH (2010) Complex I disorders: causes, mechanisms, and development of treatment strategies at the cellular level. Dev Disabil Res Rev 16:175–182
21. Schon EA (2000) Mitochondrial genetics and disease. Trends Biochem Sci 25:555–560
22. Hirst J, Carroll J, Fearnley IM, Shannon RJ, Walker JE (2003) The nuclear encoded subunits of complex I from bovine heart mitochondria. Biochim Biophys Acta 1604: 135–150
23. Wiedemann N, Frazier AE, Pfanner N (2004) The protein import machinery of mitochondria. J Biol Chem 279: 14473–14476
24. Kuffner R, Rohr A, Schmiede A, Krull C, Schulte U (1998) Involvement of two novel chaperones in the assembly of mitochondrial NADH:ubiquinone oxidoreductase (complex I). J Mol Biol 283: 409–417
25. Dunning CJ, McKenzie M, Sugiana C et al (2007) Human CIA30 is involved in the early assembly of mitochondrial complex I and mutations in its gene cause disease. EMBO J 26:3227–3237
26. Ogilvie I, Kennaway NG, Shoubridge EA (2005) A molecular chaperone for mitochondrial complex I assembly is mutated in a progressive encephalopathy. J Clin Invest 115:2784–27842
27. Pagliarini DJ, Calvo SE, Chang B et al (2008) Mootha, A mitochondrial protein compendium elucidates complex I disease biology. Cell 134:112–123
28. Mimaki M, Wang X, McKenzie M, Thorburn DR, Ryan MT (2012) Understanding mitochondrial complex I assembly in health and disease. Biochim Biophys Acta 1817(6):851–862 (PMID: 21924235)
29. McKenzie M, Tucker EJ, Compton AG et al (2011) Mutations in the gene encoding C8orf38 block complex I assembly by inhibiting production of the mitochondria-encoded subunit ND1. J Mol Biol 414(3):413–462 (PMID: 22019594)
30. Saada A, Edvardson S, Rapoport M et al (2008) C6ORF66 is an assembly factor of mitochondrial complex I. Am J Hum Genet 82:32–38
31. Saada A, Vogel RO, Hoefs SJ et al (2009) Mutations in NDUFAF3 (C3ORF60), encoding an NDUFAF4 (C6ORF66)-interacting complex I assembly protein, cause fatal neonatal mitochondrial disease. Am J Hum Genet 84:718–727
32. Sugiana C, Pagliarini DJ, McKenzie M et al (2008) Mutation of C20orf7 disrupts complex I assembly and causes lethal neonatal mitochondrial disease. Am J Hum Genet 83:468–478
33. Nouws J, Nijtmans L, Houten SM et al (2010) Acyl-CoA dehydrogenase 9 is required for the biogenesis of oxidative phosphorylation complex I. Cell Metab 12:283–294
34. Calvo SE, Tucker EJ, Compton AG et al (2010) High-throughput, pooled sequencing identifies mutations in NUBPL and FOXRED1 in human complex I deficiency. Nat Genet 42:851–858
35. Fassone E, Duncan AJ, Taanman JW et al (2010) FOXRED1, encoding an FAD-dependent oxidoreductase complex-I-specific molecular chaperone, is mutated in infantile-onset mitochondrial encephalopathy. Hum Mol Genet 19:4837–4847
36. Vogel RO, Janssen RJ, van den Brand MA et al (2007) Cytosolic signaling protein Ecsit also localizes to mitochondria where it interacts with chaperone NDUFAF1 and functions in complex I assembly. Genes Dev 21:615–624
37. Ugalde C, Janssen RJ, van den Heuvel LP, Smeitink JA, Nijtmans LG (2004) Differences in assembly or stability of complex I and other mitochondrial OXPHOS complexes in inherited complex I deficiency. Hum Mol Genet 13:659–667
38. Vogel RO, Dieteren CE, van den Heuvel LP et al (2007) Identification of mitochondrial complex I assembly intermediates by tracing tagged NDUFS3 demonstrates the entry point of mitochondrial subunits. J Biol Chem 282:7582–7590
39. Lazarou M, McKenzie M, Ohtake A, Thorburn DR, Ryan MT (2007) Analysis of the assembly profiles for mitochondrial- and nuclear-DNA-encoded subunits into complex I. Mol Cell Biol 27:4228–42237
40. Dieteren CE, Willems PH, Vogel RO et al (2008) Subunits of mitochondrial complex I exist as part of matrix- and membrane-associated subcomplexes in living cells. J Biol Chem 283:34753–34761

41. Wessels HJ, Vogel RO, van den Heuvel L et al (2009) LC-MS/MS as an alternative for SDS-PAGE in blue native analysis of protein complexes. Proteomics 9:4221–4228
42. Perales-Clemente E, Fernández-Vizarra E et al (2010) Five entry points of the mitochondrially encoded subunits in mammalian complex I assembly. Mol Cell Biol 30:3038–3047
43. Sheftel D, Stehling O, Pierik AJ et al (2009) Human Ind1, an iron–sulfur cluster assembly factor for respiratory complex I. Mol Cell Biol 29: 6059–6073
44. Fontanesi F, Soto IC, Horn D, Barrientos A (2006) Assembly of mitochondrial cytochrome c-oxidase, a complicated and highly regulated cellular process. Am J Physiol Cell Physiol 291: 1129–1147
45. Saada A, Edvardson S, Shaag A et al (2012) Combined OXPHOS complex I and IV defect, due to mutated complex I assembly factor C20ORF7. J Inherit Metab Dis 35(1):125–131 (PMID: 21607760)
46. Wegrzyn J, Potla R, Chwae YJ et al (2009) Function of mitochondrial Stat3 in cellular respiration. Science 323:793–797
47. Martinvalet D, Dykxhoorn DM, Ferrini R, Lieberman J (2008) Granzyme A cleaves a mitochondrial complex I protein to initiate caspase-independent cell death. Cell 133:681–692
48. Vahsen N, Candé C, Brière JJ et al (2004) AIF deficiency compromises oxidative phosphorylation. EMBO J 23:4679–4689
49. Ghezzi D, Sevrioukova I, Invernizzi F et al (2010) Severe X-linked mitochondrial encephalomyopathy associated with a mutation in apoptosis-inducing factor. Am J Hum Genet 86:639–649
50. Liu S, Lee YF, Chou S et al (2011) Mice lacking TR4 nuclear receptor develop mitochondrial myopathy with deficiency in complex I. Mol Endocrinol 25:1301–1310
51. Schaefer AM, Taylor RW, Turnbull DM, Chinnery PF (2004) The epidemiology of mitochondrial disorders—past, present and future. Biochim Biophys Acta 1659: 115–120
52. Skladal D, Halliday J, Thorburn DR (2003) Minimum birth prevalence of mitochondrial respiratory chain disorders in children. Brain 126:1905–1912
53. Morgan-Hughes JA, Darveniza P, Landon DN, Land JM, Clark JB (1979) A mitochondrial myopathy with a deficiency of respiratory chain NADH-CoQ reductase activity. J Neurol Sci 43: 27–46
54. Valsecchi F, Koopman WJ, Manjeri GR, Rodenburg RJ, Smeitink JA, Willems PH (2010) Complex I disorders: causes, mechanisms, and development of treatment strategies at the cellular level. Dev Disabil Res Rev 16:175–182
55. Pitkanen S, Feigenbaum A, Laframboise R, Robinson BH (1996) NADH-coenzyme Q reductase (complex I) deficiency: heterogeneity in phenotype and biochemical findings. J Inherit Metab Dis 19: 675–686
56. Kirby DM, Crawford M, Cleary MA, Dahl HH, Dennett X, Thorburn DR (1999) Respiratory chain complex I deficiency: an underdiagnosed energy generation disorder. Neurology 52: 1255–1264
57. Vilain C, Rens C, Aeby A et al (2011) A novel NDUFV1 gene mutation in complex I deficiency in consanguineous siblings with brainstem lesions and Leigh syndrome. Clin Genet. doi:10.1111
58. Rustin P, Chretien D, Bourgeron T et al (1994) Biochemical and molecular investigations in respiratory chain deficiencies. Clin Chim Acta 228:35–51
59. Janssen AJ, Smeitink JAM, van den Heuvel LWPJ (2003) Some practical aspects of providing a diagnostic service for respiratory chain defects. Ann Clin Biochem 40:3–8
60. Saada A, Bar-Meir M, Belaiche C, Miller C, Elpeleg O (2004) Evaluation of enzymatic assays and compounds affecting ATP production in mitochondrial respiratory chain complex I deficiency. Anal Biochem 335:66–72
61. Jonckheere AI, Huigsloot M, Janssen AJ et al (2010) High-throughput assay to measure oxygen consumption in digitonin-permeabilized cells of patients with mitochondrial disorders. Clin Chem 56:424–431
62. Suomalainen A, Elo JM, Pietiläinen KH et al (2011) FGF-21 as a biomarker for muscle-manifesting mitochondrial respiratory chain deficiencies: a diagnostic study. Lancet Neurol 10(9):806–818

63. Lebre AS, Rio M, Faivre d'Arcier L, Vernerey D, Landrieu P, Slama A et al (2011) A common pattern of brain MRI imaging in mitochondrial diseases with complex I deficiency. Med Genet 48:16–23
64. Wallace DC (1994) Mitochondrial DNA mutations in diseases of energy metabolism. J Bioenerg Biomembr 26:2412–2450
65. Mitchell AL, Elson JL, Howell N, Taylor RW, Turnbull DM (2006) Sequence variation in mitochondrial complex I genes: mutation or polymorphism? J Med Genet 43, 175–179
66. Loeffen J, Smeitink J, Triepels R et al (1998) The first nuclear-encoded complex I mutation in a patient with Leigh syndrome. Am J Hum Genet 63:1598–1608
67. Schuelke M, Smeitink J, Mariman E et al (1999) Mutant NDUFV1 subunit of mitochondrial complex I causes leukodystrophy and myoclonic epilepsy. Nat Genet 21:260–261
68. Bénit P, Chretien D, Kadhom N et al (2001) Large-scale deletion and point mutations of the nuclear NDUFV1 and NDUFS1 genes in mitochondrial complex I deficiency. Am J Hum Genet 68:1344–1352
69. Laugel V, This-Bernd V, Cormier-Daire V, Speeg-Schatz C, de Saint-Martin A, Fischbach M (2007) Early-onset ophthalmoplegia in Leigh-like syndrome due to NDUFV1 mutations. Pediatr Neurol 36:54–57
70. Bénit P, Beugnot R, Chretien D et al (2003) Mutant NDUFV2 subunit of mitochondrial complex I causes early onset hypertrophic cardiomyopathy and encephalopathy. Hum Mutat 21:582–586
71. Pagniez-Mammeri H, Lombes A, Brivet M et al (2009) Rapid screening for nuclear genes mutations in isolated respiratory chain complex I defects. Mol Genet Metab 96:196–200
72. Martín MA, Blázquez A, Gutierrez-Solana LG et al (2005) Leigh syndrome associated with mitochondrial complex I deficiency due to a novel mutation in the NDUFS1 gene. Arch Neurol 62:659–661
73. Bugiani M, Invernizzi F, Alberio S et al (2004) Clinical and molecular findings in children with complex I deficiency. Biochim Biophys Acta 1659:136–147
74. Hoefs SJ, Skjeldal OH, Rodenburg RJ et al (2010) Novel mutations in the NDUFS1 gene cause low residual activities in human complex I deficiencies. Mol Genet Metab 100:251–256
75. Pagniez-Mammeri H, Landrieu P, Legrand A, Slama A (2010) Leukoencephalopathy with vanishing white matter caused by compound heterozygous mutations in mitochondrial complex I NDUFS1 subunit. Mol Genet Metab 101:297–298
76. Ferreira M, Torraco A, Rizza T et al (2011) Progressive cavitating leukoencephalopathy associated with respiratory chain complex I deficiency and a novel mutation in NDUFS1. Neurogenetics 12:9–17
77. Loeffen J, Elpeleg O, Smeitink J el al (2001) Mutations in the complex I NDUFS2 gene of patients with cardiomyopathy and encephalomyopathy. Ann Neurol 49:195–201
78. Ngu LH, Nijtmans LG, Distelmaier F et al (2012) A catalytic defect in mitochondrial respiratory chain complex I due to a mutation in NDUFS2 in a patient with Leigh syndrome. Biochim Biophys Acta 1822(2):169–175 (PMID: 22036843)
79. Tuppen HA, Hogan VE, He L et al (2010) The p.M292T NDUFS2 mutation causes complex I-deficient Leigh syndrome in multiple families. Brain 133:2952–2963
80. Bénit P, Slama A, Cartault F et al (2004) Mutant NDUFS3 subunit of mitochondrial complex I causes Leigh syndrome. J Med Genet 41:14–7
81. Triepels RH, van den Heuvel LP, Loeffen JL et al (1999) Leigh syndrome associated with a mutation in the NDUFS7 (PSST) nuclear encoded subunit of complex I. Ann Neurol 45:787–790
82. Lebon S, Rodriguez D, Bridoux D et al (2007) A novel mutation in the human complex I NDUFS7 subunit associated with Leigh syndrome. Mol Genet Metab 90:379–382
83. Procaccio V, Wallace DC (2004) Late-onset Leigh syndrome in a patient with mitochondrial complex I NDUFS8 mutations. Neurol 62:1899–1901
84. van den Heuvel L, Ruitenbeek W, Smeets R et al (1998) Demonstration of a new pathogenic mutation in human complex I deficiency: a 5-bp duplication in the nuclear gene encoding the 18-kD (AQDQ) subunit. Am J Hum Genet 62:262–268

85. Petruzzella V, Vergari R, Puzziferri I et al (2001) A nonsense mutation in the NDUFS4 gene encoding the 18 kDa (AQDQ) subunit of complex I abolishes assembly and activity of the complex in a patient with Leigh-like syndrome. Hum Mol Genet 10:529–535
86. Budde SM, van den Heuvel LP, Janssen AJ et al (2000) Combined enzymatic complex I and III deficiency associated with mutations in the nuclear encoded NDUFS4 gene. Biochem Biophys Res Commun 275:63–68
87. Bénit P, Steffann J, Lebon S, Chretien D et al (2003) Genotyping microsatellite DNA markers at putative disease loci in inbred/multiplex families with respiratory chain complex I deficiency allows rapid identification of a novel nonsense mutation (IVS1nt −1) in the NDUFS4 gene in Leigh syndrome. Hum Genet 112:563–566
88. Anderson SL, Chung WK, Frezzo J et al (2008) A novel mutation in NDUFS4 causes Leigh syndrome in an Ashkenazi Jewish family. J Inherit Metab Dis 2(31 Suppl):461–467
89. Leshinsky-Silver E, Lebre AS, Minai L et al (2009) NDUFS4 mutations cause Leigh syndrome with predominant brainstem involvement. Mol Genet Metab 97:185–189
90. Kirby DM, Salemi R, Sugiana C et al (2004) NDUFS6 mutations are a novel cause of lethal neonatal mitochondrial complex I deficiency. J Clin Invest 114:837–845
91. Spiegel R, Shaag A, Mandel H, Reich D et al (2009) Mutated NDUFS6 is the cause of fatal neonatal lactic acidemia in Caucasus Jews. Eur J Hum Genet 17:1200–1203
92. Fernandez-Moreira D, Ugalde C, Smeets R et al (2007) X-linked NDUFA1 gene mutations associated with mitochondrial encephalomyopathy. Ann Neurol 61: 73–83
93. Mayr JA, Bodamer O, Haack TB et al (2011) Heterozygous mutation in the X chromosomal NDUFA1 gene in a girl with complex I deficiency. Mol Genet Metab 103:358–261
94. Potluri P, Davila A, Ruiz-Pesini E et al (2009) A novel NDUFA1 mutation leads to a progressive mitochondrial complex I-specific neurodegenerative disease. Mol Genet Metab 96:189–195
95. Hoefs SJ, Dieteren CE, Distelmaier F et al (2008) NDUFA2 complex I mutation leads to Leigh disease. Am J Hum Genet 82(6):1306–1315
96. Hoefs SJ, van Spronsen FJ, Lenssen EW et al (2011) NDUFA10 mutations cause complex I deficiency in a patient with Leigh disease. Eur J Hum Genet 19:270–274
97. Berger I, Hershkovitz E, Shaag A, Edvardson S, Saada A, Elpeleg O (2008) Mitochondrial complex I deficiency caused by a deleterious NDUFA11 mutation. Ann Neurol. 63:405–408
98. Ostergaard E, Rodenburg RJ, van den Brand M et al (2011) Respiratory chain complex I deficiency due to NDUFA12 mutations as a new cause of Leigh syndrome. Med Genet 48:737–740
99. Pagniez-Mammeri H, Rak M, Legrand A, Benit, P, Rustin, P, Slama, A (2012) Mitochondrial complex I deficiency of nuclear origin. II. Non-structural genes. Mol Genet Metab 105(2):173–179. doi:10.1016/j.ymgme.2011.10.001
100. Fassone E, Taanman JW, Hargreaves IP et al (2011) Mutations in the mitochondrial complex I assembly factor NDUFAF1 cause fatal infantile hypertrophic cardiomyopathy. J Med Genet 48:691–697
101. Herzer M, Koch J, Prokisch H, Rodenburg R et al (2010) Leigh disease with brainstem involvement in complex I deficiency due to assembly factor NDUFAF2 defect. Neuropediatr 41:30–34
102. Barghuti F, Elian K, Gomori JM et al (2008) The unique neuroradiology of complex I deficiency due to NDUFA12L defect. Mol Genet Metab 94:78–82
103. Gerards M, Sluiter W, van den Bosch BJ et al (2010) Defective complex I assembly due to C20orf7 mutations as a new cause of Leigh syndrome. J Med Genet 507–512
104. Haack TB, Danhauser K, Haberberger B et al (2010) Exome sequencing identifies ACAD9 mutations as a cause of complex I deficiency. Nat Genet 42:1131–1134
105. Gerards M, van den Bosch BJ, Danhauser K et al (2011) Riboflavin-responsive oxidative phosphorylation complex I deficiency caused by defective ACAD9: new function for an old gene. Brain 134:210–219
106. Berger I, Ben-Neriah Z, Dor-Wolman T et al (2011) Early prenatal ventriculomegaly due to an AIFM1 mutation identified by linkage analysis and whole exome sequencing. Mol Genet Metab 104(4):517–520 (PMID: 22019070)

107. Murphy MP (2009) How mitochondria produce reactive oxygen species. Biochem J 417:1–13
108. Esteitie N, Hinttala R, Wibom R et al (2005) Secondary metabolic effects in complex I deficiency. Ann Neurol 58:544–552
109. Schapira AH (2010) Complex I: inhibitors, inhibition and neurodegeneration. Exp Neurol 224:331–335
110. Coskun P, Wyrembak J, Schriner S et al (2012) A mitochondrial etiology of Alzheimer and Parkinson disease. Biochim Biophys Acta 1820(5):553–564 (PMID: 21871538)
111. Rosenfeld M, Brenner-Lavie H, Ari SG, Kavushansky A, Ben-Shachar D (2011) Perturbation in mitochondrial network dynamics and in complex I dependent cellular respiration in schizophrenia. Biol Psychiatry 69:980–988
112. DiMauro S, Mancuso M (2007) Mitochondrial diseases: Therapeutic approaches. Biosci Rep 27:125–137
113. Klopstock T, Yu-Wai-Man P, Dimitriadis K et al (2011) A randomized placebo-controlled trial of idebenone in Leber's hereditary optic neuropathy. Brain 134:2677–2686
114. Snow BJ, Rolfe FL, Lockhart MM et al A double-blind, placebo-controlled study to assess the mitochondria-targeted antioxidant MitoQ as a disease-modifying therapy in Parkinson's disease. Mov Disord 25:1670–1674
115. Koopman WJ, Verkaart S, van Emst-de Vries SE et al (2008) Mitigation of NADH: ubiquinone oxidoreductase deficiency by chronic Trolox treatment. Biochim Biophys Acta 1777:853–859
116. Bastin J, Aubey F, Rötig A, Munnich A, Djouadi F (2008) Activation of peroxisome proliferator-activated receptor pathway stimulates the mitochondrial respiratory chain and can correct deficiencies in patients' cells lacking its components. J Clin Endocrinol Metab 93:1433–1441
117. Golubitzky A, Dan P, Weissman S, Link G, Wikstrom JD, Saada A (2011) Screening for active small molecules in mitochondrial complex I deficient patient's fibroblasts, reveals AICAR as the most beneficial compound. PLoS ONE 6: E26883
118. Grad LI, Lemire BD (2004) Mitochondrial complex I mutations in Caenorhabditis elegans produce cytochrome c oxidase deficiency, oxidative stress and vitamin-responsive lactic acidosis. Hum Mol Genet 13:303–314
119. Kruse SE, Watt WC, Marcinek DJ et al (2008) Mice with mitochondrial complex I deficiency develop a fatal encephalomyopathy. Cell Metab 7:312–320
120. Qi X, Lewin AS, Sun L, Hauswirth WW, Guy J (2004) SOD2 gene transfer protects against optic neuropathy induced by deficiency of complex I. Ann Neurol 182–191
121. Bénit P, Goncalves S, Dassa EP, Brière JJ, Rustin P(2008) The variability of the harlequin mouse phenotype resembles that of human mitochondrial-complex I-deficiency syndromes. PLoS ONE 3: E3208

Chapter 13
Mitochondrial Respiratory Chain Complex II

Jaya Ganesh, Lee-Jun C. Wong and Elizabeth B. Gorman

SDH and Mitochondrial Respiratory Chain Disorders

Structure of the SDH Complex and its Genes

The SDH enzyme complex is a heterotetramer with two catalytic subunits, SDHA and SDHB, anchored to the inner mitochondrial membrane by two hydrophobic subunits, SDHC and SDHD [1]. The catalytic subunits, SDHA (a flavoprotein) and SDHB (an iron-sulfur protein), are hydrophilic and extend into the mitochondrial matrix [2]. The ubiquinone-binding site is located in between amino acid residues derived from SDHB, SDHC, and SDHD. The interface of SDHC and SDHD contains a bound heme *b* moiety [3]. SDH is the only membrane bound enzyme of the tricarboxylic acid cycle (TCA).

SDHA

SDHA contains a covalently attached flavin adenine dinucleotide (FAD) cofactor and the succinate-binding site. Flavination is important for succinate oxidation [4].

E. B. Gorman (✉)
Medical Genetics Laboratories, Baylor College of Medicine, One Baylor Plaza, NAB 2015, Houston, TX 77030, USA
e-mail: egorman@bcm.edu

J. Ganesh
Clinical Pediatrics, Section of Metabolic Diseases, The Children's Hospital of Philadelphia, Philadelphia, PA, USA

L.-J. C. Wong
Department of Molecular and Human Genetics, Baylor College of Medicine, One Baylor Plaza, NAB 2015, Houston, TX 77030, USA
e-mail: ljwong@bcm.edu

SDHB

SDHB contains three iron–sulfur (Fe–S) clusters: [2Fe–2S], [4Fe–4S], and [3Fe–4S]. Proper insertion and/or stabilization of these Fe–S centers may be important for the assembly and stability of SDH.

SDH Genes

The four subunits are entirely encoded by four nuclear genes; *SDHA* (15 exons), *SDHB* (8 exons), *SDHC* (6 exons), and *SDH D* (4 exons), mapping to chromosomes 5p15, 1p36.1-p35 1q21, and 11q23, respectively [5]. The SDHA and SDHB proteins are conserved among species but the amino acid sequences of SDHC and SDHD vary. Pseudogenes have been reported for *SDHA* (3q29), *SDHC* (17p13.3 and 10q23.32) and *SDHD* (1p36.11, 2q32.1, 18p11.21, 7q32.1, 3q26.3, and 3p21.3).

SDH Assembly

The assembly of SDH in the mitochondrial membrane shows little similarity among species [6]. Recently two, apparently dedicated, SDH assembly factors (SDHFA1 and 2) were described in which disease-causing mutations have been identified.

SDHAF1

The *SDHAF1* gene transcript is ubiquitously expressed. SDHAF1 is a hydrophilic protein with no predicted transmembrane domain, and is located in the mitochondrial matrix. SDHAF1 is not physically associated with complex II in vivo but is essential for SDH biogenesis. SDHAF1 contains a LYR tripeptide motif, which is present in proteins involved in Fe–S metabolism. SDHAF1 could be important for the insertion or maintenance of the Fe–S centers within complex II. Abnormalities in Fe–S incorporation in complex II may affect its formation and stability. Other proteins involved in Fe–S biosynthesis and mitochondrial chaperone proteins may be important for normal activity of complex II but, SDHAF1 is the only protein so far identified with a specific role in complex II assembly [7].

SDHAF2 (SDH5)

In the yeast, deletion of *SDH5* abolishes attachment of the FAD cofactor to SDHA although SDHA was still present. The residual SDHA level was high in the *SDH5* mutants but complex II activity, as measured by blue native polyacrylamide gel electrophoresis (BN-PAGE), was lost. These data suggest that SDHA assembles in the absence of SDH5 but is not stably bound in complex II. As a result, complex II is susceptible to degradation, which results in the reduction of activity of all four subunits [8]. Given the crucial role of SDH in cellular respiration additional factors are likely involved in complex II assembly and their discovery will explain the molecular basis of disease phenotypes associated with complex II deficiency [3].

Function of SDH

The SDH enzyme couples the oxidation of succinate to fumarate involving the 2-electron transfer to reduce quinone to quinol. The former is part of the Krebs cycle while the latter is part of the respiratory chain [2]. SDH is the only mitochondrial respiratory complex that does not pump protons across the inner mitochondrial membrane [3].

When succinate is converted to fumarate FAD is converted to $FADH_2$. The electrons are then transferred from $FADH_2$ through the Fe–S clusters. The Fe–S clusters are single-electron carriers, hence electron transfer steps occur successively and in tandem transferring electrons from $FADH_2$ to the 2Fe–2S cluster, the 4Fe–4S cluster, and finally, to the 3Fe–4S cluster [3]. From the 3Fe–4S cluster, the two electrons are transferred to ubiquinone reducing it to ubiquinol.

SDH can also catalyze the reverse reaction involving quinol oxidation and the reduction of fumarate. However, in human tissues this reverse reaction is negligible under standard conditions and this is likely a protective mechanism to decrease the production of reactive oxygen species (ROS) during aerobic respiration [9, 10].

SDH and Mitochondrial Respiratory Chain Disorders

Leigh syndrome, also known as subacute necrotizing encephalomyelopathy, is a genetically heterogeneous disorder of mitochondrial dysfunction. Leigh and Leigh-like syndromes have been associated with pyruvate dehydrogenase deficiency and defects involving the subunits of the electron transport chain (ETC) complexes I–V, that are due to mitochondrial DNA mutations, as well as mutations in the nuclear-encoded assembly genes of various ETC complexes. SDH dysfunction can be an underlying molecular cause of Leigh syndrome.

Clinically, Leigh syndrome (LS) is an early-onset progressive neurodegenerative disorder [11]. Affected patients present with symptoms of neurological function of varying degrees of severity including developmental delay, psychomotor retardation, ataxia, dystonia, seizures, cranial nerve palsies, nystagmus, ophthalmoparesis, optic atrophy, vomiting, failure to thrive with feeding difficulties, and respiratory impairment. The symptoms are often recognized in early infancy and may follow a varying period of normal growth and development. An intercurrent illness may be identified as the trigger for the sometimes rapid psychomotor regression. Biochemical abnormalities such as lactic acidosis may be present but serum chemistries may be entirely normal. Histology of the brain reveals focal, bilaterally symmetric, spongiform, necrotic lesions associated with demyelination, vascular proliferation, and gliosis in the brainstem, diencephalon, basal ganglia, cerebellum, or cerebral white matter.

Complex II deficiency in mitochondrial disease is very rare and thought to account for only 2–4 % of respiratory chain deficiencies [3]. Mutations in *SDHA* and *SDHAF1* have now been linked to mitochondrial disease.

The first described cases of SDH deficiency where the molecular defect was confirmed as an *SDHA* mutation presented with a neurodegenerative phenotype. *SDHA* mutations have since been associated with LS, mitochondrial encephalopathy, and optic atrophy. Defects that interfere with SDHA activity likely reduce overall complex II activity as the flavin moiety remains oxidized. This may lead to the decreased growth rate *in vitro* and *in vivo* that has been seen in experimental models of SDHA deficiency [12].

Mutations in *SDHAF1* also cause an isolated deficiency of complex II. Disruption of the *YDR379C-A* gene, a yeast ortholog of *SDHAF1*, resulted in a marginal reduction in the Km value for succinate of in the null mutant compared with wild type. This suggests that the defective SDH activity could be caused by a reduction in the numbers of SDH enzymes and not by alterations in complex II itself [7].

Reported Cases of SDH Deficiency

The first patient with isolated SDH deficiency was described by Rivner et al. [13] and was noted to have features of Kearns–Sayre syndrome including conduction defects. Reichman and Angelini [14] reported two additional SDH deficiency patients with isolated hypertrophic cardiomyopathy and hypertrophic cardiomyopathy with skeletal myopathy, respectively, but in these early cases the molecular defect was not identified.

Bourgeron et al. [15] identified complex II deficiency in two sisters of consanguineous parents, presenting with LS. In the family reported by Bourgeron et al. [15], homozygosity for a recessively inherited mutation, p.R554W, in *SDHA* was identified in these siblings. This was the first case of a nuclear gene defect causing respiratory chain deficiency in humans.

Parfait et al. [16] described an additional case of complex II deficiency in a 9-month-old girl, also of consanguineous parents, presenting with psychomotor delay and cerebellar ataxia after an initial period of normal development. There was mild elevation of lactate in the blood and CSF and bilateral necrotic lesions in the basal ganglia were seen on brain magnetic resonance imaging (MRI). Enzyme analysis on lymphoblastoid cell lines revealed reductions in SDH and complex II+III activities. This infant was found to be compound heterozygous for mutations in *SDHA* [16]. Van Coster et al. reported a 5-month-old girl, also born to consanguineous parents, who was failing to thrive. She became acutely ill during a respiratory illness and was found to have cardiomegaly and hepatosplenomegaly. She died from an apneic event. Laboratory parameters showed metabolic acidosis, ketotic hypoglycemia, elevated hepatic transaminases, lactic acidosis, low CSF glucose level, and excretion of Krebs cycle intermediaries in the urine including succinate and to a lesser extent malate. Spectrophotometric analysis and BN-PAGE separation of the respiratory complexes followed by catalytic staining confirmed complex II deficiency in skin fibroblasts and skeletal muscle. Imunoblotting with antibodies to SDHA and SDHB showed a marked decrease in the cross-reacting material (CRM). Beta oxidation of fatty acids was normal in the fibroblasts of this patient. The authors postulate that the severe hypoglycemia seen in this instance is likely due to SDH deficiency inducing a reduction of oxaloacetate which is a distal intermediary in the Krebs cycle. Oxaloacetate diffuses from the mitochondrial matrix to the cytosol to form phosphoenolpyruvate which is an important step in gluconeogenesis. A homozygous c.1664G>A substitution affecting codon p.G555E (of *SDHA*) in the patient and its heterozygous state in both parents was confirmed. Of note, in the patients described by Bourgeron et al. [15], the mutation affects codon 554 of the *SDHA* subunit suggesting that mutations in this region of SDHA likely interfere with SDHA–SDHB interactions and affect the overall stability, thus, predisposing the mutant complex to degradation as indicated by the decreased CRM [17].

An additional patient with compound heterozygosity for *SDHA* mutations and complex II deficiency with psychomotor regression was studied by Horvath et al. [18]. However, in six other patients with complex II deficiency described in this study who presented with various symptoms including neurocognitive delay, hypotonia, cardiomyopathy, and cerebellar ataxia, no mutations in *SDHA* were found.

Taylor et al. [19] described two sisters who presented with late onset optic atrophy, myopathy, and ataxia, with partial reduction (50 %) of complex II activity in muscle and platelet mitochondria but this reduction was not observed in immortalized lymphocytes or skin fibroblasts from these patients. A C to T transition resulting in p.R408C change in a highly conserved region affecting flavin binding in one allele of the *SDHA* gene was subsequently identified in this family [20]. This mutation was found in only one allele in each of the two siblings leading the authors to speculate that this mutation is inherited in an autosomal-dominant fashion, which suggests that the late-onset disease might be due to the about 50 % reduction in enzyme activity observed in these siblings.

Bugiani et al. [21] described the onset of psychomotor decline in a cohort of three Italian children who had evidence of Complex II deficiency. Two of these children were second cousins and the third was from the same village and likely had common ancestry with the other two. In this cohort, a 10-month-old girl had neurocognitive decline that appeared to be triggered by a febrile illness. She developed severe spastic quadriplegia and growth failure. The second was a male infant who presented with deafness, vomiting, and slow psychomotor regression with onset at 6 months. This child also experienced severe spastic quadriplegia and growth retardation (less than the 3 percentile). The last child was also male, and at 11 months had onset of acute psychomotor regression triggered by fever. He developed severe spastic quadriplegia, and severe irritability, but his growth was normal. Sequencing of all four SDH subunits did not reveal any mutations but the SDHA and SDHB activities were reduced in muscle tissue. The children responded to riboflavin supplementation. Ghezzi et al. [7] subsequently identified a p.G57R mutation in the *SDHAF1* gene in this family.

Brockmann et al. [22] described two female children from a large multiconsanguineous Turkish kindred with onset at 10 and 9 months of acute psychomotor regression, respectively. The former patient's regression was triggered by febrile illness. They both developed severe spastic quadriplegia, moderate cognitive impairment, and their growth was less than the 3 percentile. A p.R55P mutation in the *SDHAF1* gene was identified in both these children and two additional female children from this kindred. The latter two patients had onset of acute psychomotor regression at 10 months and developed severe spastic quadriplegia and severe irritability. One had growth delay, while the other had growth less than the 3 percentile [7].

Utility of Brain-Imaging Studies in Patients with Complex II Deficiency

Brain MRI is an important tool in the evaluation of any patient presenting with neurologic symptoms. Magnetic resonance imaging (MRI) detects the energy exchange between an external magnetic field and specific molecules within a tissue and projects the information as a radiofrequency signal, which is then translated into the anatomic image by computerized software. Its utility is even greater in the pediatric population given its noninvasive nature.

Though MRI findings in mitochondrial disease can be nonspecific, certain MRI findings can be suggestive of mitochondrial disease. The most common specific MRI findings are symmetrical signal abnormalities of deep gray matter seen as hyperintensities on T2 and FLAIR images, and hypointensities on T1 images. The lesions can be patchy or homogeneous. Cerebral and cerebellar atrophy may be present [23]. The specific signature of LS and Leigh-like changes in the brain can be seen in complex II deficiency and include defective myelination and T2-weighted hyperintensities in the basal ganglia involving the thalamus, globus pallidus, substantia nigra, nucleus ruber, and caudate nucleus [18]. Changes in the brain stem and cerebellum may also be seen. In some instances, the gray matter, basal ganglia, and the ventricles may be normal but degenerative white matter changes have been reported in complex II deficiency. T2-weighted hyperintensities may be seen in the supratentorial white matter with predominance in the frontal and occipital lobes with sparing of the subcortical U fibers, periventricular white matter, corpus callosum, and pons.

In a study that looked at MRI changes in mitochondrial disease, two pediatric patients with isolated complex II deficiency showed abnormal signal intensities in deep white matter with progressive degradation and replacement of neural tissue by

CSF, leading to cystic changes. This finding was similar to that seen in leukoencephalopathies affecting primary myelin synthesis such as childhood ataxia with diffuse central nervous system hypomyelination (CACH, or vanishing white matter disorder). The findings were unlike leukodystrophies seen in peroxisomal and lysosomal disorders. Such cystic degeneration of the affected white matter is suggestive of mitochondrial diseases. Defective oxidative phosphorylation in cerebral microvessels that are dependent on adequate mitochondrial function may impair vascular permeability leading to tissue swelling, parenchymal thinning and finally cyst formation. In the pediatric population, cerebral white matter in the developing brain is an actively myelinating tissue and thus may be especially vulnerable to mitochondrial dysfunction [24].

Magnetic resonance spectroscopy (MRS) is a noninvasive technique that can be used to measure the concentrations of different chemical components within tissues. The information produced by MRS is displayed spectrum with peaks consistent with the various chemicals that are detected. In the evaluation of a child with suspected mitochondrial disease, typically an MRI image of the brain is first obtained, and then MRS is performed targeting various areas of interest in the brain including the basal ganglia, brain parenchyma and CSF spaces. MRS can be performed using modified MRI equipment. The main findings of MRS in pateints with mitochondrial disease are elevated lactate that may be associated with decrease in N-acetyl aspartate (NAA) peaks compared with controls. These changes reflect neuronal damage and loss [23].

In a cohort of three pediatric patients, two sisters and an unrelated boy; MRS at 13 and 17 months in one sister, at 12 months in the other sister, and at 50 months in the boy revealed accumulation of succinate in cerebral and cerebellar white matter. Biochemical confirmation of SDH deficiency was made in two of these patients. Succinate is undetectable in normal brain but has been seen in MRS studies of brain abscesses, certain cerebral infections, and radiation-induced brain injuries. However, in the absence of any evidence of these specific diagnoses and in the setting of a leukoencephalopathy, the succinate peak in MRS is a strong evidence for complex II deficiency [22].

SDH and Cancer

Most of the genes involved in nonsyndromic inherited paragangliomas (PGLs) are related to SDH. These include *SDHA*, *SDHB*, *SDHC*, *SDHD*, and *SDHAF2*. How mutations in *SDH* genes lead to cancer in general and PGLs specifically is the subject of this section. Even though the dysregulation of metabolism is a common feature of cancers, why mutations in a "house-keeping" protein such as SDH would lead to PGL susceptibility is unknown. Accumulation of succinate and ROS due to SDH malfunction (in the TCA cycle or the ETC, respectively) both seem to be significant. The build-up of either can lead to pseudohypoxia and the induction of more than 100 hypoxia-inducible genes involving angiogenesis, proliferation, and cell survival; key genes in tumorigenesis. But even once the general link between *SDH* mutations and cancer has been established, the fascinating question of why these particular cancers are formed when *SDH* is mutated remains. The tendency of PGLs to form near oxygen-sensing organs may provide the first hint of an answer.

Paragangliomas and SDH

Paraganglia compose a diffuse neuroendocrine system extending from the middle ear to the pelvic floor containing components of the autonomic nervous system (both parasympathetic and sympathetic). Paraganglia are important in protection from hypoxia, bleeding, cold, and hypoglycemia [25]. They work to maintain homeostasis, either as chemical sensors or by secreting catecholamines in response to stress [26].

Paragangliomas (PGLs) are tumors arising from the paraganglia. These include intraadrenal PGLs (formerly pheochromocytomas, PCCs) and extraadrenal PGLs (formerly extraadrenal PCCs or sympathetic tumors and parasympathetic tumors). Sympathetic tumors secrete catecholamine, while parasympathetic tumors are usually nonsecreting. PGLs are rare and mostly benign but they can cause injury due to their compressive nature, complications from removing them, or from the consequences of catecholamine secretion (e.g., hypertensive crisis).

PGLs can occur sporadically, where tumors form de novo. Sporadic PGLs can be caused by the low oxygen concentrations present at high altitudes. PGLs can also be syndromic, where the tumors are part of the clinical manifestations of a larger syndrome, such as multiple endocrine neoplasia (MEN), von Hippel Lindau (VHL) disease, or neurofibromatosis type 1 (NF1). Alternatively, PGLs can be hereditary nonsyndromic due to a genetic predisposition. About 30 % of all PGLs fall into this latter category that is often referred to as familial PGL. The familial PGLs share an autosomal-dominant inheritance pattern at the pedigree level, accompanied by loss of heterozygosity (LOH) of the nonmutant allele at the cellular level in the tumors.

Four familial PGL susceptibility loci were described before the responsible genes were identified. These were PGL1 [27] and PGL2 [28] on chromosome 11 (q23 and q13) and, PGL3 [29] and PGL4 on chromosome 1 (q21). PGL4 was not finely mapped until the responsible gene was known. Although mutations in these loci share many clinical features, they are also unique in their presentation. PGL1 tumors tend to be located in the head and neck region, often with multiple tumors, but are seldom malignant. PGL1 tumors are almost exclusively paternally inherited. Similar to PGL1 tumors, PGL2 tumors are also paternally inherited, often present with multiple tumors, and primarily occur in the head and neck region. However, PGL2 tumors tend to present at a younger age. PGL3 results in isolated tumors, with a low rate of recurrence, and a low risk of malignancy. Similar to PGL3, PGL4 tumors are often isolated but they tend to be associated with a high rate of recurrence and a high risk of malignancy [3, 30, 31]. In addition, PGL4 tumors are more likely to secrete catecholamines [32].

Intriguingly, in 2000, the gene responsible for PGL1 was identified as *succinate dehydrogenase subunit D* (*SDHD*) [33]. That same year, PGL3 was identified as

SDHC [34], while in the following year, PGL4 was identified as *SDHB* [35]. It was only recently (2009); however, that PGL2 has been identified as *SDH assembly factor 2* (*SDHAF2*). The gene of the largest subunit of SDH, *SDHA,* was also recently found to lead to PGLs when mutated [36]. That PGLs appear to be the result of SDH dysfunction provides a fascinating opportunity to learn more about the fundamental nature of cancer.

Possible Mechanisms Linking SDH Function to Cancer

How mutations in *SDH* could mechanistically result in cancer is intriguing. Some unique characteristics of PGLs may suggest possible mechanisms of cancer pathogenesis. Although no definitive theory has been established, succinate accumulation, ROS accumulation, and decreased apoptosis likely all play important roles.

Hypoxia Mechanism

When humans are exposed to low oxygen conditions (hypoxia), both systemic and cellular changes occur to help promote survival and restore sufficient oxygen. Hypoxia-inducible factor (HIF) is an important cellular oxygen sensor that responds to decreases in available oxygen. Its response is the transcription of genes involved in a wide variety of processes including angiogenesis, apoptosis, cell proliferation, and energy metabolism [37], many of which can contribute to tumorigenesis.

The oxygen-sensing capacity of HIF is both elegant and complex. HIF is composed of an α and a β subunit. HIF α and HIF β are both constitutively expressed, but under normal oxygen (normoxic) conditions, HIF α is rapidly degraded via the ubiquination pathway. More specifically, under normoxic conditions, HIF α is hydroxylated by prolyl hydroxlase domain (PHD) proteins (primarily PHD1), increasing HIF α's affinity for von Hippen Lindau tumor-suppressor protein (VHL) which marks HIF α for 26S proteasome degradation. The PHD hydroxylation step is oxygen sensitive as the PHD proteins require oxygen, as well as iron and α—ketoglutarate, for their function. During hypoxia, HIF α is stabilized, allowing it to translocate to the nucleus, where the HIF-regulated transcriptsome complex forms [38]. In the nucleus, the HIF complex can then regulate the expression of hypoxia-inducible genes containing a hypoxia responsive element (5′-ACGTG-3′) in their promoter.

Over expression of hypoxia-inducible genes in PGLs suggests that SDH malfunction mimics hypoxic stimulation, thus leading to adaptive proliferation of the paraganglia (i.e., tumorigenesis) [39]. Hence, an alternative route through which the induction of the hypoxic response genes can occur has been suggested. In this alternative route, pseudohypoxia occurs when HIF is induced by the inhibition of the PHDs under normoxic conditions. Therefore, when succinate accumulates in the mitochondria due to impaired SHD function, some succinate leaks into the cytosol where it competitively inhibits PHD by competing with α—ketoglutarate for substrate binding [40].

Reactive Oxygen Species (ROS) Mechanism

Mitochondria generate ROS as a toxic by product of ETC. This occurs when carriers are not fully reduced during electron passage through the ETC. These ROS molecules can oxidize virtually any biomolecule including nucleic acids, proteins, and lipids. ROS damage to DNA can lead to deleterious mutations and general genomic instability, thus, setting the stage for tumorigenesis [41].

In addition, ROS production increases when the electron flux through the ETC is reduced [42]. Such a reduction in electron flux can occur when SDH function is compromised. Excess ROS generation can lead to pseudo-hypoxia. In this case HIF α stabilization occurs because ROS can oxidize the iron that is required for HIF α hydroxylation by PHD1 [43].

ROS damage may generate even more ROS. Fe–S centers are particularly vulnerable to ROS damage and SDHB has three Fe–S centers [41]. Interestingly, mutations in *SDHB* also seem to lead to the most severe PGLs—with the highest recurrence rate and the highest risk of malignancy.

Compromised Intrinsic Apoptosis Mechanism

Succinate accumulation can lead to inhibition of PHD proteins other than PHD1. Specifically, PHD3, which has a potential role in developmental apoptosis, is also inhibited by excessive succinate. Thus, succinate accumulation due to SDH deficiency would allow precursor neuronal cells to escape developmentally scheduled apoptosis and predispose individuals with *SDH* mutations to the development of PGLs [44].

In addition, HIF may be more directly involved in tumorigenesis. A subset of the HIF-inducible genes are glucose transporters and glycolytic enzymes. These genes are crucial for the glycolytic shift that is typical of cancer cells and known as the Warburg effect. Warburg originally observed that many cancer cells produce energy via aerobic glycolysis where pyruvate is converted to lactic acid instead of being thoroughly oxidized in the TCA. It has been proposed that this glycolytic shift makes tumors less susceptible to apoptosis [41].

Effect of SDH Subunits and Assembly Factors

Although most SDH subunits and assembly factors have now had known mutations linked to PGL susceptibility, not all have. Mutations in *SDHA* typically display LS but recently PGL cases have been attributed to *SDHA* mutations [36, 45]. Some authors now consider *SDHA* located in chromosome 5, p15, as a fifth loci for familial PGL (PGL5). Conversely, mutations in *SHDAF1* present with infantile leukoencephalopathy [7] and no link to PGLs has been established.

That these different clinical features exist for mutations in different subunits of SDH is intriguing since all of the *SDH* mutations that have been linked to PGLs exhibit the features of tumor suppressors. They show loss of the normal allele in the tumor in conjunction with a germline mutation. This results in the dysfunction of the particular subunit involved, destabilization of the SDH complex as a whole, and compromised enzymatic function [46]. Of particular interest should be the fact that *SDHA* mutations usually result in LS and only rarely result in cancer, while mutations in *SDHAF2*, the assembly factor responsible for SDHA flavination do result in cancer and do not seem to play a role in LS. Similarly, *SDHB* mutations usually result in cancer, but mutations in the assembly factor believed to be involved with the assembly or stability of SDHB's Fe–S centers result in infantile leukoencephalopathy [7].

SDH and Oxygen Sensing

An early and intriguing association between exposure to chronic hypoxia (for example, living at a high altitude) and a higher prevalence of PGLs suggests that their etiology might involve the biological pathways triggered by environmental hypoxia [47]. High altitude appears to increase the frequency of existing familial *SDH* mutations resulting in tumorigenesis and leads to sporadic tumors in those without any family history of *SDH* mutations. Complementary to these observations is the high occurrence of nonfamilial founder mutations in the very low-altitude Netherlands [48].

If oxygen-sensing defects are crucial to the formation of PGLs, mutations in the actual oxygen sensor, the PHD proteins, would be expected to result in PGLs. However, in the current largest screen of PGLs for PHD mutations, none have been found [49]. Based on the dearth of PHD mutations associated with PGL, Baysal made the intriguing hypothesis that maybe SDH does not simply cause a dysregulation of other oxygen sensors in paraganglionic tissues, but perhaps plays a major role in oxygen sensing itself [50].

Paternal Inheritance of SDHD and SDHAF2

Since there is no evidence for parent-specific disease transmission of *SDHA*, *SDHB*, or *SDHC* mutations, the largely paternal transmission of *SDHD* and *SDHAF2* mutations suggests an epigenetic mechanism for the inheritance of these genes rather than a simple lack of SDH complex function [26]. For a long time it was believed that *SDHD* had to be maternally imprinted as originally all the cases described showed paternal inheritance. Nevertheless, *SDHD* shows bi-allelic expression in normal tissues with no promoter hypermethylation in normal tissues or tumors [51]. Furthermore, the region of chromosome 11 that *SDHD* and *SDHAF2* are located on is not typically imprinted. In addition, maternal inheritance of *SDHD*, although extremely rare, has been described [52].

Another possibility which was suggested by Hensen et al. [53] was that since *SDHD* and *SDHAF2* are both on chromosome 11, that a tumor-suppressor gene in the imprinted region of chromosome 11 may be lost concurrently. Thus, when this hypothetical imprinted tumor-suppressor gene and the normal maternal *SDHD* gene are lost through maternal whole chromosome loss, the mutant paternal *SDHD* gene is expressed. In the reverse case, if the normal paternal copy is lost, it takes another change to lose the hypothetical imprinted tumor-suppressor gene [50]. Furthermore, loss of the maternal allele appears to be a prerequisite for PGL formation. However, if the maternal *SDHD* allele is already inactivated, the loss of the maternal allele would be redundant.

Given the role of SDH in the induction of hypoxia-inducible genes, it has been suggested by Baysal [50] that genetic or epigenetic suppression of *SDHD* in specific normal tissues might facilitate early detection of hypoxia. Thus, in a paraganglia specific genomic imprinting model, the *SDHD* gene might be normally suppressed in the paraganglionic tissues in a reversible fashion during female gametogenesis [50]. But there are currently no data to support imprinting of *SDHD* in paraganglia.

In a recent article, Muller [32] hypothesizes that partial imprinting of the *SDHD* gene may be the answer. In this scenario, some residual SDH activity would remain in paraganglioma cells with a paternal mutation. But over time ROS and succinate would accumulate, eventually resulting in pseudo-hypoxia. Hypoxia favors nondisjunction, leading to the loss of some or all of maternal chromosome 11. With this loss, the residual SDH activity is lost and tumorigenesis occurs. However, no mechanism is known for partial imprinting and no methylation has been observed in the *SDHD* promoter region [32].

Despite the wealth of information that has been generated in the last decade linking PGLs to SDH dysfunction, much remains to be discovered. A definitive molecular link between the inactivation of SDH and the induction of hypoxia inducible genes has been established, but the precise role of the key players remains in question. Is succinate accumulation, ROS accumulation, or reduced apoptosis the major contributor or does this molecular link require the complex interplay of all three? What exactly is the nature of the almost exclusive paternal inheritance of *SDHD* and *SDHAF2* mutations linked to PGLs? Why are the progenitor PGL cells more sensitive to *SDH* mutations? Finding the answers to these questions may be challenging, but will no doubt enrich our understanding of "house-keeping" genes such as *SDH*, oxygen sensing, and cancer. In addition, the new understanding gained by searching for these answers will provide new ideas for treatments, cures, and the prevention of many devastating diseases.

Fig. 13.1 Number of unique variants reported in *SDH* genes [54]

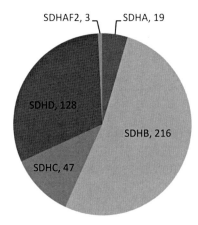

Fig. 13.2 Number of reported cases with *SDH* variants [54]

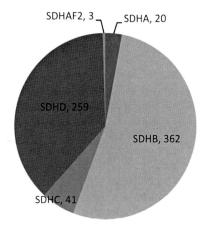

Conclusion

SDH is an important component of both the mitochondrial respiratory chain and the Krebs cycle. Mutations in all four *SDH* subunit genes and their two assembly factors have now been reported. *SDHA* and *SDHFA1* mutations primarily cause encephalomyopathy in childhood while mutations in the other three subunits and *SDHFA2* have been primarily associated with tumor formation (Figs. 13.1 and 13.2). This difference in the phenotypes associated with SDH deficiency likely has its origin in the dual function of SDH in the mitochondrial respiratory chain and the Krebs cycle. By discovering how SDH deficiency caused by different *SDH* mutations leads to distinct phenotypes will shed light on the fundamental processes of respiration and the Krebs cycle. This knowledge should provide new ideas for treatments, cures, and the prevention of many devastating diseases.

Acknowledgments To all the patients and their families who inspire us every day.

References

1. Sun F et al (2005) Crystal structure of mitochondrial respiratory membrane protein complex II. Cell 121(7):1043–1057
2. Lancaster CR (2002) Succinate:Quinone oxidoreductases: an overview. Biochim Biophys Acta 1553(1–2):1–6
3. Rutter J Winge DR, Schiffman JD (2010) Succinate dehydrogenase—assembly, regulation and role in human disease. Mitochondrion 10(4):393–401
4. Robinson KM et al (1994) The covalent attachment of FAD to the flavoprotein of Saccharomyces cerevisiae succinate dehydrogenase is not necessary for import and assembly into mitochondria. Eur J Biochem 222(3):983–990
5. Rustin P, Rotig A (2002) Inborn errors of complex II—unusual human mitochondrial diseases. Biochim Biophys Acta 1553(1–2):117–22
6. Schafer G, Anemuller S, Moll R (2002) Archaeal complex II: 'Classical' and 'non-classical' succinate:quinone reductases with unusual features. Biochim Biophys Acta 1553(1–2):57–73
7. Ghezzi D et al (2009) SDHAF1, encoding a LYR complex-II specific assembly factor, is mutated in SDH-defective infantile leukoencephalopathy. Nat Genet 41(6):654–656
8. Hao HX et al (2009) SDH5, a gene required for flavination of succinate dehydrogenase, is mutated in paraganglioma. Science 325(5944):1139–1142
9. Sucheta A et al (1992) Diode-like behaviour of a mitochondrial electron-transport enzyme. Nature 356(6367):361–362
10. Yankovskaya V et al (2003) Architecture of succinate dehydrogenase and reactive oxygen species generation. Science 299(5607):700–704
11. Finsterer J (2008) Leigh and Leigh-like syndrome in children and adults. Pediatr Neurol 39(4):223–235
12. Guzy RD et al (2008) Loss of the SdhB, but not the SdhA, subunit of complex II triggers reactive oxygen species-dependent hypoxia-inducible factor activation and tumorigenesis. Mol Cell Biol 28(2):718–731.
13. Rivner MH et al (1989) Kearns-Sayre syndrome and complex II deficiency. Neurology 39(5):693–696
14. Reichmann H, Angelini C (1994) Single muscle fibre analyses in 2 brothers with succinate dehydrogenase deficiency. Eur Neurol 34(2):95–98
15. Bourgeron T et al (1995) Mutation of a nuclear succinate dehydrogenase gene results in mitochondrial respiratory chain deficiency. Nat Genet 11(2):144–149
16. Parfait B et al (2000) Compound heterozygous mutations in the flavoprotein gene of the respiratory chain complex II in a patient with Leigh syndrome. Hum Genet 106(2):236–243
17. Van Coster R et al (2003) Homozygous Gly555Glu mutation in the nuclear-encoded 70 kDa flavoprotein gene causes instability of the respiratory chain complex II. Am J Med Genet A 120A(1):13–18
18. Horvath R et al (2006) Leigh syndrome caused by mutations in the flavoprotein (Fp) subunit of succinate dehydrogenase (SDHA). J Neurol Neurosurg Psychiatry 77(1):74–76
19. Taylor RW et al (1996) Deficiency of complex II of the mitochondrial respiratory chain in late-onset optic atrophy and ataxia. Ann Neurol 39(2):224–232
20. Birch-Machin MA et al (2000) Late-onset optic atrophy, ataxia, and myopathy associated with a mutation of a complex II gene. Ann Neurol 48(3):330–335
21. Bugiani M et al (2006) Effects of riboflavin in children with complex II deficiency. Brain Dev 28(9):576–581
22. Brockmann K et al (2002) Succinate in dystrophic white matter: a proton magnetic resonance spectroscopy finding characteristic for complex II deficiency. Ann Neurol 52(1):38–46
23. Saneto RP, Friedman SD, Shaw DW (2008) Neuroimaging of mitochondrial disease. Mitochondrion 8(5–6):396–413
24. Moroni I et al (2002) Cerebral white matter involvement in children with mitochondrial encephalopathies. Neuropediatrics 33(2):79–85

25. Pasini B, Stratakis CA (2009) SDH mutations in tumorigenesis and inherited endocrine tumours: lesson from the phaeochromocytoma-paraganglioma syndromes. J Intern Med 266(1):19–42
26. Baysal BE (2002) Hereditary paraganglioma targets diverse paraganglia. J Med Genet 39(9):617–622
27. Heutink P et al (1992) A gene subject to genomic imprinting and responsible for hereditary paragangliomas maps to chromosome 11q23-qter. Hum Mol Genet 1(1):7–10
28. Mariman EC et al (1995) Fine mapping of a putatively imprinted gene for familial non-chromaffin paragangliomas to chromosome 11q13.1: evidence for genetic heterogeneity. Hum Genet 95(1):56–62
29. Niemann S et al (2001) Assignment of PGL3 to chromosome 1 (q21-q23) in a family with autosomal dominant non-chromaffin paraganglioma. Am J Med Genet 98(1):32–36
30. Welander J, Soderkvist P, Gimm O (2011) Genetics and clinical characteristics of hereditary pheochromocytomas and paragangliomas. Endocr Relat Cancer 18(6): R253–R276
31. Bardella C, Pollard PJ, Tomlinson I (2011) SDH mutations in cancer. Biochim Biophys Acta 1807(11):1432–1443
32. Muller U (2011) Pathological mechanisms and parent-of-origin effects in hereditary paraganglioma/pheochromocytoma (PGL/PCC). Neurogenetics 12(3):175–181
33. Baysal BE et al (2000) Mutations in SDHD, a mitochondrial complex II gene, in hereditary paraganglioma. Science 287(5454):848–851
34. Niemann S, Muller U (2000) Mutations in SDHC cause autosomal dominant paraganglioma, type 3. Nat Genet 26(3):268–270
35. Astuti D et al (2001) Gene mutations in the succinate dehydrogenase subunit SDHB cause susceptibility to familial pheochromocytoma and to familial paraganglioma. Am J Hum Genet 69(1):49–54
36. Burnichon N et al (2010) SDHA is a tumor suppressor gene causing paraganglioma. Hum Mol Genet 19(15):3011–3020
37. Semenza GL (2004) Hydroxylation of HIF-1: oxygen sensing at the molecular level. Physiology (Bethesda) 19:176–182
38. Smith TG, Robbins PA, Ratcliffe PJ (2008) The human side of hypoxia-inducible factor. Br J Haematol 141(3):325–334
39. Baysal BE, Myers EN (2002) Etiopathogenesis and clinical presentation of carotid body tumors. Microsc Res Tech 59(3):256–261
40. King A, Selak MA, Gottlieb E (2006) Succinate dehydrogenase and fumarate hydratase: linking mitochondrial dysfunction and cancer. Oncogene 25(34):4675–82
41. Fogg VC, Lanning NJ, Mackeigan JP (2011) Mitochondria in cancer: at the crossroads of life and death. Chin J Cancer 30(8):526–539
42. Wallace DC (2005) A mitochondrial paradigm of metabolic and degenerative diseases, aging, and cancer: a dawn for evolutionary medicine. Annu Rev Genet 39:359–407
43. Gottlieb E, Tomlinson IP (2005) Mitochondrial tumour suppressors: a genetic and biochemical update. Nat Rev Cancer 5(11):857–66
44. Lee S et al (2005) Neuronal apoptosis linked to EglN3 prolyl hydroxylase and familial pheochromocytoma genes: developmental culling and cancer. Cancer Cell 8(2):155–167
45. Korpershoek E et al (2011) SDHA immunohistochemistry detects germline SDHA gene mutations in apparently sporadic paragangliomas and pheochromocytomas. J Clin Endocrinol Metab 96(9):E1472–E1476
46. Douwes Dekker PB et al (2003) SDHD mutations in head and neck paragangliomas result in destabilization of complex II in the mitochondrial respiratory chain with loss of enzymatic activity and abnormal mitochondrial morphology. J Pathol 201(3):480–486
47. Saldana MJ, Salem LE, Travezan R (1973) High altitude hypoxia and chemodectomas. Hum Pathol 4(2):251–263
48. Astrom K et al (2003) Altitude is a phenotypic modifier in hereditary paraganglioma type 1: evidence for an oxygen-sensing defect. Hum Genet 113(3):228–237

49. Astuti D et al (2011) Mutation analysis of HIF prolyl hydroxylases (PHD/EGLN) in individuals with features of phaeochromocytoma and renal cell carcinoma susceptibility. Endocr Relat Cancer 18(1):73–83
50. Baysal BE (2008) Clinical and molecular progress in hereditary paraganglioma. J Med Genet 45(11):689–694
51. Cascon A et al (2004) Genetic and epigenetic profile of sporadic pheochromocytomas. J Med Genet 41(3):e30
52. Pigny P et al (2008) Paraganglioma after maternal transmission of a succinate dehydrogenase gene mutation. J Clin Endocrinol Metab 93(5):1609–1615
53. Hensen EF et al (2004) Somatic loss of maternal chromosome 11 causes parent-of-origin-dependent inheritance in SDHD-linked paraganglioma and phaeochromocytoma families. Oncogene 23(23):4076–4083
54. Bayley JP, Devilee P, Taschner PE (2005) The SDH mutation database: an online resource for succinate dehydrogenase sequence variants involved in pheochromocytoma, paraganglioma and mitochondrial complex II deficiency. BMC Med Genet 6:39

Chapter 14
Mitochondrial Complex III Deficiency of Nuclear Origin: Molecular Basis, Pathophysiological Mechanisms, and Mouse Models

Alberto Blázquez, Lorena Marín-Buera, María Morán,
Alberto García-Bartolomé, Joaquín Arenas, Miguel A. Martín
and Cristina Ugalde

Introduction

Mitochondrial respiratory chain complex III (CIII, ubiquinol-cytochrome *c* oxidoreductase or cytochrome bc1 complex, E.C.1.10.2.2) is a multiprotein enzyme complex embedded in the mitochondrial inner membrane. As a middle component of the oxidative phosphorylation (OXPHOS) system, complex III catalyzes the transfer of electrons from reduced coenzyme Q to cytochrome *c* with a concomitant pump of protons to the intermembrane space [1]. The crystallized bovine complex consists of a symmetric homo-dimer with a combined molecular mass of ~450 kDa [2, 3]. Each monomer is composed of 11 subunits, ten encoded in the nucleus and one (cytochrome *b*) in the mitochondrial genome. Three subunits, namely cytochrome *b* (MTCYB), cytochrome *c*1 (CYC1), and the Rieske iron-sulphur protein (UQCRFS1 or RISP), form the catalytic core that contains four active redox prosthetic groups which are responsible for electron transfer. The remaining eight subunits are structural components of unknown specific function [3].

Isolated mitochondrial complex III enzyme deficiency [MIM 124000] is a fairly infrequently diagnosed defect of the OXPHOS system that is associated with a wide variety of visceral, muscular, and neurological disorders in childhood and adulthood, and presents with a number of clinical manifestations of variable severity [4–6]. Due to its dual genetic origin, complex III enzyme defects can be induced by mutations located either in mitochondrial or nuclear genes. Mitochondrial DNA (mtDNA) mutations in the cytochrome *b* *(MT-CYB)* gene [OMIM 516020] constitute a major cause of complex III deficiency that underlies a wide range of neuromuscular disorders where exercise intolerance is the predominant symptom [4, 5, 7, 8]. However,

C. Ugalde (✉) · A. Blázquez · L. Marín-Buera · M. Morán · A. García-Bartolomé ·
J. Arenas · M. A. Martín
Instituto de Investigación, Hospital Universitario 12 de Octubre, Avda. de Córdoba s/n,
28041 Madrid, Spain
e-mail: cugalde@h12o.es

Centro de Investigación Biomédica en Red de Enfermedades Raras (CIBERER),
46010 Valencia, Spain

MT-CYB mutations have only been associated with a small percentage of the total number of complex III-deficient patients. In the last 10 years, interest has shifted toward Mendelian genetics in mitochondrial complex III-associated diseases, not only because the rest of the complex III subunits are encoded by the nuclear genome, but also because an increasing number of nuclear regulatory proteins and assembly factors (such as Bcs1p/BCS1L, TTC19, and yeast Cbp3p, Cbp4p, Cbp6p, Bca1, and Mzm1) have been recently described to be involved in complex III biosynthesis and function [9–16]. Mutations in these genes may also contribute to complex III deficiency.

Nuclear Gene Defects Leading to Complex III Deficiency

As in other mitochondrial disorders, symptoms indicative of neurodegeneration are frequent in primary complex III deficiencies [4, 5, 8]. In addition, mitochondrial complex III dysfunction of nuclear origin is becoming an important cause for neonatal liver disease [17, 18]. A clinical-genetic classification can be proposed for nuclear defects that affect the biogenesis and function of complex III, as follows: (1) disorders due to mutations in nuclear genes encoding complex III structural components and (2) disorders due to mutations in complex III assembly factors. The nuclear-encoded proteins involved in complex III biogenesis and the clinical phenotypes associated with their genetic defects are shown in Tables 14.1 and 14.2.

Mutations in Complex III Nuclear Structural Genes

Mutations in nuclear structural genes are rare, and only the genes *UQCRQ* [OMIM 609653] and *UQCRB* [OMIM 609653] have been associated with complex III deficiency.

UQCRB Mutations

A homozygous 4-bp deletion was first detected in the *UQCRB* gene in a Turkish girl, born to consanguineous parents, who displayed isolated complex III enzyme deficiency associated with recurrent hypoglycaemic crises and hepatomegaly at 8 months of age [19]. She followed a benign disease course, and at the age of 2.5 years she was essentially normal with sporadic hypoglycaemic episodes. The 4-bp deletion spanned nucleotides 338–341 of the *UQCRB* cDNA, predicting both a change in the last seven amino acids and an addition of a stretch of 14 amino acids at the C-terminus of the protein. *UQCRB* encodes complex III subunit VI or subunit 7 (the nomenclature varies the subunit number from Roman to Arabic numerals), a putative ubiquinone-binding protein of 13.4 kDa that is thought to participate in the transfer

Table 14.1 Clinical/biochemical findings and mutations found in patients with complex III deficiency of nuclear genetic origin

	UQCRB Turkish[a] [19][b]	UQCRQ Israeli Bedouin[a] [23][b]	TTC19 Italian[a] [11][b]				BCS1L GRACILE Finnish[a] [31][b]	Björnstad Finnish[a] [34][b]
Sex	F	M/F	F	F	F	M	M/F	F/M
Gestational age (weeks)	Term						37.6	
IUGR	–	–	–	–	–	–	+	–
Age at onset	8 months	1 month	5 years	10 years	5 years	42 years	Birth	Childhood
Age at death	Alive 4 years	Alive	Alive 37 years	Alive 26 years	Alive 19 years	45 years	< 4 months	Alive
Hepatopathy	+	–	–	–	–	–	+	–
Tubulopathy	–	–	–	–	–	–	+	–
Neurology								
Hypothonia	–	+	–	–	–	–	–	–
Developmental delay	–	+	–	–	–	–	+	–
Seizures	–	–	–	–	–	–	–	–
Encephalopathy	–	–	+	+	+	+	–	–
Microcephaly	–	–	–	–	–	–	+	–
Deafness	–	–	–	+	–	–	–	+
Abnormal hair	–	–	–	–	–	–	–	+
Other symptoms		Brisk tendon reflexes Ataxia, mental retardation	Mental retardation Ataxia, PN	Cognitive regression Ataxia	Cognitive regression Ataxia, PN	Bradykinesia Paraparesis, PN		

Table 14.1 (continued)

	UQCRB Turkish[a] [19][b]	UQCRQ Israeli Bedouin[a] [23][b]	TTC19 Italian[a] [11][b]			BCS1L GRACILE Finnish[a] [31][b]	Björnstad Finnish[a] [34][b]
		Dystonia	Dysarthria Dysphonia Dysphagia	Dysarthia Dysphonia Comatose status	Comatose status		
					Dystonia Dysarthria		
Laboratory data							
Lactic acidosis	+	+	+	ND	+	+	–
Hypoglycaemia	+	–	–	–	–	–	–
Iron overload	ND	ND	ND	ND	ND	+	–
Complex III defect	Severe (L, Fb)	Mild (Mu)	Severe (Mu)	Severe (Mu)	Severe (Mu)	Normal (Mu)	Mild (Ly)
Brain MRI	ND	Abnormal findings in basal ganglia	Necrotic lesions, cerebellar atrophy, leukodystrophy	Necrotic lesions,	Necrotic lesions, cerebellar atrophy	Normal	ND
					cortical atrophy		
Mutations	c.338-341 del AAAA	S45P	L219X	L219X	Q173X	S78G	See Fig. 14.1

Respiratory chain enzyme activities refer to the percentages of the residual enzyme activities relative to mean control values: *Severe* <50 % control values, *Mild* >50 % mean control values, and *ND* not determined
+ present, – absent, *M* male, *F* female, *LS* Leigh syndrome, *PN* peripheral neuropathy, *Mu* muscle, *L* liver, *Fb* fibroblasts, *Ly* lymphocytes
[a]Descent
[b]Reference number

14 Mitochondrial Complex III Deficiency of Nuclear Origin: Molecular Basis...

BCS1L

	Complex III deficiency Turkish[a] [30][b]		Turkish[a] [30][b]		British[a] [31][b]		British[a] [31][b]		Spanish[a] [33][b]		Italian[a] [32][b]	
Sex	F	F	M	F	M	M	F	F	F	M	F
Gestational age (weeks)	38	39	Term	41	39	38	39	38	Term	38	30
IUGR	–	–	+	+	–	+	+	+	+	+	+
Age at onset	Birth	Birth	Birth	Birth	Birth	Birth	Birth	Birth	Birth	Birth	3 months
Age at death	3 months	Alive 9 years	6 months	2 years	Alive 5 months	2 days	42 days	105 days	3 weeks	3 months	4 years
Feeding difficulties	+	–	–	–	–	–	+	+	–	–	+
Hepatopathy	+	+	+	+	+	–	+	+	+	+	–
Tubulopathy	+	+	+	+	–	+	+	+	–	+	–
Neurology											
Hypothonia	–	–	+	–	–	+	+	+	+	+	+
Developmental delay	–	+	+	+	–	+	+	+	–	+	+
Seizures	–	–	–	–	–	+	–	–	–	–	+
Encephalopathy	+	+	+	+	–	+	+	+	–	+	+
Microcephaly	+	+	–	–	–	+	+	+	+	–	–
Deafness	+	–	–	–	–	–	–	–	–	–	–
Abnormal hair	–	–	–	–	–	–	–	–	–	–	+
Other symptoms	Blindness		Brisk tendon reflexes		Acute myoglobinuria					Ptosis	Dysmorphic features
											Spastic quadriparesis
Laboratory data											
Lactic acidosis	+	+	+	+	–	+	+	+	+	+	+
Hypoglycaemia	–	–	–	–	+	–	–	–	+	+	–
Iron overload	ND	ND	ND	ND	ND	ND	ND	ND	+	+	–
CIII defect	Severe (L)	Yes (Mu)	Mild (Fb)	Severe (L)	Severe (L,Mu), mild (Fb)	Mild (Mu)	Severe (Mu)	Severe (Mu)	Mild (Fb)	Severe (L)	Severe (Fb)
Brain MRI	ND	Atrophy	LS	LS	ND	ND	ND	ND	Normal	Delayed myelination	Cerebral atrophy
Mutations	S277N	S277N	P99L	P99L	R155P,V353M	R56X, V327A	g.1433 T > A, IVS3+1G > T	S78G, R144Q	R45C, R56X	R45C, R56X	R73C, F368I

Table 14.2 (continued)

	BCS1L								
	Complex III deficiency								
	Moroccan[a] [32][b]	Finnish[a] [34][b]	Spanish[a] [37][b]	Spanish[a] [36][b]	Spanish[a] [38][b]	Turkish[a] [29][b]	Spanish[a] [29][b]	Kenyan[a] [39][b]
Sex	F	F	M	M	F	F	F	F
Gestational age (weeks)	36			35	39			Term
IUGR	–		+			+	–	+
Age at onset	9 months		Birth	Birth	Birth	Birth	19 months	Birth
Age at death	Alive 4 years	Alive 4 years	Alive 4 years	11 months	6 months	7 months	Alive 5 years	Alive 20 years
Feeding difficulties	–				+			+
Hepatopathy	–	–	+	+	+	+	+	–
Tubulopathy	–	–	–	+	+	+	–	–
Neurology								
Hypothonia	+	+	+	+	+	+	–	+
Developmental delay	+	+	+	+	+		+	
Seizures	–		–	–	+	–	+	+
Encephalopathy	+		+	–	+	+	+	–
Microcephaly	–		+	+	+		–	–
Deafness	+	+	+	–	–	–	–	–
Abnormal hair	+	+	–	–	–	–	–	–
Other symptoms		Abnormal subcutaneous fat distribution	Dysmorphic features Hypertrichosis	Anaemia	Cataracts		Spasticity Hyperreflexia	Optic atrophy Muscle weakness
Laboratory data								
Lactic acidosis	+	–	+	+	+	+	+	–
Hypoglycaemia	–	–	+	+	+	+	–	–
Iron overload	–	ND	–	–	–	+	ND	ND
CIII defect	Severe (Mu)	Severe (Mu)	Severe (M) Mild (Fb)	Severe (Mu), Severe (Fb)	Yes (Mu)	Severe (Fb)	Severe (Mu)	Severe (Mu)
Brain MRI	Atrophy	ND	Normal	Normal	Normal	ND	Abnormal	Normal
Mutations	R183C, R184C	G35R, R184C	T50A	R56X g.1181A > G, g.1164C > G	R56X, R45C	P99L	R184C	G129R

Table legend as in Table 14.1

14 Mitochondrial Complex III Deficiency of Nuclear Origin: Molecular Basis...

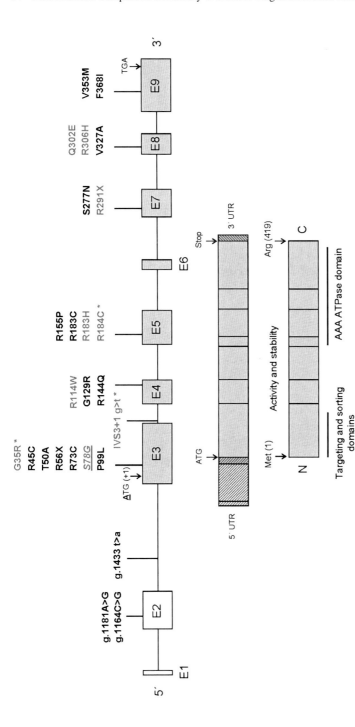

Fig. 14.1 Known mutations in the *BCS1L* gene. Mutations causing Björnstad syndrome are shown in *gray*. The p.S78G mutation causing GRACILE syndrome is indicated in gray (*underlined in italics*), and mutations causing complex III deficiency are shown in *black*. *Asterisks* indicate mutations associated with Björnstad syndrome and complex III deficiency. Coding exons representative of the BCS1L open-reading frame are indicated as *E1–E7*. The translation initiation site is indicated with an *arrow* in a schematic representation of the *BCS1L* cDNA (GenBank reference sequence NM_004328.4). Nucleotide 1 corresponds to nucleotide 2021 of the *BCS1L* genomic sequence (GenBank accesion no. AF516670). The BCS1L protein functional domains, deduced by homology with yeast Bcs1p, are indicated

of electrons when ubiquinone is bound [20, 21]. Recent functional experiments have demonstrated that this protein plays a key role in the oxygen-sensing mechanisms that regulate cellular responses to hypoxia [22].

UQCRQ Mutations

The first missense mutation in the *UQCRQ* gene was later reported in a large Israeli Bedouin kindred presenting with severe psychomotor retardation and extrapyramidal signs [23]. The mutation consisted of a homozygous c.208C > T transition in exon 2 of the *UQCRQ* gene that resulted in a serine to phenylalanine substitution at amino acid residue 45 (p.S45F). The *UQCRQ* gene encodes complex III subunit VII or subunit 8, another hypothetical ubiquinone-binding protein of 9.5 kDa whose exact function remains unknown [24, 25].

Mutations in Complex III Assembly Factors

Human disorder-associated mutations have been recently identified in two genes encoding proteins required for the proper assembly and functioning of complex III, namely *BCS1L* [OMIM 603647] and *TTC19* [OMIM 613814].

BCS1L Mutations

BCS1L is a mitochondrial inner membrane chaperone belonging to the AAA$^+$ family of proteins that plays an essential role in the biogenesis of respiratory chain complex III [9, 10, 26, 27]. The vast majority of nuclear mutations leading to complex III deficiency have been discovered in the *BCS1L* gene [27–29]. In 2001, de Lonlay et al. [30] reported the first *BCS1L* mutations in six patients from four unrelated families suffering from metabolic acidosis at birth, neonatal proximal tubulopathy, liver failure, and encephalopathy; functional complementation analyses in yeast confirmed the deleterious effects of such genetic alterations. Since then, more than 25 different pathogenic mutations that affect BCS1L function have been reported (Fig. 14.1) [31–39].

Human BCS1L protein consists of 419 amino acids, and shares functional homology and 50 % identity with yeast Bcs1p [9, 40]. In yeast, Bcs1p was first described as an ATP-dependent chaperone that maintained a preassembled complex III in a competent state for the subsequent incorporation of the Rieske and Qcr10 subunits [10]. The Bcs1p protein was demonstrated to interact with complex III, but a substantial fraction of the protein was present in a high molecular-weight complex in which no complex III subunits were found [10, 26, 41]. Although these findings suggested a transient interaction between BCS1L and complex III, this physical association has not been yet confirmed in humans [27]. More recently, the role of Bcs1p has been redefined as a protein translocase involved in the export of the folded Fe–S domain of Rieske subunit across the inner membrane and the insertion of its transmembrane segment into an assembly intermediate of the cytochrome *bc1* complex [42].

Consequently, the absence of Bcs1p prevents the assembly of Rieske within complex III and leads to the accumulation of a partially assembled precomplex that lacks respiratory enzyme activity [10, 32, 42, 43]. Bcs1p spans three different domains: (1) the N-terminal domain comprises amino acid residues 1–126, which contain all the required information for the mitochondrial targeting and protein sorting [40, 44]; (2) a Bcs1p-specific domain contains amino acid residues that are important for Bcs1p activity and stability [45]; and (3) the C-terminal region contains a ~ 200 amino acids domain that encompasses two motifs corresponding to nucleotide binding sites, a characteristic feature of the AAA^+ (for ATPases associated with various cellular activities) protein superfamily [46]. Pathogenic mutations have been described in the three different domains of human BCS1L, although the pathophysiological mechanisms that contribute to the different clinical manifestations of the disease remain poorly understood.

Mutations in *BCS1L* lead to three main clinical phenotypes: (1) The most benign is the Björnstad syndrome [OMIM 262000], an autosomal-recessive disorder characterized by sensorineural hearing loss and *pili torti* [34]; (2) The most severe clinical consequence of *BCS1L* mutations is the GRACILE syndrome [OMIM 603358], a Finnish-heritage disease caused by the homozygous c.232A > G (p.S78G) mutation, which is characterized by fetal growth retardation, aminoaciduria, cholestasis, iron overload, lactic acidosis, and early death [31, 47, 48]. Affected infants are small for gestational age due to intrauterine growth retardation, and they usually develop Fanconi-type aminoaciduria, cholestasis with progressive liver dysfunction, and iron overload with liver hemosiderosis. All patients die within weeks of birth. The disease is dominated by liver failure and there are no dysmorphic or neurological signs probably due to the early onset and fast evolution of the disease; and (3) An intermediate clinical phenotype is represented by complex III enzyme deficiency in neonates, infants, or adults who may present with encephalopathy, alone or in combination with other neurological signs, congenital metabolic acidosis, iron overload, neonatal proximal tubulopathy, and liver failure [30, 32, 33, 36–39]. In this regard, the GRACILE syndrome and the most severe presentation of complex III deficiency are only subtly different. Other clinical traits may include microcephaly, hypotonia, eye defects, dismorphic features, deafness, and abnormal hair.

It has been hypothesized that the physiological impact of the *BCS1L* mutations might account for the clinical spectrum of the disease [32, 34, 38]. For instance, GRACILE segregates with the Finnish disease heritage mutation c.232A > G (p.S78G), which leads to the most severe clinical phenotype (50 % die before 12 days and 50 % before 4 months), complex III deficiency-associated mutations are relatively less severe (children live from 3 months to 5 years), and Björnstad syndrome mutations are the mildest (normal life span). However, in patients presenting with GRACILE syndrome complex III activity was either normal or mildly decreased despite low BCS1L levels [31, 35, 48], while in some patients who harbored *BCS1L* mutations associated with Björnstad syndrome or complex III deficiency, an assembly defect of the RISP subunit led to a more severe mitochondrial complex III enzyme defect [32, 34, 37]. Iron overload has been reported in patients with *BCS1L* mutations leading to GRACILE syndrome or complex III deficiency, but not Björnstad Syndrome, suggesting an involvement of BCS1L in iron metabolism [29, 33, 47]. It has also been proposed that the pathophysiology of the *BCS1L* mutations might be regulated by reactive oxygen species (ROS) production in a mutation and disease-dependent manner, most severely by the complex III deficiency-associated mutations [29, 34]. Although BCS1L seems to be ubiquitously expressed, observations during mouse embryonic development support tissue-specific differences in its expression that might account for the organ-specific manifestations of disease [49]. It is therefore likely that phenotypic variability of the *BCS1L* mutations depends on the nature of the mutation—the most severe mutations seem to be located in the N-terminus of BCS1L and affect the protein sorting, while the mutations that produce the least severe phenotype are usually placed in C-terminal residues that affect the external chaperone surface—in combination with the tissue-specific expression of the gene [34, 38, 49].

TTC19 Mutations

Recently, Ghezzi et al. [11] identified the first mutations in the *TTC19* gene, which encodes a tetratricopeptide repeat (TPR) motif-containing protein, in four Italian patients presenting with severe isolated complex III deficiency who lacked alterations in either complex III subunits or in *BCS1L*. Two Italian siblings, who presented with a slowly progressive neurodegenerative disorder with onset in late infancy, carried a homozygous nucleotide change c.656T > G that predicted a truncated protein, p.L219X. Brain MRI of one of these patients showed hyperintense lesions in the olives, putamen, and substantia nigra, as well as cerebral and cerebellar atrophy. A third unrelated patient from the same country region had a similar disease course, and haplotype analysis indicated a founder effect. A fourth independent patient carried the homozygous nonsense mutation c.517C > T, which predicted the truncated protein p.Q173X. He presented with adult onset of subacute, rapidly progressive neurologic failure resulting in death. Consistent with these genetic and biochemical features, decreased *TTC19* mRNA or protein levels were found in patients' muscle tissue or fibroblasts. The authors showed that TTC19 physically interacts with complex III at the mitochondrial inner membrane, and immunofluorescence studies indicated a defective assembly of respiratory chain complex III in the patients' tissues. A *Drosophila melanogaster* knockout model for *TTC19* also presented with neurological abnormalities linked to complex III deficiency.

Cellular Pathophysiological Consequences of Complex III Deficiency

Respirasome Assembly and Activity Defects

The most obvious consequence of the genetic defects in complex III structural subunits or assembly factors is an altered assembly of this complex either due to the malfunctioning of essential chaperones, or to conformational changes or a complete lack of the affected subunits [27, 28]. Despite of the vital importance of a correctly functioning respiratory chain complex III for any organism, the contribution of isolated complex III deficiency to the overall respiratory chain enzyme dysfunction in human patients is significantly lower than that of complex I or complex IV deficiencies, and it only accounts for approximately 10–15 % of the overall respiratory chain defects [5, 8, 50]. The reason probably relies in the fact that OXPHOS complexes are associated in higher order assemblies called supercomplexes or respirasomes [51–57]. These structures are thought to offer structural or functional advantages to the system, such as the prevention of destabilization and degradation of OXPHOS complexes, the enhancement of electron-transport efficiency and substrate channeling, or the decrease of electron or proton leakage [58]. As a consequence, there is an interdependence between the individual OXPHOS complexes [59–62], with major implications for the diagnostics of mitochondrial disorders since some combined deficiencies can be attributed to the genetic defect of a single complex. Mitochondrial complex I is functionally associated with the supercomplexes, and complex III plays an essential role in the stabilization of complex I within these structures [59, 63, 64]. For this reason, structural alterations that severely impair complex III biosynthesis may induce pleiotropic deleterious effects on the assembly and enzyme activity of complex I that will probably be translated into combined respiratory deficiencies of both complexes. Accordingly, pathogenic mutations in complex III structural or assembly genes have been described in some patients presenting with combined enzyme deficiencies of complexes I, III, and even complex IV [19, 23, 29, 32, 65]. However, this is not a general feature of complex III deficiency and, contrary to what is commonly accepted, there is no such clear correlation between the severity of the complex III enzyme defect, the complex III assembly impairment and the pleiotropic complex I assembly and activity defects [23]. Exceptions can be observed for instance in some patients harboring mutations in *BCS1L* and *TTC19*, who showed normal complex I activity levels despite a dramatic loss of fully assembled complex III in a variety of tissues [11, 32]. Such differences may be attributed to the nature of the mutation or the functional role of the complex III-mutated gene, and suggest that not all complex III genes are equally necessary to maintain complex I stability. Further research is necessary in order to gain more insight in the mechanisms that regulate the interdependences between OXPHOS complexes.

Bioenergetic Defects and Increased ROS Production

As a result of respiratory chain assembly defects, disturbances in the electron transport and proton pumping occur, which in turn may lead to decreased mitochondrial membrane potential and decreased OXPHOS-derived ATP synthesis [66–69]. Besides the energetic defects, an altered electron transport often induces increased rates of superoxide anion production. Once superoxide is generated, it can be eliminated by cellular antioxidants, but in some situations an increased superoxide production can induce the appearance of other free radicals and precursors of free radicals, known as reactive oxygen species (ROS). These ROS are highly reactive molecules that oxidize macromolecules such as lipids, DNA, and proteins. If they are not efficiently eliminated by the cellular antioxidants, the cellular components become damaged by oxidation, which in turn leads to oxidative stress, cellular dysfunction, and even cell death.

Mitochondrial respiratory chain complexes I and III constitute the two main sources of mitochondrial ROS [70]. Two studies have analyzed the impact of complex III deficiency on ROS production due to mutations in the BCS1L assembly factor [29, 34]. One of these studies revealed an increased ROS production in isolated mitochondria of lymphocytes from patients with either complex III deficiency or Björnstad syndrome [34]. The second study was performed with skin fibroblasts from patients with complex III deficiency associated with *BCS1L* mutations, and only four out of the six analyzed cell lines showed an increase in H_2O_2 levels and unbalanced expression of the cellular antioxidant defences [29]. All these complex III-deficient samples with increased ROS levels showed alterations in the assembly of respiratory chain complexes I and III or complexes III and IV, suggesting that a correlation can be established between the severity of the respiratory chain assembly defect and increased ROS production. In agreement, complexes I, IV, and supercomplexes were decreased in Rieske-deficient mouse cells in a ROS-regulated manner [43]. However, the stable silencing of the catalytic Rieske protein in 143B cells and mice showed low superoxide induction and normal hydrogen peroxide levels despite a significant loss of complex III activity and a general down-regulation of OXPHOS genes expression, which could be indicative of a lowering of mitochondrial biogenesis and a shift toward anaerobic energy metabolism to improve cellular survival [71, 72]. In agreement with these results, the *C. elegans* Rieske knocked-down (*isp-1*) mutants displayed low oxygen consumption and an elevated generation of superoxide (but not of overall ROS) that specifically increased their longevity compared with wild-type worms, suggesting a role of superoxide as a protective signal [73, 74]. The lack of studies analyzing the potential sites of superoxide production in "mutated respiratory chains", only allows speculation with the general idea that alterations in the assembly of the complexes due to conformational changes in the mutated subunits or a complete lack of them would increase electron leakage throughout the respiratory chain.

Fragmentation of the Mitochondrial Network

Technical improvements in the microscopy field have allowed researchers to realize that the "old static" mitochondria were able to move continuously, fuse with each other creating mitochondrial networks, and also to become solitary units after fission events. The overall mitochondrial morphology in each cell line is determined by the equilibrium between fusion and fission events, rendering mitochondrial networks that may range from mainly tubular to highly fragmented [75–78]. Alterations in mitochondrial morphology have been frequently reported in cultured cells from patients with OXPHOS disorders [68, 79–82], but studies relative to complex III deficiency are scarce. Antimycin treatment induced complex III deficiency and a partial mitochondrial fragmentation in control fibroblasts; however, cells from a patient presenting with complex III deficiency of unknown genetic origin showed essentially normal mitochondrial morphology with a high proportion of long filamentous mitochondria [69]. In HeLa cells, *BCS1L* knockdown caused not only the disassembly of the respiratory chain, but also the down regulation of LETM1, a protein that maintains the mitochondrial tubular shape and is implicated in the human Wolf–Hirschhorn syndrome [83]. Loss of BCS1L also led to morphological alterations in the mitochondrial network in a LETM1-independent fashion, suggesting a different role for these two proteins in the maintenance of mitochondrial morphology [83]. In order to unveil the cell biological consequences of *BCS1L* genetic defects, our group performed a comparative series of cellular and biochemical studies in skin fibroblasts from six complex III-deficient patients harboring different mutations in the *BCS1L* gene [29]. Our results showed a mitochondrial network fragmentation in all patients' fibroblasts that neither correlated with other pathophysiological parameters such as respiratory chain assembly defects, enzyme deficiencies, or ROS production, nor with the severity of the clinical manifestations of the patients. A link between mitochondrial dynamics and ROS production has been described [78, 84]; however, some of our complex III-deficient patients' fibroblasts showed normal respiratory chain activities and ROS levels but mitochondrial network derangements [29], suggesting that the above-mentioned association is not straightforward. It is worth noting that all the studies regarding mitochondrial dynamics have been performed in cultured cells, and data about human tissues are lacking. Therefore, the dysfunctions due to alterations in mitochondrial dynamics that occur in tissues from human patients remain unexplored.

Mice Models of Complex III Deficiency

Targeted mutagenesis of the mouse-germ line has proved to be a powerful approach to understand the *in vivo* function of mitochondrial genes. Conventional knockouts usually result in embryonic lethality [85, 86], which can be circumvented by the generation of conditional knockouts that have led to mouse models of complex I and complex IV deficiencies [87–91]. However, mammalian models of specific mitochondrial diseases caused by nuclear mutations are scarce. Till date, three transgenic models have been developed to understand the functional role of complex III nuclear gene products in health and disease [72, 92, 93].

The RISP Mice

As previously mentioned, RISP is one of the three essential catalytic subunits of complex III [94]. It contains an iron–sulfur cluster [2Fe–2S] that has been suggested to function as a proton-exiting gate within the complex [95]. A mouse with a *knock-in* RISP protein was first developed by Garcia et al. [93]. A targeting vector consisting of the full-length *RISP* intact gene plus a neomycin/thymidine kinase selection cassette in the 3' UTR was inserted in mouse embryonic stem cells by homologous recombination. The retention of the neomycin resistance cassette in the 3' UTR did not disrupt the splicing or reading frame of RISP, but resulted in a larger transcript than the endogenous *RISP*. Homozygous knocked-in mice showed reduced RISP expression specifically in the skin melanocytes, with normal RISP expression levels and complex III enzyme activity in isolated fibroblasts. Interestingly, pigmentation changes occurred in approximately 4–7 months old mutant mice.

A second group recently generated a *knock-in* mouse with a loss-of-function RISP mutation (p.P224S) that was homozygous lethal [72]. However, heterozygous mice were viable, with unaffected performance and fertility, despite decreased RISP expression levels and a mild impairment of complex III enzyme activity. This decrease was sufficient to impair mitochondrial respiration and to decrease overall metabolic rates in males, but surprisingly not in females, suggesting a sex-dependent response to respiratory chain dysfunction. Biomarkers of oxidative stress such as aconitase activity or protein carbonylation were unaffected in both young and aged animals regardless their sex. Despite this, the average lifespan of male heterozygous mice was shortened and their health deteriorated more rapidly than females. These results showed that mild perturbations of the mitochondrial respiratory chain may have significant physiological effects in mammals, and that the severity of those effects can be sex-dependent.

The GRACILE Mouse

Leveen et al. [92] recently presented the first viable model of OXPHOS deficiency mimicking a human mitochondrial disorder. This group developed a mouse model for GRACILE syndrome, caused by the homozygous *BCS1L* mutation c.232A > G (p.S78G). After 3 weeks of age, homozygous-mutant mice developed remarkable similarities to the human disorder, such as growth retardation, hepatic glycogen depletion, steatosis, fibrosis, cirrhosis, tubulopathy, lactic acidosis, and short lifespan. BCS1L expression levels were decreased in liver mitochondria of all mutants, and symptomatic animals displayed complex III deficiency due to RISP misincorporation into complex III. Strikingly, RISP was correctly assembled in complex III in young homozygotes, which suggests the necessity of another complex III assembly factor during early ontogenesis. Supporting this, a new complex III assembly factor involved in RISP assembly has been recently described in yeast [16]. Electron flux kinetics through complex III demonstrated that complex III deficiency was remarkably evident in liver compared with other tissues of the symptomatic mutants, reaching only 20 % of the control enzyme activity. In high-resolution respirometry, complex III dysfunction resulted in decreased electron-transport capacity through the respiratory chain under maximum substrate input. However, complex I function was unaffected, in agreement with the previously mentioned reports from patients harboring mutations in *BCS1L* or in the assembly factor *TTC19* [11, 32]. Due to the early onset of symptoms, the convenient time range for disease progress and the reproducible pathology, this model constitutes a valuable tool for the investigation of the clinical manifestations, pathophysiology, and treatment of the GRACILE syndrome and mitochondrial complex III deficiency, as well as of other OXPHOS disorders.

Concluding Remarks

In the last 10 years, interest has moved toward Mendelian genetics of mitochondrial complex III-associated disorders, not only because ten out of the 11 complex III subunits are nuclear-encoded, but also because a growing number of nuclear regulatory proteins and assembly factors have been recently described to be involved in complex III biosynthesis and function. Of these, only four genes have been identified so far as disease-responsible causes of mitochondrial complex III deficiency, but an increasing number of nuclear genetic defects that affect complex III biogenesis have been described. It is likely that yet unknown regulatory proteins involved in the translation, chaperoning, and assembly of complex III components will emerge as new genetic causes of complex III deficiency. In complex III-related disorders, as for other pathologies that involve mitochondrial dysfunction, a clear relationship between the alterations in the biosynthesis of the respiratory chain complexes, and their specific pathophysiological consequences, has not been fully elucidated yet. The considerable variability in the disease manifestations and organ differences involved in complex III deficiency constitute an unsolved matter. Significant effort must still be invested to better understand how a malfunctioning respiratory chain may affect different cellular processes, which will lead to new therapeutic approaches. The development of new transgenic models that mimic OXPHOS disorders will be invaluable.

Conflict of Interests There are no actual or potential conflicts of interest with other people or organizations.

Acknowledgments This study was supported by Instituto de Investigación Hospital Universitario 12 de Octubre (I+12), and Instituto de Salud Carlos III (ISCIII)/Ministry of Science and Innovation (MCINN) to C.U. (grant numbers PI08-0021 and PI11-00182), to M.M. (CP11-00151) and to M.A.M. (PI09-01359). M.A.M. was the recipient of a "Intensificación de la Actividad Investigadora" action from ISCIII and Comunidad Autónoma de Madrid (CAM), Spain.

References

1. Baum H, Rieske JS, Silman HI, Lipton SH (1967) On the mechanism of electron transfer in complex III of the electron transfer chain. Proc Natl Acad Sci U S A 57:798–805

2. Xia D, Yu CA, Kim H et al (1997) Crystal structure of the cytochrome bc1 complex from bovine heart mitochondria. Science 277:60–66
3. Iwata S, Lee JW, Okada K et al (1998) Complete structure of the 11-subunit bovine mitochondrial cytochrome bc1 complex. Science 281:64–71
4. Borisov VB (2002) Defects in mitochondrial respiratory complexes III and IV, and human pathologies. Mol Aspects Med 23:385–412
5. Benit P, Lebon S, Rustin P (2009) Respiratory-chain diseases related to complex III deficiency. Biochim Biophys Acta 1793:181–185
6. Scaglia F, Towbin JA, Craigen WJ et al (2004) Clinical spectrum, morbidity, and mortality in 113 pediatric patients with mitochondrial disease. Pediatrics 114:925–931
7. Andreu AL, Hanna MG, Reichmann H et al (1999) Exercise intolerance due to mutations in the cytochrome b gene of mitochondrial DNA. N Engl J Med 341:1037–1044
8. Mourmans J, Wendel U, Bentlage HA et al (1997) Clinical heterogeneity in respiratory chain complex III deficiency in childhood. J Neurol Sci 149:111–117
9. Nobrega FG, Nobrega MP, Tzagoloff A (1992) BCS1, a novel gene required for the expression of functional Rieske iron-sulfur protein in Saccharomyces cerevisiae. EMBO J 11:3821–3829
10. Cruciat CM, Hell K, Folsch H, Neupert W, Stuart RA (1999) Bcs1p, an AAA-family member, is a chaperone for the assembly of the cytochrome bc(1) complex. EMBO J 18:5226–5233
11. Ghezzi D, Arzuffi P, Zordan M et al (2011) Mutations in TTC19 cause mitochondrial complex III deficiency and neurological impairment in humans and flies. Nat Genet 43:259–263
12. Shi G, Crivellone MD, Edderkaoui B (2001) Identification of functional regions of Cbp3p, an enzyme-specific chaperone required for the assembly of ubiquinol-cytochrome c reductase in yeast mitochondria. Biochim Biophys Acta 1506:103–116
13. Kronekova Z, Rodel G (2005) Organization of assembly factors Cbp3p and Cbp4p and their effect on bc(1) complex assembly in Saccharomyces cerevisiae. Curr Genet 47:203–212
14. Dieckmann CL, Tzagoloff A (1985) Assembly of the mitochondrial membrane system. CBP6, a yeast nuclear gene necessary for synthesis of cytochrome b. J Biol Chem 260:1513–1520
15. Mathieu L, Marsy S, Saint-Georges Y, Jacq C, Dujardin G (2011) A transcriptome screen in yeast identifies a novel assembly factor for the mitochondrial complex III. Mitochondrion 11:391–396
16. Atkinson A, Smith P, Fox JL, Cui TZ, Khalimonchuk O, Winge DR (2011) The LYR protein Mzm1 functions in the insertion of the Rieske Fe/S protein in yeast mitochondria. Mol Cell Biol 31:3988–3996
17. Fellman V, Kotarsky H (2011) Mitochondrial hepatopathies in the newborn period. Semin Fetal Neonatal Med 16:222–228
18. Dimauro S, Garone C (2011) Metabolic disorders of fetal life: glycogenoses and mitochondrial defects of the mitochondrial respiratory chain. Semin Fetal Neonatal Med 16:181–189
19. Haut S, Brivet M, Touati G et al (2003) A deletion in the human QP-C gene causes a complex III deficiency resulting in hypoglycaemia and lactic acidosis. Hum Genet 113:118–122
20. Wakabayashi S, Takao T, Shimonishi Y et al (1985) Complete amino acid sequence of the ubiquinone binding protein (QP-C), a protein similar to the 14,000-dalton subunit of the yeast ubiquinol-cytochrome c reductase complex. J Biol Chem 260:337–343
21. von Jagow G, Link TA, Ohnishi T (1986) Organization and function of cytochrome b and ubiquinone in the cristae membrane of beef heart mitochondria. J Bioenerg Biomembr 18:157–179
22. Jung HJ, Shim JS, Lee J et al (2010) Terpestacin inhibits tumor angiogenesis by targeting UQCRB of mitochondrial complex III and suppressing hypoxia-induced reactive oxygen species production and cellular oxygen sensing. J Biol Chem 285:11584–11595
23. Barel O, Shorer Z, Flusser H et al (2008) Mitochondrial complex III deficiency associated with a homozygous mutation in UQCRQ. Am J Hum Genet 82:1211–1216
24. Usui S, Yu L, Yu CA (1990) The small molecular mass ubiquinone-binding protein (QPc-9.5 kDa) in mitochondrial ubiquinol-cytochrome c reductase: isolation, ubiquinone-binding domain, and immunoinhibition. Biochemistry 29:4618–4626

25. Yu L, Deng K, Yu CA (1995) Cloning, gene sequencing, and expression of the small molecular mass ubiquinone-binding protein of mitochondrial ubiquinol-cytochrome c reductase. J Biol Chem 270:25634–25638
26. Conte L, Trumpower BL, Zara V (2011) Bcs1p can rescue a large and productive cytochrome bc(1) complex assembly intermediate in the inner membrane of yeast mitochondria. Biochim Biophys Acta 1813:91–101
27. Fernandez-Vizarra E, Tiranti V, Zeviani M (2009) Assembly of the oxidative phosphorylation system in humans: what we have learned by studying its defects. Biochim Biophys Acta 1793:200–211
28. Diaz F, Kotarsky H, Fellman V, Moraes CT (2011) Mitochondrial disorders caused by mutations in respiratory chain assembly factors. Semin Fetal Neonatal Med 16:197–204
29. Moran M, Marin-Buera L, Gil-Borlado MC et al (2010) Cellular pathophysiological consequences of BCS1L mutations in mitochondrial complex III enzyme deficiency. Hum Mutat 31:930–941
30. de Lonlay P, Valnot I, Barrientos A et al (2001) A mutant mitochondrial respiratory chain assembly protein causes complex III deficiency in patients with tubulopathy, encephalopathy and liver failure. Nat Genet 29:57–60
31. Visapaa I, Fellman V, Vesa J et al (2002) GRACILE syndrome, a lethal metabolic disorder with iron overload, is caused by a point mutation in BCS1L. Am J Hum Genet 71:863–876
32. Fernandez-Vizarra E, Bugiani M, Goffrini P et al (2007) Impaired complex III assembly associated with BCS1L gene mutations in isolated mitochondrial encephalopathy. Hum Mol Genet 16:1241–1252
33. De Meirleir L, Seneca S, Damis E et al (2003) Clinical and diagnostic characteristics of complex III deficiency due to mutations in the BCS1L gene. Am J Med Genet A 121A:126–131
34. Hinson JT, Fantin VR, Schonberger J et al (2007) Missense mutations in the BCS1L gene as a cause of the Bjornstad syndrome. N Engl J Med 356:809–819
35. Fellman V, Lemmela S, Sajantila A, Pihko H, Jarvela I (2008) Screening of BCS1L mutations in severe neonatal disorders suspicious for mitochondrial cause. J Hum Genet 53:554–558
36. Gil-Borlado MC, Gonzalez-Hoyuela M, Blazquez A et al (2009) Pathogenic mutations in the 5′ untranslated region of BCS1L mRNA in mitochondrial complex III deficiency. Mitochondrion 9:299–305
37. Blazquez A, Gil-Borlado MC, Moran M et al (2009) Infantile mitochondrial encephalomyopathy with unusual phenotype caused by a novel BCS1L mutation in an isolated complex III-deficient patient. Neuromuscul Disord 19:143–146
38. Ramos-Arroyo MA, Hualde J, Ayechu A et al (2009) Clinical and biochemical spectrum of mitochondrial complex III deficiency caused by mutations in the BCS1L gene. Clin Genet 75:585–587
39. Tuppen HA, Fehmi J, Czermin B et al (2010) Long-term survival of neonatal mitochondrial complex III deficiency associated with a novel BCS1L gene mutation. Mol Genet Metab 100:345–348
40. Petruzzella V, Tiranti V, Fernandez P, Ianna P, Carrozzo R, Zeviani M (1998) Identification and characterization of human cDNAs specific to BCS1, PET112, SCO1, COX15, and COX11, five genes involved in the formation and function of the mitochondrial respiratory chain. Genomics 54:494–504
41. Zara V, Conte L, Trumpower BL (2007) Identification and characterization of cytochrome bc(1) subcomplexes in mitochondria from yeast with single and double deletions of genes encoding cytochrome bc(1) subunits. FEBS J 274:4526–4539
42. Wagener N, Ackermann M, Funes S, Neupert W (2011) A Pathway of protein translocation in mitochondria mediated by the AAA-ATPase Bcs1. Mol Cell 44:191–202
43. Diaz F, Enriquez JA, Moraes CT (2011) Cells lacking Rieske iron sulfur protein have a ROS-associated decrease in respiratory complexes I and IV. Mol Cell Biol 32(2):415–429. doi:10.1128/MCB.06051-11
44. Folsch H, Guiard B, Neupert W, Stuart RA (1996) Internal targeting signal of the BCS1 protein: A novel mechanism of import into mitochondria. EMBO J 15:479–487

45. Nouet C, Truan G, Mathieu L, Dujardin G (2009) Functional analysis of yeast bcs1 mutants highlights the role of Bcs1p-specific amino acids in the AAA domain. J Mol Biol 388:252–261
46. Frickey T, Lupas AN (2004) Phylogenetic analysis of AAA proteins. J Struct Biol 146:2–10
47. Fellman V (2002) The GRACILE syndrome, a neonatal lethal metabolic disorder with iron overload. Blood Cells Mol Dis 29:444–450
48. Kotarsky H, Karikoski R, Morgelin M et al (2010) Characterization of complex III deficiency and liver dysfunction in GRACILE syndrome caused by a BCS1L mutation. Mitochondrion 10:497–509
49. Kotarsky H, Tabasum I, Mannisto S, Heikinheimo M, Hansson S, Fellman V (2007) BCS1L is expressed in critical regions for neural development during ontogenesis in mice. Gene Expr Patterns 7:266–273
50. von Kleist-Retzow JC, Cormier-Daire V, de Lonlay P et al (1998) A high rate (20 %–30 %) of parental consanguinity in cytochrome-oxidase deficiency. Am J Hum Genet 63:428–435
51. Acin-Perez R, Fernandez-Silva P, Peleato ML, Perez-Martos A, Enriquez JA (2008) Respiratory active mitochondrial supercomplexes. Mol Cell 32:529–539
52. Bianchi C, Genova ML, Parenti Castelli G, Lenaz G (2004) The mitochondrial respiratory chain is partially organized in a supercomplex assembly: kinetic evidence using flux control analysis. J Biol Chem 279:36562–36569
53. Boekema EJ, Braun HP (2007) Supramolecular structure of the mitochondrial oxidative phosphorylation system. J Biol Chem 282:1–4
54. Schagger H (2002) Respiratory chain supercomplexes of mitochondria and bacteria. Biochim Biophys Acta 1555:154–159
55. Schagger H (2001) Respiratory chain supercomplexes. IUBMB Life 52:119–128
56. Schagger H, Pfeiffer K (2001) The ratio of oxidative phosphorylation complexes I-V in bovine heart mitochondria and the composition of respiratory chain supercomplexes. J Biol Chem 276:37861–37867
57. Schagger H, Pfeiffer K (2000) Supercomplexes in the respiratory chains of yeast and mammalian mitochondria. EMBO J 19:1777–1783
58. Lenaz G, Genova ML (2010) Structure and organization of mitochondrial respiratory complexes: a new understanding of an old subject. Antioxid Redox Signal 12:961–1008
59. Acin-Perez R, Bayona-Bafaluy MP, Fernandez-Silva P et al (2004) Respiratory complex III is required to maintain complex I in mammalian mitochondria. Mol Cell 13:805–815
60. Li Y, D'Aurelio M, Deng JH et al (2007) An assembled complex IV maintains the stability and activity of complex I in mammalian mitochondria. J Biol Chem 282:17557–17562
61. Soto IC, Fontanesi F, Valledor M, Horn D, Singh R, Barrientos A (2009) Synthesis of cytochrome c oxidase subunit 1 is translationally downregulated in the absence of functional F1F0-ATP synthase. Biochim Biophys Acta 1793:1776–1786
62. Vempati UD, Han X, Moraes CT (2009) Lack of cytochrome c in mouse fibroblasts disrupts assembly/stability of respiratory complexes I and IV. J Biol Chem 284:4383–4391
63. Schagger H, de Coo R, Bauer MF, Hofmann S, Godinot C, Brandt U(2004) Significance of respirasomes for the assembly/stability of human respiratory chain complex I. J Biol Chem 279:36349–36353
64. Calvaruso MA, Willems P, van den Brand M et al (2011) Mitochondrial complex III stabilizes complex I in the absence of NDUFS4 to provide partial activity. Hum Mol Genet 21(1):115–120. doi:10.1093/hmg/ddr446
65. Lamantea E, Carrara F, Mariotti C, Morandi L, Tiranti V, Zeviani M (2002) A novel nonsense mutation (Q352X) in the mitochondrial cytochrome b gene associated with a combined deficiency of complexes I and III. Neuromuscul Disord 12:49–52
66. Distelmaier F, Visch HJ, Smeitink JA, Mayatepek E, Koopman WJ, Willems PH (2009) The antioxidant Trolox restores mitochondrial membrane potential and Ca2+-stimulated ATP production in human complex I deficiency. J Mol Med 87:515–522
67. Iuso A, Scacco S, Piccoli C et al (2006) Dysfunctions of cellular oxidative metabolism in patients with mutations in the NDUFS1 and NDUFS4 genes of complex I. J Biol Chem 281:10374–10380

68. Moran M, Rivera H, Sanchez-Arago M et al (2010) Mitochondrial bioenergetics and dynamics interplay in complex I-deficient fibroblasts. Biochim Biophys Acta 1802:443–453
69. Guillery O, Malka F, Frachon P, Milea D, Rojo M, Lombes A (2008) Modulation of mitochondrial morphology by bioenergetics defects in primary human fibroblasts. Neuromuscul Disord 18:319–330
70. Murphy MP (2009) How mitochondria produce reactive oxygen species. Biochem J 417:1–13
71. Levanets O, Reinecke F, Louw R et al (2011) Mitochondrial DNA replication and OXPHOS gene transcription show varied responsiveness to Rieske protein knockdown in 143B cells. Biochimie 93:758–765
72. Hughes BG, Hekimi S (2011) A mild impairment of mitochondrial electron transport has sex-specific effects on lifespan and aging in mice. PLoS One 6:e26116
73. Yang W, Hekimi S (2010) A mitochondrial superoxide signal triggers increased longevity in Caenorhabditis elegans. PLoS Biol 8:e1000556
74. Feng J, Bussiere F, Hekimi S (2001) Mitochondrial electron transport is a key determinant of life span in Caenorhabditis elegans. Dev Cell 1:633–644
75. Chan DC (2006) Mitochondria: dynamic organelles in disease, aging, and development. Cell 125:1241–1252
76. Campello S, Scorrano L (2010) Mitochondrial shape changes: orchestrating cell pathophysiology. EMBO Rep 11:678–684
77. Liesa M, Palacin M, Zorzano A (2009) Mitochondrial dynamics in mammalian health and disease. Physiol Rev 89:799–845
78. Benard G, Bellance N, James D et al (2007) Mitochondrial bioenergetics and structural network organization. J Cell Sci 120:838–848
79. Sauvanet C, Duvezin-Caubet S, di Rago JP, Rojo M (2010) Energetic requirements and bioenergetic modulation of mitochondrial morphology and dynamics. Semin Cell Dev Biol 21:558–565
80. Benard G, Karbowski M (2009) Mitochondrial fusion and division: Regulation and role in cell viability. Semin Cell Dev Biol 20:365–374
81. Koopman WJ, Verkaart S, Visch HJ et al (2007) Human NADH: ubiquinone oxidoreductase deficiency: radical changes in mitochondrial morphology? Am J Physiol Cell Physiol 293:C22–C29
82. Pham NA, Richardson T, Cameron J, Chue B, Robinson BH (2004) Altered mitochondrial structure and motion dynamics in living cells with energy metabolism defects revealed by real time microscope imaging. Microsc Microanal 10:247–260
83. Tamai S, Iida H, Yokota S et al (2008) Characterization of the mitochondrial protein LETM1, which maintains the mitochondrial tubular shapes and interacts with the AAA-ATPase BCS1L. J Cell Sci 121:2588–2600
84. Yu T, Robotham JL, Yoon Y (2006) Increased production of reactive oxygen species in hyperglycemic conditions requires dynamic change of mitochondrial morphology. Proc Natl Acad Sci U S A 103:2653–2658
85. Vempati UD, Torraco A, Moraes CT (2008) Mouse models of oxidative phosphorylation dysfunction and disease. Methods 46:241–247
86. Wallace DC, Fan W (2009) The pathophysiology of mitochondrial disease as modeled in the mouse. Genes Dev 23:1714–1736
87. Kruse SE, Watt WC, Marcinek DJ, Kapur RP, Schenkman KA, Palmiter RD (2008) Mice with mitochondrial complex I deficiency develop a fatal encephalomyopathy. Cell Metab 7:312–320
88. Pospisilik JA, Knauf C, Joza N et al (2007) Targeted deletion of AIF decreases mitochondrial oxidative phosphorylation and protects from obesity and diabetes. Cell 131:476–491
89. Diaz F, Garcia S, Hernandez D et al (2008) Pathophysiology and fate of hepatocytes in a mouse model of mitochondrial hepatopathies. Gut 57:232–242
90. Diaz F, Thomas CK, Garcia S, Hernandez D, Moraes CT (2005) Mice lacking COX10 in skeletal muscle recapitulate the phenotype of progressive mitochondrial myopathies associated with cytochrome c oxidase deficiency. Hum Mol Genet 14:2737–2748

91. Fukui H, Diaz F, Garcia S, Moraes CT (2007) Cytochrome c oxidase deficiency in neurons decreases both oxidative stress and amyloid formation in a mouse model of Alzheimer's disease. Proc Natl Acad Sci U S A 104:14163–14168
92. Leveen P, Kotarsky H, Morgelin M, Karikoski R, Elmer E, Fellman V (2011) The GRACILE mutation introduced into Bcs11 causes postnatal complex III deficiency: a viable mouse model for mitochondrial hepatopathy. Hepatology 53:437–447
93. Garcia S, Diaz F, Moraes CT (2008) A 3′ UTR modification of the mitochondrial rieske iron sulfur protein in mice produces a specific skin pigmentation phenotype. J Invest Dermatol 128:2343–2345
94. Brandt U, Haase U, Schagger H, von Jagow G (1991) Significance of the "Rieske" iron-sulfur protein for formation and function of the ubiquinol-oxidation pocket of mitochondrial cytochrome c reductase (bc1 complex). J Biol Chem 266:19958–19964
95. Gurung B, Yu L, Xia D, Yu CA (2005) The iron-sulfur cluster of the Rieske iron-sulfur protein functions as a proton-exiting gate in the cytochrome bc(1) complex. J Biol Chem 280:24895–24902

Chapter 15
Mitochondrial Cytochrome *c* Oxidase Assembly in Health and Human Diseases

Flavia Fontanesi and Antoni Barrientos

Introduction

Defects in the assembly and function of cytochrome *c* oxidase (COX) or mitochondrial respiratory chain complex IV are a frequent cause of oxidative phosphorylation defects in humans. Patients suffering from the associated mitochondrial diseases present heterogeneous clinical phenotypes ranging from encephalomyopathy, hypertrophic cardiomyopathy and liver disease to Leigh syndrome [1–4]. Over the last 25 years, the progress made on the characterization of COX structure and function [5–7], and the process of COX biogenesis [8–10] have allowed new insight into the molecular basis of these human disorders.

COX is the terminal oxidase of the mitochondrial respiratory chain (MRC), which catalyzes the reduction of dioxygen to water by ferrocytochrome *c* [11]. This enzymatic reaction serves as one of the coupling sites of the respiratory chain where the energy of the redox reaction is converted in an electrochemical proton gradient, subsequently used by the mitochondrial ATPase to drive the synthesis of ATP [12]. Human COX is formed by 13 subunits of dual genetic origin. The mitochondrial genome encodes the three catalytic core subunits (subunits 1, 2, and 3) while the nuclear genome encodes the remaining subunits that form a protective shield around the catalytic core. The catalytic subunits 1 and 2 (COX1 and COX2) coordinate metal prosthetic groups essential for electron transfer. Subunit 1 contains one redox center formed by a low-spin heme *a* and a second center formed by a high-spin heme a_3 and a Cu(I) atom that forms the Cu_B site. The third redox center of the enzyme

---

A. Barrientos (✉)
Department of Neurology, Biochemistry and Molecular Biology,
University of Miami Miller School of Medicine, 1600 NW 10th Ave.,
RMSB # 2067, Miami, FL 33136, USA
e-mail: abarrientos@med.miami.edu

F. Fontanesi
Department of Neurology, University of Miami Miller School of Medicine,
1600 NW 10th Ave., RMSB # 2067, Miami, FL 33136, USA
e-mail: ffontanesi@med.miami.edu

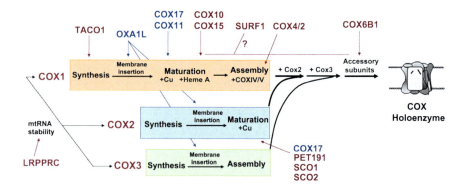

Fig. 15.1 Human cytochrome *c* oxidase biogenesis in health and disease. The cartoon depicts the assembly lines of the three catalytic core subunits and their final formation of the holoenzyme. In *red* are marked the assembly factors and assembly subunits so far found mutated in patients suffering from mitochondrial diseases associated with COX deficiencies

resides in subunit 2 and is formed by two copper ions of mixed valences that form the Cu_A site. Electrons are transferred from ferrocytochrome *c* to the Cu_A active site and subsequently to the heme *a* site. From heme *a*, electrons are intramolecularly transferred to the heme a_3-Cu_B binuclear center where molecular oxygen binds and is reduced to water [13]. Subunit 3 (COX3) does not contain prosthetic groups and it could be involved in the assembly/stability of subunits 1 and 2. It has also been suggested, it could modulate the access of oxygen to the binuclear center [14, 15] or proton translocation through the catalytic subunits [16].

COX biogenesis is believed to be a linear process with the different subunits and cofactors being added in an ordered manner. Based on analyses of assembly intermediates by native electrophoresis, it is known that assembly is initiated around a seed formed by subunit 1 and proceeds with the formation of subassemblies with progressively higher complexity [17]. Maturation of subunit 1 by insertion of its metal cofactors seems to be the first step in the process and does occur once COX1 has been inserted into the membrane [18]. The assembly proceeds with the addition of subunits COX4 and COX5 to subunit 1 and the subsequent addition to subunit 2, 3, and the remaining nuclear-encoded subunits with the exception of subunits COX6a and COX7a/b, which are finally added to complete the holoenzyme [17]. The incorporation of subunit 2 into the second assembly intermediate probably occurs upon its copper metallation because the intermediate COX1-COX4-COX5 accumulates in cells deficient in this function [19, 20]. Heme A biosynthesis and insertion into the apoenzyme, mitochondrial copper homeostasis, and copper delivery to COX subunits are crucial pathways that converge into the initial steps of the linear "subunit-incorporation" assembly process (Fig. 15.1).

In addition to the structural subunits, a large number of nuclear-encoded factors are required for all the steps of the COX assembly process and are generally known as COX assembly factors. Many of these proteins were first identified by the systematic analysis of *Saccharomyces cerevisiae* mutants defective in COX assembly [21, 22].

15 Mitochondrial Cytochrome c Oxidase Assembly in Health and Human Diseases   241

Table 15.1 Human disorders presenting with COX deficiency caused by mutations in COX structural subunits and genes involved in their expression

Gene	Clinical manifestations	References
Mitochondrial-encoded subunits		
COX1	Myopathy, multisystemic mitochondrial disorder, motor neuron-like degeneration, idiopathic sideroblastic anemia, exercise-induced recurrent myoglobinuria, epilepsia partialis continua, rhabdomyolisis, and MELAS-like syndrome	[43–49]
COX2	Myopathy, encephalomyopathy, multisystemic mitochondrial disorder, and LHON	[50–53]
COX3	Encephalopathy, proximal myopathy, myoglobinuria, MELAS, LHON, and Leigh-like syndrome	[54–59]
Nuclear-encoded subunits		
COX6B1	Early-onset leukodystrophic encephalomyopathy	[28]
COX4I2	Exocrine pancreatic insufficiency, dyseythropoietic anemia, calvarial hyperostosis	[29]
Expression of mitochondrial-encoded subunits		
LRPPRC	French-Canadian type Leigh's syndrome	[38, 60, 61]
TACO1	Leigh's syndrome	[39, 62]

Among the COX ancillary factors, several are involved in the expression of the core subunits, copper homeostasis and insertion into subunits 1 and 2, heme biosynthesis and incorporation into subunit 1, and in the formation of assembly intermediates [23, 24]. More recently, the analysis of samples and cultured cell lines derived from the patients suffering from COX deficiencies has allowed for the extensive characterization of some of the human homologues [25, 26] and for the identification of mammalian-specific COX assembly factors [27].

Several cases of maternal inherited isolated COX deficiencies with a very broad spectrum of clinical manifestations have been reported (Table 15.1), associated with mutations in the mitochondrial DNA (mtDNA)-encoded COX subunits 1, 2 and 3 (for a complete list of pathogenic mtDNA mutations see MITOMAP at www.mitomap.org). In contrast, till date only two cases of Mendelian disorders presenting isolated COX deficiency have been assigned to mutations in COX structural subunits, an infantile encephalomyopathy caused by a mutation in *COX6B1* [28] and an exocrine pancreatic insufficiency caused by a mutation in the *COX4I2* gene [29] (Table 15.1). All the remaining cases described so far are due to mutations in ancillary factors. Specifically, mutations have been found in *SURF1* [30, 31] and *PET191* [32] required for the formation of early assembly intermediates, *SCO1* and *SCO2* required for COX copper metallation [33, 34], *COX10* and *COX15*, essential for heme A biosynthesis [35–37] and finally in *LRPPRC*[38] and *TACO1* [39], required for the expression of COX subunits (Tables 15.1–4).

In this chapter, we will briefly review the process of COX assembly, the specific functions of the proteins involved and the molecular basis of human COX deficiencies resulting from mutations in these proteins.

Synthesis of Mitochondrial DNA-Encoded COX Core Subunits in Health and Disease

The expression of the three mtDNA-encoded COX subunits forming the core of the enzyme is far from been understood. In part, this can be explained because given the qualitative differences between the mitochondrial DNA and mRNA of *S. cerevisiae* and mammals, the yeast model has not provided extensive clues concerning human mitochondrial gene expression. For example, splicing factors are not required in humans because *COX1*, as any other gene in the human mitochondrial genome, lacks introns. Mammalian mtDNA transcription is polycistronic and the structural genes are flanked by tRNAs which serve as marks for processing by specific RNAses to free the individual transcripts. In mammalian mitochondria, the mRNAs are subsequently polyadenylated. Detailed information about mammalian mitochondrial transcription can be found in recent reviews [40–42].

Concerning translation, the existence of mRNA-specific translational factors in mammalian mitochondria has long been a subject of speculation. Mammalian mitochondrial mRNAs lack 5′UTRs entirely. Thus, a Shine–Dalgarno interaction between the mRNA and the 12S rRNA is not used during mitochondrial translation. The yeast mitochondrial mRNAs also lack a typical Shine–Dalgarno element. However, in yeast, the mRNA-specific translational activators could be involved in the localization of the small ribosomal subunit near the translational start codon [63]. It has been proposed that such factors, if they exist in mammals, they should interact with the coding regions of mammalian mitochondrial mRNAs. Unfortunately, these translation factors are poorly conserved at the primary sequence level, making their orthologues difficult to be identified in mammalian genomes. However, the human homologue of the yeast nuclear gene *PET309* [64], encoding a COX1-specific translational activator, was recently identified. Mutations in this gene, termed *LRPPRC* (leucine-rich pentatricopeptide repeat cassette) are responsible for the French-Canadian form of Leigh syndrome (LSFC), a neurodegenerative disorder characterized by COX deficiency in brain and liver [38, 60]. LSFC is a common disorder (1:2000) in the Saguenary-Lac St-Jean region of Quebec and the majority of the affected individuals carry the homozygous c.1061C>T nucleotide substitution (1119C>T genomic DNA substitution in exon 9), missense mutation resulting in the p.Ala354Val substitution [38].

LRPPRC is part of the pentatricopeptide repeat (PPR) protein family, a group of RNA-binding proteins characterized by a structural motif of degenerate 35-aminoacids tandem repeats [38]. Although LRPPRC was first identified as a nuclear protein [65], several groups have shown, by subcellular fractionation and immunofluorescence analysis, that most LRPPRC protein localizes to mitochondria [61, 66, 67], where it binds polyadenylated RNAs [66]. In vitro, however, LRPPRC binds preferentially to poly(U) tracts and less efficiently to poly(C) and poly(G) RNAs, while it does not associate with poly(A) RNA homopolymers [66]. The LRPPRC RNA-binding activity was mapped in the C-terminus domain of the protein [66], but the exact LRPPRC target sequences in mitochondrial mRNAs remain to be identified. In a first report by Xu et al. [67], a specific decrease in the steady-state levels of mitochondrial *COX1* and *COX3* mRNA was observed in *LRPPRC*-mutant cells. However, a general decrease in mitochondrial mRNAs, but not rRNAs and tRNAs, associated with mutations in *LRPPRC* or *LRPPRC* depletion by siRNA has been described in several subsequent studies [61, 68, 69]. In particular, Sasarman et al. [61] reported a global reduction in the levels of mitochondrial mRNA, with the exception of *ND3* and *ND6*, encoding for MRC complex I subunits, in LSFC fibroblasts, but unaltered levels of mitochondrial polycistronic transcripts, suggesting that LRPPRC function is limited to mature mRNAs. Moreover, the residual amount of mitochondrial mRNAs correlated with the decrease in LRPPRC level, with COX mRNAs been the most affected. As a result of the mRNA instability, mitochondrial translation is also impaired in mutant cells. In addition, the synthesis of COX subunits, in particular COX1 and COX2, is disproportionately reduced in patient fibroblasts [61]. Currently, it remains unclear why the LRPPRC-dependent translation is poorer or why the mitochondrial COX mRNAs are more unstable. Recently it has been shown that LRPPRC interacts with SLIRP, a stem-loop RNA-binding protein, to form a high molecular weight complex of about 250 kDa that also contains mature mitochondrial mRNAs [61]. Taken together, these data support a post transcriptional role of LRPPRC in mitochondria. Nevertheless, while it is clear that LRPPRC is required for the stability of mature mitochondrial mRNAs, its involvement in mitochondrial translation remains an open question.

A new COX assembly factor has been recently identified by genome-wide linkage analysis in a patient affected by Leigh syndrome associated with an isolated COX deficiency [39]. The product of the gene *CCDC44*, renamed TACO1 for translational activator of COX1, is a soluble protein localized to the mitochondrial matrix. Deletion of the yeast homologue, *YGR021W*, does not produce any respiratory-deficient phenotypes, suggesting a specific role for TACO1 in mammalian mitochondria. Fibroblasts from a patient carrying a TACO1 frameshift mutation present normal levels of *COX1* mRNA but approximately 65% reduction in newly synthesized COX1 protein. Weraarpachai et al. [39] speculated that TACO1 could act by securing an accurate initiation of *COX1* mRNA translation or by stabilizing the elongation polypeptide and ensuring completion of its translation.

The identification of mammalian-specific mitochondrial COX gene translational activators and the characterization of their mechanism of action are expected to provide crucial information concerning how translation of mammalian mitochondrial mRNAs is activated.

Mitochondrial Copper Metabolism and Insertion into COX in Health and Disease

Copper metallation of mitochondrial COX subunits 1 and 2 is required for COX catalytic function and also for its biogenesis, assembly and stability (reviewed in [70]).

Mitochondria have their own copper trafficking and distribution network to ensure the metallation of several copper metalloenzymes, including COX and of the small portion of the Cu-Zn superoxide dismutase (SOD1) located in the intermembrane space. The power of yeast genetics has enabled the identification of several copper metallochaperones required for COX assembly. However, the understanding of how copper is delivered to mitochondria remains very limited.

Once in the IMS, copper is inserted into COX by the action of at least three key proteins conserved from yeast to human. COX copper metallation involves the copper chaperone COX17, a small hydrophilic protein that binds copper ions and also contains a twin CX_9C structural motif [71]. COX17 transfers copper ions to two additional chaperones [72] that facilitate copper insertion into the COX Cu_A and Cu_B active sites, respectively SCO1 [73] and COX11 [74, 75]. These proteins are anchored to the mitochondrial inner membrane through a transmembrane α-helix and expose their copper-binding sides in the IMS where copper transfer occurs [76, 77]. It remains unclear, however, how copper is transferred from COX17 to these proteins because physical interactions among them have not been detected [72].

The human COX17 homologue shares 48 % sequence identity with its yeast counterpart. Depletion of *COX17* by siRNA in HeLa cells cause the accumulation of a COX1-containing COX assembly intermediates devoid of COX2 [78]. This observation suggested a role for human COX17 in the maturation of COX2 but not COX1. Nevertheless, it is possible that COX1 protein is more stable than COX2 in the absence of copper.

SCO1 was originally identified as essential for COX assembly in yeast [79] and subsequently as a multicopy suppressor of a *COX17* null mutant strain [80]. Yeast SCO1 transfers copper from COX17 to COX2 and directly interacts with COX2 [81]. SCO1 has a metal-binding thioredoxin-like CX_3C motif analogous to the copper-binding motif of COX2 and this motif is essential for its function as demonstrated by site-direct mutagenesis [82]. Given the structural similarity of SCO1 with the protein family of di sulfide reductases it has been suggested that SCO1 could be involved in the reduction of cysteines in the COX2 copper-binding site [83], which is necessary for the cofactor incorporation [84, 85]. SCO1 has the ability to form homodimeric complexes [81], which could facilitate the performance of both COX assembly and cysteine reduction by the collaborative action of each monomer.

Yeast *SCO1* has a highly conserved homologue, *SCO2* [86], whose deletion does not affect COX assembly [80]. Although *SCO2* over-expression does not suppress the COX assembly defect of a *SCO1* null mutant strain, it is able to partially rescue a *SCO1* point mutant [80]. In addition, *SCO2* over-expression also suppresses *COX17* mutations, although less efficiently than *SCO1*, and only in the presence of exogenous copper [80]. These data were interpreted to indicate that yeast SCO1 and SCO2 have overlapping but nonidentical functions [80].

Humans also have two SCO proteins, SCO1 and SCO2, both homologues of yeast SCO1 and that probably originated from a duplication that occurred separately in the two organisms [33]. Both SCO proteins are essential for COX assembly and mutations in *SCO1* and *SCO2* genes result in severe cardioencephalomyopathies. In contrast with their yeast homologues, human SCO1 and SCO2 have been shown to perform independent, but cooperative functions [87, 88]. As a result, overexpression of each SCO protein in fibroblasts from patients with mutation in the other SCO protein results in a dominant negative phenotype [87]. SCO proteins contain a CX_3C copper-binding motif, shown to be essential for their function in COX2 biogenesis [72, 88]. Furthermore, both SCO1 and SCO2 proteins have a higher affinity for copper than COX17 [89], that allow for the quantitative transfer of Cu(I) from COX17 to SCO1 and SCO2 [71, 90]. SCO1 protein exists as a mixed population of oxidized and reduced thiols, the proportion of which depends upon the presence of a functional SCO2 protein [88]. These data, together with the observation that SCO proteins contain a highly conserved thioredoxin domain [91], has brought Leary et al. [88] to propose a thiol-disulphide oxidoreductase function for SCO2. Following the proposed model, after COX2 metallation by SCO2 and SCO1-dependent simultaneous or sequential copper insertion, SCO2 reoxidize SCO1 cysteines, a reaction that allows to reset both proteins for further round of COX2 biogenesis [88]. While both SCO proteins are required for copper transfer to COX2, SCO2 is also necessary for COX2 synthesis, since SCO2 depletion decrease the accumulation of newly synthesized COX2 in culture cells [88]. Finally, both SCO1 and SCO2 play additional roles in the maintenance of cellular copper homeostasis [92]. Recently, the mitochondrial copper pool of *SCO1* and *SCO2*-mutant fibroblasts has been specifically estimated using a targetable fluorescent sensor and no alteration in the mitochondrial copper content of mutant cells has been observed compare with control cells despite a global cellular copper deficiency, suggesting a tight regulation of mitochondrial copper homeostasis [93]. In contrast, reduced amounts of mitochondrial copper, but not total copper content in brain, liver, muscle, and heart, was reported in mice models homozygous or heterozygous for the pathogenic *SCO2* E129K allele [94].

Defects in mitochondrial copper transport and delivery to COX have been found to be responsible for several different clinical presentations of COX deficiency (Table 15.2). Specifically, mutations in the genes involved in copper insertion into the Cu_A site, either *SCO1* or *SCO2,* lead to decreased COX activity and early death. Both proteins are ubiquitously expressed in all tissues [33, 95]. However, mutations in each gene lead to significantly different phenotypes. Mutations in the *SCO1* gene, located in human chromosome 17p13.1, lead to fatal infantile hepato-encephalomyopathy [34], and the phenotype resulting from mutations in the *SCO2* gene, located on human chromosome 22q13, causes encephalocardiomyopathy, hyperthropic cardiomyopathy and in rare cases spinal muscular atrophy (Table 15.2). The reason for the different tissue involvement in the two disorders is currently unknown.

Mutations in *SCO1* have been found in a single family with two affected children who died within the first two months of life [34]. Mutation screening revealed compound heterozygosity for *SCO1* gene mutations in the patients [34]. The two alleles consisted of a 2 bp frameshift deletion resulting in a premature stop codon and a change of a highly conserved proline 174 to leucine (p.P174L) in a position adjacent to the CxxxC copper-binding domain of the protein, and thus probably affecting its three-dimensional conformation. Immortalized fibroblasts from a different *SCO1*-deficient patient showed a severe decrease in COX activity that was partially rescued by overexpression of p.P174L SCO1 [109], suggesting that the mutated protein retains some residual activity. Although the mutant protein retained the ability to bind

Table 15.2 Human disorders presenting with COX deficiency caused by mutations in conserved COX copper delivery genes

Gene	Clinical manifestations	References
SCO1	Hepatoencephalomyopathy	[34]
SCO2	Fatal infantile cardioencephalomyopathy, hypertrophic cardiomyopathy, delayed infantile onset of cardiomyopathy and neuropathy, spinal muscular atrophy type I-like, encephalomyopathy, SMA-like presentation	[33, 96–101]; [102–108]

Cu(I) and Cu(II) normally when expressed in bacteria, the COX17-mediated copper transfer was severely compromised both *in vitro* and in a yeast cytoplasmic assay [109]. A yeast *SCO1* allele carrying the corresponding p.P153L substitution was able to complement the respiratory defect of a strain carrying a *SCO1* null mutation; however, it failed to suppress the phenotype of cells harboring a C57Y allele of *COX17*, which severely affects COX17 function [109]. These results were interpreted to indicate that the p.P174L mutation attenuates a transient interaction with COX17 that is necessary for copper transfer [109]. The studies in yeast suggest that alterations in COX17-mediated copper metallation of SCO1, as well as the subsequent failure of Cu$_A$ site maturation, is the basis for the inefficient COX assembly in SCO1-mutant patients [109].

Isolated COX deficiency caused by mutations in *SCO2* seems to be more frequent than mutations in *SCO1*. More than 50 cases have been described in the literature and most of them are listed in Table 15.2. Interestingly, all but one patient [106] with *SCO2* mutations reported till date carry at least one allele with the same common mutation: c.418G>A (1541G>A genomic DNA substitution), converting glutamic acid at amino acid position to lysine (p.E140K) [33, 96–99, 101–105, 107, 108, 110]. Although the disease is fatal for all affected children, the rate of progression of the clinical course of the disease may differ depending on the genotype. Patients carrying the prevalent p.E140K mutation in homozygosity start within the first 6 months of life with progressive encephalomyopathy and delayed onset (8–18 months) of hypertrophic cardiomyopathy [97, 101, 108]. In contrast, the phenotype of patients compound heterozygous for the p.E140K are generally characterized by rapidly progressive hypertrophic cardiomyopathy accompanied by severe neurological involvement [33, 96, 99, 101–105, 107, 108], and death usually occurs within the first 6 months of life.

There is no clear explanation for the observation of a common allele in all reported *SCO2* patients. The pathogenicity of the p.E140K mutation is probably related to its proximity to the predicted CxxxC copper-binding domain of the protein [33, 96], thereby disrupting the copper-binding ability of SCO2 [111]. However, the homologous p.E140K mutation in yeast *SCO1* does not produce a respiratory defect [112]. This suggests that, although the SCO proteins of yeast and human could work on slightly different copper-delivery pathways, the p.E140K is a comparatively mild mutation. The mutated human protein probably retains some residual function, which could explain why patients homozygous for this mutation have a later onset and more prolonged course of disease. In addition, the mutation results in a significant reduction in protein content in fibroblasts [87], which may be more relevant than the effect of the mutation on copper binding [105].

Heme A Biosynthesis and Insertion into COX Subunit 1 in Health and Disease

Insertion of heme A into COX subunit 1 is an essential requirement for COX activity and assembly (reviewed in [8, 24, 113]). Heme A is a unique heme compound present exclusively in COX. It differs from protoheme (heme B) because it has a farnesyl instead of a vinyl group at carbon C_2 and a formyl instead of a methyl group at carbon C_8 [114]. In yeast, the conversion of heme B into heme A seems to occur in three steps. The first step of the heme A biosynthetic pathway is the conversion of heme B in heme O catalyzed by the farnesyl-transferase COX10 [115]. The subsequent oxidation of heme O to heme A occurs in two discrete monooxygenase steps. The first consists of a monooxygenase-catalyzed hydroxylation of the methyl group at C_8, resulting in an alcohol that is then further oxidized to an aldehyde by a dehydrogenase. This first step is catalyzed by COX15, in concert with ferredoxin (Yah1) and the putative ferredoxin reductase Arh1, which are probably necessary for the electron supply to the oxygenase COX15 [116–118]. The identity of the putative gene product involved in the oxidation of the alcohol resulting from the COX15 action to the corresponding aldehyde to yield heme A remains unclear. However, an alternative model suggested that COX15 may utilize two successive monooxygenase reactions to generate a germinal diol, which could then spontaneously be dehydrated [117]. Studies in *B. subtillis* have provided strong evidence indicating that COX15 oxidizes heme O to heme A via successive monooxygenase reactions [117].

Human orthologues of *COX10* and *COX15* have been described. There is no doubt that they are functionally equivalent to their yeast counterparts. Human *COX10* was actually identified in a functional complementation assay of a yeast *COX10* mutant with a human cDNA library [119]. Overexpression of human *COX15* is able to rescue the respiratory defect of a yeast *COX15* mutant (Tzagoloff A, personal communication). Till date, however, mammalian homologues of Yah1 and Arh1 have not been identified.

Mutations in *COX10* and *COX15* have been shown to underlie a few cases of human COX deficiencies (Table 15.3). The range of phenotypes in which these mutations manifest is broad and heterogeneous, mainly involving the nervous system and heart.

The first case of mitochondrial disorder caused by a mutation in *COX10* gene was described in affected members of a large consanguineous family presenting ataxia and tubulopathy associated with COX deficiency [35]. By linkage analysis, the mutated gene was mapped in a region on chromosome 17 containing both *SCO1* and *COX10* genes. Sequencing of the seven coding exons of *COX10* revealed a homozygous c. 612C>A mutation in exon 4, resulting in the change of a conserved asparagine at amino acid position 240 to lysine (p.N204K). To further test whether this mutation

Table 15.3 Human disorders presenting with COX deficiency caused by mutations in conserved heme a biosynthesis genes

Gene	Clinical manifestations	References
COX10	Tubulopathy, leukodystrophy, Leigh's syndrome, Leigh-like syndrome, and hypertrophic cardiomyopathy	[35, 36, 120, 121]
COX15	Hypertrophic cardiomyopathy, Leigh's syndrome, atypical long survival Leigh's syndrome, and encephalopathy	[37, 122–124]

was responsible for the COX deficiency, Valnot et al. [35] took advantage of the functional equivalence of the yeast and human genes to test the ability of the $COX10^{N204K}$ mutant to complement the respiratory defect of a yeast *COX10* null mutant. While the wild-type *COX10* human gene was able to partially rescue the respiratory-deficient phenotype of a *COX10* null mutant strain either in single copy or in high-copy plasmid, the mutant $COX10^{N204K}$ failed to do it when expressed in low or single-copy plasmids. The mutant protein, however, retained some residual activity and was able to partially restore the respiratory defect of the *COX10* null mutant when expressed in a multicopy plasmid. The results obtained from these functional complementation assays in yeast strongly support the conclusion that the COX deficiency in those patients was a result of the *COX10* mutation. Till date, four other patients with isolated COX deficiency and different clinical presentations (Table 15.3) carrying either compound heterozygous or homozygous missense mutations in conserved positions of the COX10 protein have been described.

Mutations in *COX15* present as fatal infantile hypertrophic cardiomyopathy [37, 123] and Leigh syndrome [122, 124], similar to those seen in cases with *COX10* mutations. Four patients have been described (Table 15.3), all except one who carries heterozygous *COX15* mutations. Consistently with the severe COX deficiency, a reduction in the level of heme A was measured in the heart tissue of one patient. Overexpression of *COX15* in patient fibroblasts does not fully rescue COX assembly; in particular, it increases the levels of heme A to 65 % of control. *COX15* has two splice variants; COX15.1 and COX15.2, the roles of which are unknown. Antonicka et al. [37] used COX15.1 for their complementation experiments. The results suggest that the two splicing variants are necessary for complete activity.

Analysis of COX by Blue Native-PAGE from fibroblasts obtained from patients carrying either *COX10* or *COX15* mutations revealed very low levels of fully assembled enzyme complex. In addition, the COX assembly intermediate formed by COX subunits 1, 4 and 5 was not detected [37, 125]. This result suggests that insertion of heme A in COX1 is an early event in COX biogenesis, which occurs before the formation of the COX1-COX4-COX5 subassembly. Moreover, given COX1 localization within the assembled protein buried in the inner mitochondrial membrane, it seems likely that heme A insertion occurs during translation or membrane insertion. In addition, due to heme A reactive nature it is anticipated that one or more heme-binding proteins assist the transfer of heme A into COX1. However, the mechanism for heme A incorporation into COX1 and the players involved remain largely unknown.

The high reactivity of the heme A moiety would also predict the evolution of an adaptive system to coordinate heme A biosynthesis with its incorporation into COX1.

In yeast, the biosynthesis of heme A has been shown to be regulated by downstream events in the COX assembly process [126]. A drastic reduction of steady-state levels of heme A was observed in most COX assembly mutants. The overexpression of *COX15* significantly increased the amount of heme A in these mutants including those in which COX1 is not synthesized [126]. This observation has suggested that the absence of heme A in the mutants is not due to a rapid turnover of the cofactor in the absence of COX1, but rather to a feedback regulation of the heme A synthesis when the COX assembly process is impaired. In addition to the reduced heme A level, COX mutants, with the exception of *COX10*, also show accumulation of heme O, indicating that this compound is stable. Heme O level was instead very low in a *COX15* null mutant and this phenotype was not rescued by *COX10* overexpression [126]. Taken together, these data suggest that the first step of the heme A biosynthesis is also positively regulated in a COX15 dependent manner. Notably, it has been recently reported that COX10 forms an oligomeric complex, absent in COX mutants where, as mentioned above, heme O is accumulated and COX1 fails to be assembled in early COX assembly intermediates [127]. This observation indicates that COX10 oligomerization is not required for its function as a heme O synthase. It is instead tempting to speculate that it could be involved in coupling of heme A biosynthesis and COX assembly, although the COX assembly intermediate that could drive this feedback regulation remains to be identified.

The regulation of heme A synthesis could be different in mammalian cells. In fact, contrary to what was described in the yeast *COX15* mutant strain, *COX15*-deficient fibroblasts have decreased heme A levels but an increase in heme O accumulation compared with control cells [37]. However, these cells retain some COX15 activity, as it is suggested by the presence of residual fully assembled and functional COX in fibroblasts and heart tissue from the patient [37].

Formation of COX Assembly Intermediates in Health and Disease

Leigh syndrome, also known as infantile subacute necrotizing encephalopathy, is the most frequent mitochondrial disorder in infancy [128]. It is characterized by bilaterally symmetric necrotic lesions in the subcortical regions of the brain. The most common phenotype associated with the disease is a variable but usually severe reduction of the MRC complex IV. COX deficiency is an autosomal recessive trait, and most patients belong to a single genetic complementation group. A locus for the disease has been mapped to chromosome 9p34 [30, 31]. Analysis of this region revealed the presence of a candidate gene *SURF1*, which was shown to be mutated in most of the patients affected by Leigh syndrome associated with systemic COX deficiency (Table 15.4). Additional rare clinical manifestations associated with mutations in the *SURF1* gene have also been reported. They include hypertrichosis, leukodystrophy and renal tubulopathy [129–132].

SURF1 is conserved from bacteria to human. Despite the effort of several laboratories and the growing amount of data obtained using bacterial, yeast, and cell culture

Table 15.4 Human disorders presenting with COX deficiency caused by mutations in conserved COX assembly factor genes

Gene	Clinical manifestations	References
SURF1	Leigh's syndrome, villous atrophy Hypertrichosis, peripheral neuropathy, leukodystrophy, cerebral atrophy, leigh syndrome with ragged red fibers, renal tubulopathy, late-onset Leigh-like syndrome	[30, 31, 129, 130, 139, 140, 142, 143, 147, 154]; [98, 110, 121, 155–163]; [131, 132, 145, 164, 165]
PET191 (C2orf64)	Fatal neonatal cardiomyopathy	[32]

models, the exact function of SURF1 in COX assembly remains to be fully understood. Shy1 is the yeast homolog of human SURF1 [133]. Shy1 has been proposed to be involved in either the formation or the stabilization of the heme a_3 site. This hypothesis was initially based on the observation that in a *Rhodobacter sphaeroides surf1* null mutant only 50 % of the COX1 CuB—heme a_3 binuclear center contains heme [134]. Noticeably, in this mutant, the heme a site was found to be formed. This result opens the possibility of the existence of two heme A insertases. Although there has not been any report of other candidate proteins involved in heme delivery to COX1, it is not likely that Shy1/SURF1 is the only protein involved in this process due to the 10–15 % residual COX activity detected in the yeast *shy1* null mutant and SURF1-deficient human cells. Recent data on *Paracoccus denitrificans* SURF1 isoforms, have shown that when coexpressed in *Escherichia coli* together with enzymes for heme A synthesis, they have the ability to bind heme A in a 1:1 stoichiometry with Kd values in the submicromolar range [135]. Nonetheless, these findings have yet to be confirmed *in vivo* and in eukaryotes. The bacterial study also identified a conserved histidine as a residue crucial for heme binding [135]. *COX10* is a weak multicopy suppressor of yeast *shy1* mutant cells, thus connecting Shy1 to heme biosynthesis [136]. However, mutations of either of the two conserved His residues in yeast Shy1 did not significantly affect its function [127]. Alternatively, Shy1 function may enhance the stabilization of the heme a_3 site rather than playing a direct role in heme A delivery. Moreover, studies of COX assembly intermediates in fibroblasts from human patients carrying mutations in *SURF1*, have disclosed the accumulation of COX1 alone or in an early intermediate containing human subunits 4 and 5 (yeast subunits 5a and 6) [31, 125, 137], contrary to what has been observed in fibroblasts from patients with *COX10* and *COX15* mutations [37, 125]. This observation has suggested that SURF1 could play a role in the incorporation of subunit 2 into these nascent intermediates. A role of Shy1 in incorporation of additional COX subunits into early COX1 subassemblies has not been fully discarded. For example, overexpression of *COX5a* and *COX6* significantly suppresses the respiratory defect of *shy1* mutant cells [138]. Enhanced levels of these subunits may stabilize COX1 in the *shy1* mutant, enabling progression to later stages of COX assembly. We could speculate that, for example, addition of COX5a, whose transmembrane helix is tightly packed against COX1 could contribute to the stabilization of the metal centers in COX1.

Human SURF1 is a 300 aminoacids protein anchored to the mitochondrial inner membrane by two transmembrane domains. The C-terminus transmembrane domain is necessary for protein stability, since mutant polypeptides lacking this domain are rapidly degraded [139–141]. In general, the majority of the mutations reported in patients are nonsense mutations, whereas missense mutations are rare. Moreover, in the case of the pathogenic p.G124E and p.G124R substitutions, located in the IMS loop between the two transmembrane domains of the protein [142, 143], studies in yeast have shown that the mutated full-length polypeptides are imported but fail to accumulate in mitochondria [144]. On the contrary, the protein carrying the missense mutation p.Y274D is stable and properly localized to the mitochondria inner membrane [144]. This *SURF1* allele has been associated with a mild, late-onset Leigh syndrome [145]. It has been proposed that the presence of missense mutations in the *SURF1* gene may correlate with a milder course of the disease and longer survival of Leigh syndrome patients [129, 145]. However, cases presenting late-onset or milder disease's symptoms associated with truncated or unstable SURF1 protein have been reported [146], suggesting a more complex scenario where different genetic backgrounds could influence the capacity of the cells to tolerate energy deprivation or play a role in the development of compensatory mechanisms.

Recently, a homozygous mutation in C2orf64, the human homologue of yeast *PET191* gene, has been identified in two siblings affected by fatal neonatal cardiomyopathy associated with severe COX deficiency [32]. The yeast Pet191 protein is a small protein of the mitochondrial inner membrane space containing twin CX_9C motifs. It is required for COX assembly and it has been proposed to be part of a copper-trafficking pathway. However, its specific function and the function of its human homologue still remain unknown. Accumulation of a small COX assembly intermediate containing COX1, but not COX2, COX4 or COX5 in patient fibroblasts suggests that PET191 is involved in an early step of COX assembly [32].

Concluding Remarks

Defects in mitochondrial gene expression, heme A biosynthesis, copper insertion into the cytochrome *c* oxidase apoenzyme, and COX assembly intermediate formation are associated with human diseases. Despite research efforts in many laboratories worldwide, the molecular basis of autosomal recessive COX-deficiency associated diseases remains unknown in a large number of patients [166, 167]. This suggests the involvement of new COX assembly factors remaining to be characterized. In fact, gene products required for copper import to mitochondria, copper transport across the inner mitochondrial membrane, the final step of heme A biosynthesis and heme A insertion into COX subunit 1 have yet to be elucidated. In addition, little is known about the players and the mechanisms involved in mammalian mitochondrial translation, or whether new assembly factors, yet to be identified, are required for assembly of COX into respiratory chain supercomplexes. The characterization of new factors involved in these pathways will contribute to a better understanding of

COX assembly and will yield new candidates when screening for genes responsible for human disorders associated with COX deficiency.

Acknowledgments Our research is supported by National Institutes of Health Research Grant GM071775A (to A.B.) and a Research Grant (to A.B.) and a Development Grant (to F.F.) from the Muscular Dystrophy Association.

References

1. Shoubridge EA (2001) Cytochrome c oxidase deficiency. Am J Med Genet 106:46–52
2. Solans A, Zambrano A, Barrientos A (2004) Cytochrome c oxidase deficiency: from yeast to human. Preclinica 2:336–348
3. Zee JM, Glerum DM (2006) Defects in cytochrome oxidase assembly in humans: lessons from yeast. Biochem Cell Biol 84:859–869
4. Pecina P, Houstkova H, Hansikova H et al (2004) Genetic defects of cytochrome c oxidase assembly. Physiol Res 53:213–223
5. Tsukihara T, Aoyama H, Yamashita E et al (1996) The whole structure of the 13-subunit oxidized cytochrome c oxidase at 2.8 A. Science 272:1136–1144
6. Ostermeier C, Harrenga A, Ermler U et al (1997) Structure at 2.7 A resolution of the Paracoccus denitrificans two-subunit cytochrome c oxidase complexed with an antibody FV fragment. Proc Natl Acad Sci U S A 94:10547–10553
7. Yoshikawa S, Shinzawa-Itoh K, Nakashima R et al (1998) Redox-coupled crystal structural changes in bovine heart cytochrome c oxidase. Science 280:1723–1729
8. Barrientos A, Barros MH, Valnot I et al (2002) Cytochrome oxidase in health and disease. Gene 286:53–63
9. Fontanesi F, Soto IC, Horn D et al (2006) Assembly of mitochondrial cytochrome c oxidase, a complicated and highly regulated cellular process. Am J Physiol Cell Physiol 291:C1129–1147
10. Cobine PA, Pierrel F, Winge DR (2006) Copper trafficking to the mitochondrion and assembly of copper metalloenzymes. Biochim Biophys Acta 1763:759–772
11. Saraste M (1990) Structural features of cytochrome oxidase. Q Rev Biophys 23:331–366
12. Mitchell P, Moyle J (1967) Chemiosmotic hypothesis of oxidative phosphorylation. Nature 213:137–139
13. Babcock GT, Wikstrom M (1992) Oxygen activation and the conservation of energy in cell respiration. Nature 356:301–309
14. Brunori M, Antonini G, Malatesta F et al Cytochrome-c oxidase. Subunit structure and proton pumping. Eur J Biochem 169:1–8
15. Riistama S, Puustinen A, Garcia-Horsman A et al (1996) Channelling of dioxygen into the respiratory enzyme. Biochim Biophys Acta 1275:1–4
16. Hosler JP (2004) The influence of subunit III of cytochrome c oxidase on the D pathway, the proton exit pathway and mechanism-based inactivation in subunit I. Biochim Biophys Acta 1655:332–339
17. Nijtmans LG, Taanman JW, Muijsers AO et al (1998) Assembly of cytochrome-c oxidase in cultured human cells. Eur J Biochem 254:389–394
18. Khalimonchuk O, Bestwick M, Meunier B et al (2010) Formation of the redox cofactor centers during Cox1 maturation in yeast cytochrome oxidase. Mol Cell Biol 30:1004–1017
19. Williams SL, Valnot I, Rustin P et al (2004) Cytochrome c oxidase subassemblies in fibroblast cultures from patients carrying mutations in COX10, SCO1, or SURF1. J Biol Chem 279:7462–7469

20. Stiburek L, Vesela K, Hansikova H et al (2005) Tissue-specific cytochrome c oxidase assembly defects due to mutations in SCO2 and SURF1. Biochem J 392:625–632
21. Tzagoloff A, Dieckmann CL (1990) PET genes of Saccharomyces cerevisiae. Microbiol Rev 54:211–225
22. McEwen JE, Ko C, Kloeckner-Gruissem B et al (1986) Nuclear functions required for cytochrome c oxidase biogenesis in Saccharomyces cerevisiae. Characterization of mutants in 34 complementation groups. J Biol Chem 261:11872–11879
23. Fontanesi F, Soto IC, Horn D et al (2006) Assembly of mitochondrial cytochrome c-oxidase, a complicated and highly regulated cellular process. Am J Physiol Cell Physiol 291:C1129–1147
24. Soto IC, Fontanesi F, Liu J et al (2012) Biogenesis and assembly of eukaryotic cytochrome C oxidase catalytic core. Biochim Biophys Acta 1817:883–897
25. Diaz F (2010) Cytochrome c oxidase deficiency: patients and animal models. Biochim Biophys Acta 1802:100–110
26. Barrientos A, Gouget K, Horn D et al (2009) Suppression mechanisms of COX assembly defects in yeast and human: insights into the COX assembly process. Biochim Biophys Acta 1793:97–107
27. Weraarpachai W, Antonicka H, Sasarman F et al (2009) Mutation in TACO1, encoding a translational activator of COX I, results in cytochrome c oxidase deficiency and late-onset Leigh syndrome. Nat Genet 41:833–837
28. Massa V, Fernandez-Vizarra E, Alshahwan S et al (2008) Severe infantile encephalomyopathy caused by a mutation in COX6B1, a nucleus-encoded subunit of cytochrome c oxidase. Am J Hum Genet 82:1281–1289
29. Shteyer E, Saada A, Shaag A et al (2009) Exocrine pancreatic insufficiency, dyserythropoeitic anemia, and calvarial hyperostosis are caused by a mutation in the COX4I2 gene. Am J Hum Genet 84:412–417
30. Zhu Z, Yao J, Johns T et al (1998) SURF1, encoding a factor involved in the biogenesis of cytochrome c oxidase, is mutated in Leigh syndrome. Nat Genet 20:337–343
31. Tiranti V, Hoertnagel K, Carrozzo R et al (1998) Mutations of SURF-1 in Leigh disease associated with cytochrome c oxidase deficiency. Am J Hum Genet 63:1609–1621
32. Huigsloot M, Nijtmans LG, Szklarczyk R et al (2011) A mutation in C2orf64 causes impaired cytochrome c oxidase assembly and mitochondrial cardiomyopathy. Am J Hum Genet 88:488–493
33. Papadopoulou LC, Sue CM, Davidson MM et al (1999) Fatal infantile cardioencephalomyopathy with COX deficiency and mutations in SCO2, a COX assembly gene. Nat Genet 23:333–337
34. Valnot I, Osmond S, Gigarel N et al (2000) Mutations of the SCO1 gene in mitochondrial cytochrome c oxidase deficiency with neonatal-onset hepatic failure and encephalopathy. Am J Hum Genet 67:1104–1109
35. Valnot I, von Kleist-Retzow JC, Barrientos A et al (2000) A mutation in the human heme A:farnesyltransferase gene (COX10) causes cytochrome c oxidase deficiency. Hum Mol Genet 9:1245–1249
36. Antonicka H, Leary SC, Guercin GH et al (2003) Mutations in COX10 result in a defect in mitochondrial heme A biosynthesis and account for multiple, early-onset clinical phenotypes associated with isolated COX deficiency. Hum Mol Genet 12:2693–2702
37. Antonicka H, Mattman A, Carlson CG et al (2003) Mutations in COX15 produce a defect in the mitochondrial heme biosynthetic pathway, causing early-onset fatal hypertrophic cardiomyopathy. Am J Hum Genet 72:101–114
38. Mootha VK, Lepage P, Miller K et al (2003) Identification of a gene causing human cytochrome c oxidase deficiency by integrative genomics. Proc Natl Acad Sci U S A 100:605–610
39. Weraarpachai W, Antonicka H, Sasarman F et al (2009) Mutation in TACO1, encoding a translational activator of COX I, results in cytochrome c oxidase deficiency and late-onset Leigh syndrome. Nat Genet 41:833–837

40. Scarpulla RC (2008) Transcriptional paradigms in mammalian mitochondrial biogenesis and function. Physiol Rev 88:611–638
41. Asin-Cayuela J, Gustafsson CM (2007) Mitochondrial transcription and its regulation in mammalian cells. Trends Biochem Sci 32:111–117
42. Shutt TE, Shadel GS (2010) A compendium of human mitochondrial gene expression machinery with links to disease. Environ Mol Mutagen 51:360–379
43. Bruno C, Martinuzzi A, Tang Y et al (1999) A stop-codon mutation in the human mtDNA cytochrome c oxidase I gene disrupts the functional structure of complex IV. Am J Hum Genet 65:611–620
44. Varlamov DA, Kudin AP, Vielhaber S et al (2002) Metabolic consequences of a novel missense mutation of the mtDNA CO I gene. Hum Mol Genet 11:1797–1805
45. Comi GP, Bordoni A, Salani S et al (1998) Cytochrome c oxidase subunit I microdeletion in a patient with motor neuron disease. Ann Neurol 43:110–116
46. Gattermann N, Retzlaff S, Wang YL et al (1997) Heteroplasmic point mutations of mitochondrial DNA affecting subunit I of cytochrome c oxidase in two patients with acquired idiopathic sideroblastic anemia. Blood 90:4961–4972
47. Karadimas CL, Greenstein P, Sue CM et al (2000) Recurrent myoglobinuria due to a nonsense mutation in the COX I gene of mitochondrial DNA. Neurology 55:644–649
48. Kollberg G, Moslemi AR, Lindberg C et al (2005) Mitochondrial myopathy and rhabdomyolysis associated with a novel nonsense mutation in the gene encoding cytochrome c oxidase subunit I. J Neuropathol Exp Neurol 64:123–128
49. Tam EW, Feigenbaum A, Addis JB et al (2008) A novel mitochondrial DNA mutation in COX1 leads to strokes, seizures, and lactic acidosis. Neuropediatrics 39:328–334
50. Rahman S, Taanman JW, Cooper JM et al (1999) A missense mutation of cytochrome oxidase subunit II causes defective assembly and myopathy. Am J Hum Genet 65:1030–1039
51. Campos Y, Garcia-Redondo A, Fernandez-Moreno MA et al (2001) Early-onset multisystem mitochondrial disorder caused by a nonsense mutation in the mitochondrial DNA cytochrome C oxidase II gene. Ann Neurol 50:409–413
52. Clark KM, Taylor RW, Johnson MA et al (1999) An mtDNA mutation in the initiation codon of the cytochrome C oxidase subunit II gene results in lower levels of the protein and a mitochondrial encephalomyopathy. Am J Hum Genet 64:1330–1339
53. Zhadanov SI, Atamanov VV, Zhadanov NI et al (2006) De novo COX2 mutation in a LHON family of Caucasian origin: implication for the role of mtDNA polymorphism in human pathology. J Hum Genet 51:161–170
54. Keightley JA, Hoffbuhr KC, Burton MD et al (1996) A microdeletion in cytochrome c oxidase (COX) subunit III associated with COX deficiency and recurrent myoglobinuria. Nat Genet 12:410–416
55. Manfredi G, Schon EA, Moraes CT et al (1995) A new mutation associated with MELAS is located in a mitochondrial DNA polypeptide-coding gene. Neuromuscul Disord 5:391–398
56. Hanna MG, Nelson IP, Rahman S et al (1998) Cytochrome c oxidase deficiency associated with the first stop-codon point mutation in human mtDNA. Am J Hum Genet 63:29–36
57. Choi BO, Hwang JH, Kim J et al (2008) A MELAS syndrome family harboring two mutations in mitochondrial genome. Exp Mol Med 40:354–360
58. Johns DR, Neufeld MJ (1993) Cytochrome c oxidase mutations in Leber hereditary optic neuropathy. Biochem Biophys Res Commun 196:810–815
59. Tiranti V, Corona P, Greco M et al (2000) A novel frameshift mutation of the mtDNA COIII gene leads to impaired assembly of cytochrome c oxidase in a patient affected by Leigh-like syndrome. Hum Mol Genet 9:2733–2742
60. Merante F, Petrova-Benedict R, MacKay N et al (1993) A biochemically distinct form of cytochrome oxidase (COX) deficiency in the Saguenay-Lac-Saint-Jean region of Quebec. Am J Hum Genet 53:481–487
61. Sasarman F, Brunel-Guitton C, Antonicka H et al (2010) LRPPRC and SLIRP interact in a ribonucleoprotein complex that regulates posttranscriptional gene expression in mitochondria. Mol Biol Cell 21:1315–1323

62. Seeger J, Schrank B, Pyle A et al (2010) Clinical and neuropathological findings in patients with TACO1 mutations. Neuromuscul Disord 20:720–724
63. Fox TD (1996) Genetics of mitochondrial translation. In: Hershey JWB, Matthews MB and Sonenberg N (eds) Translational control, pp. 733–758. Cold Spring Harbor Press, Cold Spring Harbor
64. Manthey GM, Przybyla-Zawislak BD, McEwen JE (1998) The Saccharomyces cerevisiae Pet309 protein is embedded in the mitochondrial inner membrane. Eur J Biochem 255:156–161
65. Mili S, Shu HJ, Zhao Y et al (2001) Distinct RNP complexes of shuttling hnRNP proteins with pre-mRNA and mRNA: candidate intermediates in formation and export of mRNA. Mol Cell Biol 21:7307–7319
66. Mili S, Pinol-Roma S (2003) LRP130, a pentatricopeptide motif protein with a noncanonical RNA-binding domain, is bound in vivo to mitochondrial and nuclear RNAs. Mol Cell Biol 23:4972–4982
67. Xu F, Morin C, Mitchell G et al (2004) The role of the LRPPRC (leucine-rich pentatricopeptide repeat cassette) gene in cytochrome oxidase assembly: mutation causes lowered levels of COX (cytochrome c oxidase) I and COX III mRNA. Biochem J 382:331–336
68. Cooper MP, Qu L, Rohas LM et al (2006) Defects in energy homeostasis in Leigh syndrome French Canadian variant through PGC-1alpha/LRP130 complex. Genes Dev 20:2996–3009
69. Cooper MP, Uldry M, Kajimura S et al (2008) Modulation of PGC-1 coactivator pathways in brown fat differentiation through LRP130. J Biol Chem 283:31960–31967
70. Horn D, Barrientos A (2008) Mitochondrial copper metabolism and delivery to cytochrome c oxidase. IUBMB Life 60:421–429
71. Banci L, Bertini I, Cavallaro G et al (2007) The functions of Sco proteins from genome-based analysis. J Proteome Res 6:1568–1579
72. Horng YC, Cobine PA, Maxfield AB et al (2004) Specific copper transfer from the Cox17 metallochaperone to both Sco1 and Cox11 in the assembly of yeast cytochrome C oxidase. J Biol Chem 279:35334–35340
73. Glerum DM, Shtanko A, Tzagoloff A (1996) SCO1 and SCO2 act as high copy suppressors of a mitochondrial copper recruitment defect in Saccharomyces cerevisiae. J Biol Chem 271:20531–20535
74. Hiser L, Di Valentin M, Hamer AG et al (2000) Cox11p is required for stable formation of the Cu(B) and magnesium centers of cytochrome c oxidase. J Biol Chem 275:619–623
75. Carr HS, George GN, Winge DR (2002) Yeast Cox11, a protein essential for cytochrome c oxidase assembly, is a Cu(I)-binding protein. J Biol Chem 277:31237–31242
76. Beers J, Glerum DM, Tzagoloff A (1997) Purification, characterization, and localization of yeast Cox17p, a mitochondrial copper shuttle. J Biol Chem 272:33191–33196
77. Carr HS, Maxfield AB, Horng YC et al (2005) Functional analysis of the domains in Cox11. J Biol Chem 280:22664–22669
78. Oswald C, Krause-Buchholz U, Rodel G (2009) Knockdown of human COX17 affects assembly and supramolecular organization of cytochrome c oxidase. J Mol Biol 389:470–479
79. Krummeck G, Rodel G (1990) Yeast SCO1 protein is required for a post-translational step in the accumulation of mitochondrial cytochrome c oxidase subunits I and II. Curr Genet 18:13–15
80. Glerum DM, Shtanko A, Tzagoloff A (1996) SCO1 and SCO2 act as high copy suppressors of a mitochondrial copper recruitment defect in Saccharomyces cerevisiae. J Biol Chem 271:20531–20535
81. Lode A, Kuschel M, Paret C et al (2000) Mitochondrial copper metabolism in yeast: interaction between Sco1p and Cox2p. FEBS Lett 485:19–24
82. Rentzsch A, Krummeck-Weiss G, Hofer A et al (1999) Mitochondrial copper metabolism in yeast: mutational analysis of Sco1p involved in the biogenesis of cytochrome c oxidase. Curr Genet 35:103–108
83. Chinenov YV (2000) Cytochrome c oxidase assembly factors with a thioredoxin fold are conserved among prokaryotes and eukaryotes. J Mol Med (Berl) 78:239–242

84. Abajian C, Rosenzweig AC (2006) Crystal structure of yeast Sco1. J Biol Inorg Chem 11:459–466
85. Ye Q, Imriskova-Sosova I, Hill BC et al (2005) Identification of a disulfide switch in BsSco, a member of the Sco family of cytochrome c oxidase assembly proteins. Biochem 44:2934–2942
86. Smits PH, De Haan M, Maat C et al (1994) The complete sequence of a 33 kb fragment on the right arm of chromosome II from Saccharomyces cerevisiae reveals 16 open reading frames, including ten new open reading frames, five previously identified genes and a homologue of the SCO1 gene. Yeast (10 Suppl A):S75–80
87. Leary SC, Kaufman BA, Pellecchia G et al (2004) Human SCO1 and SCO2 have independent, cooperative functions in copper delivery to cytochrome c oxidase. Hum Mol Genet 13:1839–1848
88. Leary SC, Sasarman F, Nishimura T et al (2009) Human SCO2 is required for the synthesis of CO II and as a thiol-disulphide oxidoreductase for SCO1. Hum Mol Genet 18:2230–2240
89. Banci L, Bertini I, Ciofi-Baffoni S et al (2010) Affinity gradients drive copper to cellular destinations. Nat 465:645–648
90. Banci L, Bertini I, Ciofi-Baffoni S et al (2008) Mitochondrial copper(I) transfer from Cox17 to Sco1 is coupled to electron transfer. Proc Natl Acad Sci U S A 105:6803–6808
91. Chinenov YV (2000) Cytochrome c oxidase assembly factors with a thioredoxin fold are conserved among prokaryotes and eukaryotes. J Mol Med (Berl) 78:239–242
92. Leary SC, Cobine PA, Kaufman BA, et al (2007) The human cytochrome c oxidase assembly factors SCO1 and SCO2 have regulatory roles in the maintenance of cellular copper homeostasis. Cell Metab 5:9–20
93. Dodani SC, Leary SC, Cobine PA et al (2011) A targetable fluorescent sensor reveals that copper-deficient SCO1 and SCO2 patient cells prioritize mitochondrial copper homeostasis. J Am Chem Soc 133:8606–8616
94. Yang H, Brosel S, Acin-Perez R et al (2010) Analysis of mouse models of cytochrome c oxidase deficiency owing to mutations in Sco2. Hum Mol Genet 19:170–180
95. Horng YC, Leary SC, Cobine PA et al (2005) Human Sco1 and Sco2 function as copper-binding proteins. J Biol Chem 280:34113–34122
96. Jaksch M, Ogilvie I, Yao J et al (2000) Mutations in SCO2 are associated with a distinct form of hypertrophic cardiomyopathy and cytochrome c oxidase deficiency. Hum Mol Genet 9:795–801
97. Jaksch M, Horvath R, Horn N et al (2001) Homozygosity (E140K) in SCO2 causes delayed infantile onset of cardiomyopathy and neuropathy. Neurol 57:1440–1446
98. Sacconi S, Salviati L, Sue CM et al (2003) Mutation screening in patients with isolated cytochrome c oxidase deficiency. Pediatr Res 53:224–230
99. Salviati L, Sacconi S, Rasalan MM et al (2002) Cytochrome c oxidase deficiency due to a novel SCO2 mutation mimics Werdnig-Hoffmann disease. Arch Neurol 59:862–865
100. Bohm M, Pronicka E, Karczmarewicz E et al (2006) Retrospective, multicentric study of 180 children with cytochrome C oxidase deficiency. Pediatr Res 59:21–26
101. Vesela K, Hansikova H, Tesarova M et al (2004) Clinical, biochemical and molecular analyses of six patients with isolated cytochrome c oxidase deficiency due to mutations in the SCO2 gene. Acta Paediatr 93:1312–1317
102. Tarnopolsky MA, Bourgeois JM, Fu MH et al (2004) Novel SCO2 mutation (G1521A) presenting as a spinal muscular atrophy type I phenotype. Am J Med Genet A 125A:310–314
103. Knuf M, Faber J, Huth RG et al (2007) Identification of a novel compound heterozygote SCO2 mutation in cytochrome c oxidase deficient fatal infantile cardioencephalomyopathy. Acta Paediatr 96:130–132
104. Verdijk RM, de Krijger R, Schoonderwoerd K et al (2008) Phenotypic consequences of a novel SCO2 gene mutation. Am J Med Genet A 146A:2822–2827
105. Leary SC, Mattman A, Wai T et al (2006) A hemizygous SCO2 mutation in an early onset rapidly progressive, fatal cardiomyopathy. Mol Genet Metab 89:129–133
106. Mobley BC, Enns GM, Wong LJ et al (2009) A novel homozygous SCO2 mutation, p.G193S, causing fatal infantile cardioencephalomyopathy. Clin Neuropathol 28:143–149

107. Joost K, Rodenburg R, Piirsoo A et al (2010) A novel mutation in the SCO2 gene in a neonate with early-onset cardioencephalomyopathy. Pediatr Neurol 42:227–230
108. Pronicki M, Kowalski P, Piekutowska-Abramczuk D et al (2010) A homozygous mutation in the SCO2 gene causes a spinal muscular atrophy like presentation with stridor and respiratory insufficiency. Eur J Paediatr Neurol 14:253–260
109. Cobine PA, Pierrel F, Leary SC et al (2006) The P174L mutation in human Sco1 severely compromises Cox17-dependent metallation but does not impair copper binding. J Biol Chem 281:12270–12276
110. Bohm M, Pronicka E, Karczmarewicz E et al (2006) Retrospective, multicentric study of 180 children with cytochrome C oxidase deficiency. Pediatr Res 59:21–26
111. Foltopoulou PF, Zachariadis GA, Politou AS et al (2004) Human recombinant mutated forms of the mitochondrial COX assembly Sco2 protein differ from wild-type in physical state and copper binding capacity. Mol Genet Metab 81:225–236
112. Dickinson EK, Adams DL, Schon EA et al (2000) A human SCO2 mutation helps define the role of Sco1p in the cytochrome oxidase assembly pathway. J Biol Chem 275:26780–26785
113. Stiburek L, Hansikova H, Tesarova M et al (2006) Biogenesis of eukaryotic cytochrome c oxidase. Physiol Res 55(Suppl 2):S27–41
114. Caughey WS, Smythe GA, O'Keeffe DH et al (1975) Heme A of cytochrome c oxicase. Structure and properties: comparisons with hemes B, C, and S and derivatives. J Biol Chem 250:7602–7622
115. Tzagoloff A, Nobrega M, Gorman N et al (1993) On the functions of the yeast COX10 and COX11 gene products. Biochem Mol Biol Int 31:593–598
116. Barros MH, Nobrega FG, Tzagoloff A (2002) Mitochondrial ferredoxin is required for heme A synthesis in Saccharomyces cerevisiae. J Biol Chem 277:9997–10002
117. Brown KR, Allan BM, Do P et al (2002) Identification of novel hemes generated by heme A synthase: evidence for two successive monooxygenase reactions. Biochem 41:10906–10913
118. Barros MH, Carlson CG, Glerum DM et al (2001) Involvement of mitochondrial ferredoxin and Cox15p in hydroxylation of heme O. FEBS Lett 492:133–138
119. Glerum DM, Tzagoloff A (1994) Isolation of a human cDNA for heme A:farnesyltransferase by functional complementation of a yeast cox10 mutant. Proc Natl Acad Sci U S A 91:8452–8456
120. Coenen MJ, Van Den Heuvel LP, Ugalde C et al (2004) Cytochrome c oxidase biogenesis in a patient with a mutation in COX10 gene. Ann Neurol 56:560–564
121. Coenen MJ, Smeitink JA, Pots JM et al (2006) Sequence analysis of the structural nuclear encoded subunits and assembly genes of cytochrome c oxidase in a cohort of 10 isolated complex IV-deficient patients revealed five mutations. J Child Neuro 21:508–511
122. Bugiani M, Tiranti V, Farina L et al (2005) Novel mutations in COX15 in a long surviving Leigh syndrome patient with cytochrome c oxidase deficiency. J Med Genet 42:e28
123. Alfadhel M, Lillquist YP, Waters PJ et al (2011) Infantile cardioencephalopathy due to a COX15 gene defect: report and review. Am J Med Genet A 155A:840–844
124. Oquendo CE, Antonicka H, Shoubridge EA et al (2004) Functional and genetic studies demonstrate that mutation in the COX15 gene can cause Leigh syndrome. J Med Genet 41:540–544
125. Williams SL, Valnot I, Rustin P et al (2004) Cytochrome c oxidase subassemblies in fibroblast cultures from patients carrying mutations in COX10, SCO1, or SURF1. J Biol Chem 279:7462–7469
126. Barros MH, Tzagoloff A (2002) Regulation of the heme A biosynthetic pathway in Saccharomyces cerevisiae. FEBS Lett 516:119–123
127. Bestwick M, Khalimonchuk O, Pierrel F et al (2010) The role of Coa2 in hemylation of yeast Cox1 revealed by its genetic interaction with Cox10. Mol Cell Biol 30:172–185
128. Leigh D (1951) Subacute necrotizing encephalomyelopathy in an infant. J Neurol Neurosurg Psychiatry 14:216–221
129. Von Kleist-Retzow JC, Yao J, Taanman JW et al (2001) Mutations in SURF1 are not specifically associated with Leigh syndrome. J Med Genet 38:109–113

130. Williams SL, Taanman JW, Hansikova H et al (2001) A novel mutation in SURF1 causes skipping of exon 8 in a patient with cytochrome c oxidase-deficient Leigh syndrome and hypertrichosis. Mol Genet Metab 73:340–343
131. Tay SK, Sacconi S, Akman HO et al (2005) Unusual clinical presentations in four cases of Leigh disease, cytochrome C oxidase deficiency, and SURF1 gene mutations. J Child Neurol 20:670–674
132. Rahman S, Brown RM, Chong WK et al (2001) A SURF1 gene mutation presenting as isolated leukodystrophy. Ann Neurol 49:797–800
133. Mashkevich G, Repetto B, Glerum DM et al (1997) SHY1, the yeast homolog of the mammalian SURF-1 gene, encodes a mitochondrial protein required for respiration. J Biol Chem 272:14356–14364
134. Smith D, Gray J, Mitchell L et al (2005) Assembly of cytochrome-c oxidase in the absence of assembly protein Surf1p leads to loss of the active site heme. J Biol Chem 280:17652–17656
135. Bundschuh FA, Hannappel A, Anderka O et al (2009) Surf1, associated with Leigh syndrome in humans, is a heme-binding protein in bacterial oxidase biogenesis. J Biol Chem 284:25735–25741
136. Pierrel F, Bestwick ML, Cobine PA et al (2007) Coa1 links the Mss51 post-translational function to Cox1 cofactor insertion in cytochrome c oxidase assembly. EMBO J 26:4335–4346
137. Stiburek L, Vesela K, Hansikova H et al (2005) Tissue-specific cytochrome c oxidase assembly defects due to mutations in SCO2 and SURF1. Biochem J 392:625–632
138. Fontanesi F, Jin C, Tzagoloff A et al (2008) Transcriptional activators HAP/NF-Y rescue a cytochrome c oxidase defect in yeast and human cells. Hum Mol Genet 17:775–788
139. Tiranti V, Jaksch M, Hofmann S et al (1999) Loss-of-function mutations of SURF-1 are specifically associated with Leigh syndrome with cytochrome c oxidase deficiency. Ann Neurol 46:161–166
140. Tiranti V, Lamantea E, Uziel G et al (1999) Leigh syndrome transmitted by uniparental disomy of chromosome 9. J Med Genet 36:927–928
141. Yao J, Shoubridge EA (1999) Expression and functional analysis of SURF1 in Leigh syndrome patients with cytochrome c oxidase deficiency. Hum Mol Genet 8:2541–2549
142. Poyau A, Buchet K, Bouzidi MF et al (2000) Missense mutations in SURF1 associated with deficient cytochrome c oxidase assembly in Leigh syndrome patients. Hum Genet 106:194–205
143. Coenen MJ, Van Den Heuvel LP, Nijtmans LG et al (1999) SURFEIT-1 gene analysis and two-dimensional blue native gel electrophoresis in cytochrome c oxidase deficiency. Biochem Biophys Res Commun 265:339–344
144. Reinhold R, Bareth B, Balleininger M et al (2011) Mimicking a SURF1 allele reveals uncoupling of cytochrome c oxidase assembly from translational regulation in yeast. Hum Mol Genet 20:2379–2393
145. Piekutowska-Abramczuk D, Magner M, Popowska E et al (2009) SURF1 missense mutations promote a mild Leigh phenotype. Clin Genet 76:195–204
146. Salviati L, Freehauf C, Sacconi S et al (2004) Novel SURF1 mutation in a child with subacute encephalopathy and without the radiological features of Leigh Syndrome. Am J Med Genet A 128A:195–198
147. Teraoka M, Yokoyama Y, Ninomiya S et al (1999) Two novel mutations of SURF1 in Leigh syndrome with cytochrome c oxidase deficiency. Hum Genet 105:560–563
148. Ogawa Y, Naito E, Ito M et al (2002) Three novel SURF-1 mutations in Japanese patients with Leigh syndrome. Pediatr Neurol 26:196–200
149. von Kleist-Retzow JC, Vial E, Chantrel-Groussard K et al (1999) Biochemical, genetic and immunoblot analyses of 17 patients with an isolated cytochrome c oxidase deficiency. Biochim Biophys Acta 1455:35–44
150. Sue CM, Karadimas C, Checcarelli N et al (2000) Differential features of patients with mutations in two COX assembly genes, SURF-1 and SCO2. Ann Neurol 47:589–595
151. Santoro L, Carrozzo R, Malandrini A et al (2000) A novel SURF1 mutation results in Leigh syndrome with peripheral neuropathy caused by cytochrome c oxidase deficiency. Neuromuscul Disord 10:450–453

152. Pequignot MO, Dey R, Zeviani M et al (2001) Mutations in the SURF1 gene associated with Leigh syndrome and cytochrome C oxidase deficiency. Hum Mutat 17:374–381
153. Pequignot MO, Desguerre I, Dey R et al (2001) New splicing-site mutations in the SURF1 gene in Leigh syndrome patients. J Biol Chem 276:15326–15329
154. Bruno C, Biancheri R, Garavaglia B et al (2002) A novel mutation in the SURF1 gene in a child with Leigh disease, peripheral neuropathy, and cytochrome-c oxidase deficiency. J Child Neurol 17:233–236
155. Capkova M, Hansikova H, Godinot C et al (2002) A new missense mutation of 574C>T in the SURF1 gene—biochemical and molecular genetic study in seven children with Leigh syndrome. Cas Lek Cesk 141:636–641
156. Rossi A, Biancheri R, Bruno C et al (2003) Leigh syndrome with COX deficiency and SURF1 gene mutations: MR imaging findings. AJNR Am J Neuroradiol 24:1188–1191
157. Pecina P, Capkova M, Chowdhury SK et al (2003) Functional alteration of cytochrome c oxidase by SURF1 mutations in Leigh syndrome. Biochim Biophys Acta 1639:53–63
158. Moslemi AR, Tulinius M, Darin N et al (2003) SURF1 gene mutations in three cases with Leigh syndrome and cytochrome c oxidase deficiency. Neurology 61:991–993
159. Darin N, Moslemi AR, Lebon S et al (2003) Genotypes and clinical phenotypes in children with cytochrome-c oxidase deficiency. Neuropediatrics 34:311–317
160. Head RA, Brown RM, Brown GK (2004) Diagnostic difficulties with common SURF1 mutations in patients with cytochrome oxidase-deficient Leigh syndrome. J Inherit Metab Dis 27:57–65
161. Monnot S, Chabrol B, Cano A et al (2005) Cytochrome c oxydase-deficient Leigh syndrome with homozygous mutation in SURF1 gene. Arch Pediatr 12:568–571
162. Ostergaard E, Bradinova I, Ravn SH et al (2005) Hypertrichosis in patients with SURF1 mutations. Am J Med Genet A 138:384–388
163. van Riesen AK, Antonicka H, Ohlenbusch A et al (2006) Maternal segmental disomy in Leigh syndrome with cytochrome c oxidase deficiency caused by homozygous SURF1 mutation. Neuropediatrics 37:88–94
164. Piekutowska-Abramczuk D, Popowska E, Pronicki M et al (2009) High prevalence of SURF1 c.845_846delCT mutation in Polish Leigh patients. Eur J Paediatr Neurol 13:146–153
165. Zhang Y, Yang YL, Sun F et al (2007) Clinical and molecular survey in 124 Chinese patients with Leigh or Leigh-like syndrome. J Inherit Metab Dis 30:265
166. Sacconi S, Salviati L, Sue CM et al (2003) Mutation screening in patients with isolated cytochrome c oxidase deficiency. Pediatr Res 53:224–230
167. Rotig A, Lebon S, Zinovieva E et al (2004) Molecular diagnostics of mitochondrial disorders. Biochim Biophys Acta 1659:129–135

Part IV
Mitochondrial Protein Translation Related Diseases

Chapter 16
Mitochondrial Aminoacyl-tRNA Synthetases

Henna Tyynismaa

Mitochondrial protein synthesis serves the purpose of transferring genetic information from the mitochondrial genome to the enzyme complexes of the oxidative phosphorylation system (OXPHOS). For mitochondrial protein synthesis, collaboration of two genomes is required: mitochondrial tRNAs and ribosomal RNAs are encoded by the mitochondrial genome whereas all protein factors involved in the translation of mitochondrial proteins are encoded by nuclear genes, translated in the cytoplasm and imported into mitochondria.

Aminoacyl-tRNA synthetases are an ancient family of enzymes that are essential in determining and accurately maintaining the genetic code [1]. In the initial step of protein synthesis, aminoacyl-tRNA synthetases perform the aminoacylation of each tRNA with the appropriate amino acid (Fig. 16.1). In principle, each of the 20 amino acids has its own aminoacyl-tRNA synthetase. The aminoacylation process requires the synthetase to specifically recognize and bind its cognate tRNA and to catalyze the transfer of the appropriate amino acid to the acceptor stem of the tRNA. The 20 synthetases are divided into two classes (class I and class II) in accordance with their active site architecture. Some synthetases have demonstrated editing activity that removes mischarged amino acids from their cognate tRNAs.

In human cells, protein synthesis takes place in the cytoplasm and in mitochondria. The aminoacylation in the two cellular compartments is largely performed by distinct sets of synthetases, which are encoded by separate nuclear genes. This separation is thought to be the consequence of evolutionary pressure of mitochondrial DNA size reduction, which has resulted in truncated tRNAs and subsequent inability of cytoplasmic aminoacyl-tRNA synthetases to recognize the mitochondrial tRNAs [2]. Two enzymes, glycyl-and lysyl-tRNA synthetase, are exceptions because the same nuclear gene (*GARS* and *KARS*, respectively) encodes both the cytoplasmic and the mitochondrial protein.

The synthetase proteins are commonly abbreviated by the three-letter amino acid code and the genes by one-letter code, with the mitochondrial protein marked with

H. Tyynismaa (✉)
Research Program of Molecular Neurology, Biomedicum Helsinki,
Haartmaninkatu 8, 00014 University of Helsinki, Finland
e-mail: henna.tyynismaa@helsinki.fi

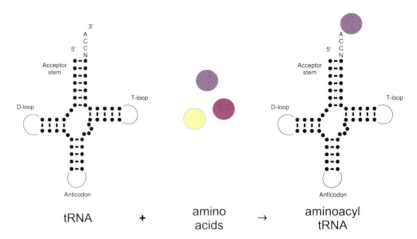

Fig. 16.1 The acceptor stem of tRNA is charged with an amino acid in aminoacylation reaction by aminoacyl-tRNA synthetase. Mitochondrial tRNAs generally have the acceptor stem and D-, T-, and anticodon loops but lack the variable loop

the prefix mt and the mitochondrial synthetase encoding genes with a suffix 2. For example, cytoplasmic and mitochondrial histidyl-tRNA synthetases are HisRS and mtHisRS for short, respectively, and the genes encoding them are abbreviated *HARS* and *HARS2*. Table 16.1 lists the genes for human aminoacyl-tRNA synthetases. Note the few exceptions: (1) Cytoplasmic ProRS is expressed as part of a multifunctional enzyme that also displays GluRS activity and is referred to as the gluprolyl-tRNA synthetase (GluProRS); (2) Cytoplasmic PheRS has two subunits that are encoded by separate genes; (3) Mitochondrial GlnRS does not exist but instead Gln-tRNA^{Gln} is synthesized indirectly via misacylation of tRNA^{Gln} with glutamic acid by GluRS to form Glu-tRNA^{Gln}, which is subsequently corrected to Gln-tRNA^{Gln} by the Glu-tRNA^{Gln} amidotransferase [3].

Aminoacyl-tRNA synthetases have been of great interest for their evolutionary role in establishing the algorithm of the genetic code but more recently also for the use of synthetases as drug targets, for their versatile noncanonical functions, and for their emerging roles in human disease. Mutations in four genes encoding cytoplasmic aminoacyl-tRNA synthetases (*AARS, YARS, GARS, KARS*) have been described in patients with Charcot-Marie-Tooth neuropathy. Notably, two of those, *GARS* and *KARS*, also encode the mitochondrial synthetase. In addition, patients with mutations in six genes (*DARS2, RARS2, YARS2, SARS2, HARS2, AARS2*) encoding mitochondrial aminoacyl-tRNA synthetases have been recently reported, making this group of enzymes a new important cause of protein synthesis defects in the mitochondria. Deficient mitochondrial translation could be expected to reduce OXPHOS capacity and lead to ATP-deficiency in all tissues, primarily affecting the high-energy demanding tissues such as brain, heart, and skeletal muscle. Surprisingly, however, the defects in mitochondrial aminoacyl-tRNA synthetases have been identified in different types of tissue-specific disorders (Table 16.2), which will be reviewed in the following table.

16 Mitochondrial Aminoacyl-tRNA Synthetases

Table 16.1 Human aminoacyl-tRNA synthetases

Aminoacyl-tRNA synthetase	Gene for the cytoplasmic	Gene for cytoplasmic and mitochondrial	Gene for mitochondrial
AlaRS	AARS		AARS2
ArgRS	RARS		RARS2
AsnRS	NARS		NARS2
AspRS	DARS		DARS2
CysRS	CARS		CARS2
GlnRS	QARS		Transamidation
GluRS	EPRS		EARS2
GlyRS		GARS	
HisRS	HARS		HARS2
IleRS	IARS		IARS2
LeuRS	LARS		LARS2
LysRS		KARS	
MetRS	MARS		MARS2
PheRS	FARSA + FARSB		FARS2
ProRS	EPRS		PARS2
SerRS	SARS		SARS2
ThrRS	TARS		TARS2
TrpRS	WARS		WARS2
TyrRS	YARS		YARS2
ValRS	VARS		VARS2

Table 16.2 Human diseases caused by defects in mitochondrial aminoacyl-tRNA synthetases

Gene	Phenotype	Main affected tissue	Age of onset or diagnosis	Method of identification	Reference
DARS2	Leukoencephalopathy with brain stem and spinal cord involvement and lactate elevation (LBSL)	Central nervous system	Childhood, adolescence	Linkage analysis and haplotype mapping	[6]
RARS2	Pontocerebellar hypoplasia type 6 (PCH6)	Central nervous system	Infancy, lived up to 16 months	Homozygosity mapping	[12]
YARS2	Myopathy, lactic acidosis, sideroblastic anemia (MLASA)	Skeletal muscle	Infancy, childhood	Homozygosity mapping	[19]
SARS2	Hyperuricemia, pulmonary hypertension, renal failure in infancy and alkalosis (HUPRA syndrome)	Kidney	Infancy, lived up to 14 months	Homozygosity mapping	[21]
HARS2	Ovarian dysgenesis and sensorineural hearing loss (Perrault syndrome)	Ovary	Childhood, adolescence	Linkage analysis	[23]
AARS2	Infantile cardiomyopathy	Heart	Infancy, lived up to 10 months	Exome sequencing	[24]

Mitochondrial Aminoacyl-tRNA Synthetases in Disease

Aspartyl-tRNA Synthetase (mtAspRS, DARS2)

Leukoencephalopathy with brain stem and spinal cord involvement and lactate elevation (LBSL) was the first identified mitochondrial aminoacyl-tRNA synthetase disease. The entity was described in 2003 with a characteristic magnetic resonance imaging pattern of cerebral white matter and selective involvement of brain stem and spinal tracts [4]. Proton magnetic resonance imaging showed increased lactate and decreased N-acetylaspartate in the abnormal white matter. The patients had a slowly progressive spastic ataxia starting in childhood or adolescence [4, 5]. Only the white matter lactate elevation suggested mitochondrial involvement since patient fibroblasts and lymphoblasts showed no defects in OXPHOS complex amount or activity [6] and patients' muscle mitochondria functioned normally [4]. Using linkage analysis and haplotype mapping in six affected sib pairs, Scheper et al. [6] localized the causative gene for this recessively inherited disorder and identified mutations in the *DARS2* gene in 38 patients. Curiously, nearly all patients were compound heterozygous for splice site mutations upstream of *DARS2* exon 3 in one allele and various missense mutations in the other allele. The effect of the identified missense mutations on mtAspRS function were measured in vitro with purified recombinant proteins and the ability of the mutant proteins to aminoacylate tRNA^{Asp} was found to be significantly reduced [6]. This deficiency was mainly due to reduced catalytic activity of the mutant proteins. The splicing defects were predicted to lead to exon 3 skipping and premature truncation of the protein by frameshift. Indeed, the aberrantly spliced cDNA variant was detected in patients' cells, although at a low concentration [6]. Later study showed that the splice site mutation was "leaky" because patient cells with one nonsense and one exon 3 splicing mutation had a reduced but detectable level of full-length mtAspRS protein [7]. Other reported *DARS2* patients also had compound heterozygous mutations with exon 3 splicing mutation in one of the alleles [5, 8, 9] except recently when two homozygous mutations were identified: the p.R609W in a benign case with exercise-induced gait ataxia [10] and the exon 3 splice site mutation in a severe case [11]. The mutation spectrum suggests that severe catalytic mutations are not tolerated as homozygotes and the leaking frameshift mutations allow the residual mtAspRS activity to adequately maintain mitochondrial translation in most tissues but not in cerebral white matter.

Expression levels of *DARS2* mRNA in tissues do not explain the selective involvement of specific white matter tracts in LBSL [6]. Van Berge et al. [7] have recently suggested that tissue-specific differences in splicing efficiency may account for the neuronal vulnerability in LBSL. They showed in cell culture studies that the acceptor site of exon 3 of *DARS2* is a weak splice site even in the absence of disease-causing mutations because exon 3 skipping was observed in all cell types tested. Neuronal cells showed more exon 3 skipping than other cells. Furthermore, the authors confirmed that the exon 3 splicing mutations increased exon 3 skipping and the effect was largest in neural cell lines [7]. The results suggest that the mechanism for the tissue-specificity in LBSL involves reduced splicing efficiency in selective neurons, thus reaching the threshold level for mtAspRS activity at which mitochondrial translation is no longer sufficient.

Arginyl-tRNA Synthetase (mtArgRS, RARS2)

Pontocerebellar hypoplasia (PCH) is a group of autosomal-recessive neurodegenerative disorders with prenatal onset. The disease is characterized by cerebellar hypoplasia with variable atrophy of the cerebellum and the ventral pons and includes variable neocortical atrophy, ventriculomegaly, and microcephaly. In 2007, Edvardson et al. [12] investigated three children born to consanguineous parents, who were hypotonic since birth and had progressive atrophy of the cerebellum, pons, cerebral cortex, and white matter by magnetic resonance imaging of the brain. The patients did not reach developmental milestones and they lived up to 16 months. Even in the end, the symptoms were confined to the central nervous system. Marked reductions of OXPHOS complex I, III, and IV activities were measured in skeletal muscle homogenates. Slightly elevated lactate was detected in plasma of one patient and in cerebrospinal fluid (CSF) of another patient. Homozygosity mapping led to the identification of a disease-causing mutation in the *RARS2* gene [12]. This was a homozygous splice-site mutation resulting in exon 2 skipping, which was predicted to cause a frameshift, abolishing mtArgRS activity entirely. Most of the *RARS2* cDNA in patient fibroblasts showed exon 2 skipping but some normal cDNA was also present, indicating leakiness of the splicing mutation. Similar to the *DARS2* exon 3 splice site, *RARS2* exon 2 splice site was predicted to be weak. The level of tRNA^{Arg} was significantly reduced in patient fibroblasts, whereas the remaining tRNA was fully acylated. This implied that some residual mtArgRS activity was present in the cells and that uncharged tRNA^{Arg} was unstable. The authors speculated that a splicing factor important for both *DARS2* and *RARS2* is present in low concentration in brain and that leakiness of the splice site mutation may allow the residual aminoacyl-tRNA synthetase activity to sustain OXPHOS function in other tissues. Pontocerebellar hypoplasia with *RARS2* mutations was classified as type 6 (PCH6).

Second study reported *RARS2* mutations in a patient with PCH6 and PEHO-like condition (progressive encephalopathy, edema, hypsarrhythmia, and optic atrophy) but without hypsarrhythmia and optic atrophy [13]. OXPHOS function was normal in muscle but transient elevations in blood and CSF lactate levels were detected. This patient had two missense mutations, one changing methionine at amino acid position 342 to valine, and the other changing glutamine at amino acid position 12 to arginine, suggesting that the mitochondrial localization signal in the N-terminus of the protein was impaired. Effects of the mutations were not experimentally tested but the latter mutation was proposed to influence the proportion of the two *RARS2* splice variants that were detected in the study, as it was located within the splice junction at the end of exon 1.

Third, a patient with compound heterozygous Q12R and exon 2 splicing mutation was reported [14]. This combination resulted in hypotonia, high CSF lactate levels, and early lethality (6 days). Postmortem neuropathological examination revealed loss of spinal anterior horn cells and diffuse gliosis, which corresponded to PCH1 phenotype.

Other types of PCH are caused by mutations in genes encoding for subunits of the tRNA splicing endonuclease (*TSEN54, TSEN2,* and *TSEN34*). The TSEN complex removes an intron that exists within the anticodon loop of 6.2 % of nuclear-encoded tRNAs [15]. TSEN is thus not linked to mitochondrial function, except that in yeast *Saccharomyces cerevisiae* the complex has been localized to the outer surface of mitochondria [16]. Kasher et al. [15] described the first animal model for mitochondrial aminoacyl-tRNA synthetase deficiencies. They generated *rars2* knockdown zebrafish embryos and observed brain hypoplasia, loss of structural definition within the brain and increased brain cell death. Similar phenotype was seen in their *tsen54* knockdown embryos, which led the authors to suggest a common disease pathway between *TSEN54-* and *RARS2*-related PCH. Further studies are required to understand the role of RARS2 in PCH.

Tyrosyl-tRNA Synthetase (mtTyrRS, YARS2)

Myopathy, lactic acidosis, and sideroblastic anemia (MLASA) is a mitochondrial disorder where the severity of progressive exercise intolerance varies between families and affected family members, sometimes including mental retardation. The onset of sideroblastic anemia may also vary from infancy to childhood. Mutations in pseudouridylate synthase 1 (*PUS1*) have been associated with MLASA, affecting protein synthesis by decreased tRNA pseudouridylation [17, 18]. Riley et al. [19] studied three MLASA patients from unrelated consanguineous families and found reduced OXPHOS complex I, III, and IV activities in muscle but not in fibroblasts. Onset of symptoms was earlier than generally reported for *PUS1* mutations. Through homozygosity mapping of two individuals, followed by candidate gene sequencing, a homozygous mutation in *YARS2*, resulting in the p.F52L amino acid change, was identified in all affected individuals [19]. Patient's myotubes showed decreased levels of mitochondrial translation products and OXPHOS complex I, III, and IV subunits. Again, patient fibroblasts had normal mitochondrial protein synthesis.

Phenylalanine 52 lies in the catalytic domain of mtTyrRS and is predicted to be important for recognizing and/or stabilizing the amino acid-accepting helix of tRNA[Tyr] [19]. Functional assays with purified recombinant proteins showed that the p.F52L mutant was catalytically active but with a reduced catalytic rate and affinity for tRNA[Tyr] substrate. The reduced aminoacylation efficiency thus resulted in defective mitochondrial protein synthesis, which manifested as OXPHOS deficiency in skeletal muscle. The connection between *PUS1* and *YARS2* may be that pseudouridylation of the second anticodon position of tRNA[Tyr] is essential for its tRNA identity [20]. However, *PUS1* also pseudouridylates cytosolic tRNAs and non-tRNA molecules and it is unclear, which of its functions contribute to MLASA. *YARS2* mutation, on the other hand, clearly affects mitochondrial protein synthesis, evidenced by the mitochondrial translation defect in cultured diseased myotubes. Whether the sideroblastic anemia phenotype also results from protein synthesis defect is not known.

Seryl-tRNA Synthetase (mtSerRS, SARS2)

HUPRA syndrome is a multiorgan disease in infancy, manifesting with hyperuricemia, hypochloremic metabolic alkalosis, progressive renal failure, and primary pulmonary hypertension. Belostotsky et al. [21] studied prematurely born HUPRA infants with feeding difficulties and developmental delay, who belonged to consanguineous kindred. High blood and CFS lactate levels and increased serum alanine were measured. Muscle biopsy demonstrated type 2 fiber atrophy with several COX-deficient fibers. OXPHOS complex I, III, and IV activities were reduced in muscle. Brain ultrasound was normal. The three described patients died by 14 months of age. Homozygosity mapping was performed with DNA samples from two patients. Direct sequencing of candidate genes resulted in the identification of a homozygous mutation in *SARS2*, resulting in p.D390G amino acid change [21]. This residue is located within one of the β -strands that assemble the active site of mtSerRS.

Two tRNA^{Ser} variants, tRNA$^{Ser}_{AGY}$ and tRNA$^{Ser}_{UCN}$, are present in mitochondria and mtSerRS aminoacylates both of them. Interestingly, these mitochondrial tRNAs lack the long variable arm that is the major determinant for recognition by SerRS in all other systems. Furthermore, the two tRNAs do not share sequence or structural similarities [22], implying that different elements of the substrate tRNAs are recognized by mtSerRS. In patient lymphocyte cultures the amount of tRNA$^{Ser}_{AGY}$ was reduced to 10–20 % of the control level and this residual tRNA pool was nonacylated whereas tRNA$^{Ser}_{UCN}$ amount was unaffected. This suggests that the mtSerRS with the p.D390G mutation only impaired the ability of the enzyme to acylate tRNA$^{Ser}_{AGY}$. The AGY codons are present in 11 out of 13 mitochondrially encoded proteins, which are thus affected by the aminoacylation defect. Why renal manifestations are prominent in patients with the *SARS2* mutation is however not clear.

Histidyl-tRNA Synthetase (mtHisRS, HARS2)

Perrault syndrome is characterized by ovarian dysgenesis in females and sensorineural hearing loss in females and males. Varying age of onset and severity have been reported as well as additional neurological manifestations. Using linkage analysis in a family with 11 children, Pierce et al. [23] identified compound heterozygous mutations in *HARS2*. Maternal mutation changed valine at amino acid position 386 to leucine and paternal mutation changed leucine at amino acid position 200 to valine. Analysis of patient cDNA showed that paternal mutant allele also created an alternative splice site, leading to an in-frame deletion of 12 amino acids (p.L200-K211del).

Protein-localization study in cultured cells suggested that the deletion mutant was unstable. The p.V386 is located at a highly conserved motif, which is involved in histidine recognition and its substitution for leucine is predicted to reduce binding of the activated histidine. In vitro, the purified recombinant mtHisRS p.V386L mutant had severely decreased activity in the pyrophosphate exchange assay, which measures the first step of the aminoacylation reaction [23]. Also, the mutant analogous to human p.V386L was unable to rescue the lethality of yeast deletion strain. The p.L200V mutant, on the other hand, was concluded to provide near wild-type activity in vivo and to rescue the yeast strain. The authors also tested the impact of reduced mtHisRS on *Caenorhabditis elegans* by RNA interference (RNAi) feeding and observed gonadal defects and reduced fertility [23]. However, in worm a single gene exists for histidyl-tRNA synthetases and therefore in their experiment, RNAi also affected the cytoplasmic HisRS. In conclusion, similar to *DARS2* and *RARS2*, one of the identified *HARS2* mutations also involved splicing, which may explain the tissue-specificity. In addition to the catalytically defective p.V386L mtHisRS, the patients' tissues had varying proportions of p.L200-K211del and p.L200V of which the former is an unstable protein and the latter is practically wild-type mtHisRS. Therefore, the splicing efficiency determines the outcome in each tissue.

Alanyl-tRNA Synthetase (mtAlaRS, AARS2)

Mitochondrial AlaRS defects were identified in a search for genetic causes of infantile mitochondrial cardiomyopathy using whole-exome sequencing [24]. The index patient died at 10 months of age of hypertrophic cardiomyopathy with combined cardiac OXPHOS complex I and IV deficiency. Blue native PAGE of autopsy tissues showed reduced OXPHOS complexes also in brain but not in liver, and COX-deficient skeletal muscle fibers were present in muscle biopsy histology. Translation assay using patient's fibroblasts was normal. The exome analysis concentrated on homozygous damaging variants in nuclear genes encoding mitochondrial proteins and resulted in identification of mtAlaRS variant p.R592W [24]. Another patient was found to be compound heterozygous with p.R592W and p.L155R amino acid changes. The second patient had died at 3 days of age of hypertrophic cardiomyopathy with near-total COX activity loss in heart.

Structural analysis of mtAlaRS predicted that the p.L155R variant severely disturbed the aminoacylation function by altering the catalytic site. The p.R592W, on the other hand, located to the surface of mtAlaRS-editing domain. Alanyl-tRNA synthetases have a recognition dilemma because they cannot distinguish serine or glycine from alanine and therefore the editing domain has developed to remove misacylated tRNAs [25]. The p.R592W mutation was predicted to interfere with mischarged tRNA recognition leading to increased mistranslation in mitochondrial proteins. The resulting OXPHOS defect was most pronounced in the heart leading to early death of patients. Other postmitotic tissues with high energy demand, such as brain and skeletal muscle, had reduced OXPHOS complex levels, suggesting that translation defect was also present in those tissues, causing progressive OXPHOS deficiency. *AARS2* mutations resulted in a phenotype close to what is expected from a mitochondrial translation defect: progressive OXPHOS deficiency in all cells but first manifesting in tissues with high energy demand. To understand if, for example high serine levels in heart caused the specific vulnerability of that tissue, amino acid levels in patient tissues were measured. No clear differences were noted in other than alanine levels, which were higher in patient heart and skeletal muscle but not in liver compared with control samples. It is unclear if the increased alanine was a compensatory consequence of the mtAlaRS defect or if the finding reflected mitochondrial dysfunction in general.

Leucyl-tRNA Synthetase (mtLeuRS, *LARS2*) in Type 2 Diabetes

Based on the inefficient aminoacylation of the mitochondrial tRNA^{Leu} in association with its A3243G mutations in maternally inherited diabetes and deafness, Hart et al. [26] studied whether genetic variation in *LARS2* also associated with type 2 diabetes. As a result, they proposed the p.H324Q mtLeuRS as a susceptibility variant for type 2 diabetes. The p.H324 residue was present at the surface of the mtLeuRS-editing domain, which deacylates mischarged Ile-tRNA^{Leu} and Met-tRNA^{Leu}. However, the authors showed in vitro using purified proteins that the mutant mtLeuRS had normal aminoacylation and editing activities. Finally, a larger replication study of the p.H324Q variant failed to confirm an association with type 2 diabetes [27].

Mitochondrial Aminoacyl-tRNA Synthetases in Charcot-Marie-Tooth Neuropathy (GlyRS, *GARS*; LysRS, *KARS*)

GARS and *KARS* are the two genes that encode both the cytoplasmic and mitochondrial aminoacyl-tRNA synthetases. Alternative splicing of the *KARS* gene results in production of the different isoforms: cytoplasmic form that needs splicing of exon 1 to exon 3, whereas the mitochondrial form includes exons 1, 2, and 3 [28]. For *GARS*, the mechanism to produce two isoforms from one gene is thought to involve alternative translation start sites but only those in yeast have been fully characterized [29].

Charcot-Marie-Tooth (CMT) is a genetically and clinically heterogeneous group of peripheral neuropathies caused by demyelination (CMT1) or axonal degeneration (CMT2), or a combination of both features. The disease is characterized by loss of muscle tissue and touch sensation in body extremities, predominantly in the feet and legs. *GARS* mutations have been found to be a frequent cause of dominant CMT2 [30] with 11 reported missense mutations. How these mutations lead to axonal damage is not clear. While some mutations caused reduced catalytic function in vitro, other mutations did not affect aminoacylation, suggesting that the mutations may have a gain-of-function consequence. Most mutations affected the formation of protein dimer and promoted conformational opening of the dimer [31]. In neuronal cells, overexpressed mutant GlyRS proteins displayed abnormal neurite distribution [32]. GlyRS may thus have a unique function in neurons. Other proposed mechanisms for *GARS* axonopathy include protein aggregation, tRNA mischarging, nucleolar dysfunction, protein toxicity, mitochondrial toxicity or dysfunction, and impaired axonal transport [33].

Recessive *KARS* mutations have recently been reported in intermediate CMT phenotype [34]. The patient also exhibited developmental delay, self-abusive behavior, and dysmorphic features. Identified mutations were compound heterozygous resulting in near-total loss of LysRS aminoacylation activity *in vitro*, which is surprising considering that the patient did not show signs of mitochondrial dysfunction.

Although *GARS* and *KARS* mutations may also impair mitochondrial tRNA aminoacylation, mitochondrial protein synthesis defects have not been largely considered to be part of the CMT pathogenesis. This reasoning is supported by the fact that two other genes encoding only cytoplasmic aminoacyl-tRNA synthetases, *AARS* and *YARS*, have also been identified as CMT disease genes [35, 36]. However, the mitochondrial aspect deserves more detailed investigation since mutations affecting MFN2, an important mitochondrial fusion protein, also underlie CMT2 [37].

Mitochondrial Aminoacyl-tRNA Synthetase Levels Influence the Outcome of mtDNA Point Mutations

Mitochondrial tRNA point mutations are a frequent cause of mitochondrial disease and the primary defect with these mutations is often inefficient aminoacylation. Recent data show that overexpression of aminoacyl-tRNA synthetases in cultured cells can correct the dysfunction caused by the mutant cognate tRNA. This has been shown for *LARS2* with tRNA$^{Leu}_{UUR}$ A3243G MELAS mutation [38], *IARS2* with tRNA^{Ile} T4277C mutation [39], and *VARS2* with tRNA^{Val} C1624T mutation [40]. Similar studies have been done in *Saccharomyces cerevisiae*. *LARS2* overexpression increased the percentage of aminoacylated tRNA$^{Leu}_{UUR}$ resulting in improved translation efficiency and respiration capacity [38]. Similarly, the remarkably low steady-state level of mutant tRNA^{Val} level was partially restored by *VARS2* overexpression, which increased the tRNA stability by charging it [40]. These findings indicate that individual variation in aminoacyl-tRNA synthetase levels may affect the outcome of mitochondrial tRNA mutations. Indeed, Perli et al. [39] reported that an affected child and his unaffected mother both had a homoplasmic tRNA^{Ile} mutation but the expression level of *IASR2* was significantly higher in mother's tissues than in the affected child's tissues, suggesting that synthetase level modulated the penetrance of the tRNA mutation. Thus, regulation of aminoacyl-tRNA synthetase expression could have therapeutic potential in mitochondrial diseases.

Conclusion

Mitochondrial aminoacyl-tRNA synthetase defects have in short time emerged as a new cause of mitochondrial disease and surprised us by their tissue-specific outcomes. Undoubtedly, the list of disease genes in this group will soon grow to cover most synthetases. All so far identified mitochondrial synthetase defects are recessively inherited and not totally inactivating, suggesting that some residual synthetase activity is required to maintain mitochondrial protein synthesis but on the other hand, a relatively low amount may be sufficient for most tissues. Splicing effects are one of the suggested reasons for the tissue-specificity (*DARS2, RARS2, HARS2*), but other explanations are also possible and to be addressed in future studies. The set of mitochondrial aminoacyl-tRNA synthetases is not well characterized and each of the proteins may even have additional functions. The cytoplasmic synthetases are known to perform widely varying noncanonical functions in, for example, tRNA export, ribosomal RNA synthesis, apoptosis, inflammation, and angiogenesis [41]. Finding additional processes where mitochondrial aminoacyl-tRNA synthetases are involved would not be surprising. Also, the effect of synthetase expression in modulating translation efficiency may be a factor in tissue-specificity of these diseases. Factors regulating synthetase gene expression have not been identified. The relative amino acid concentrations may also play a role in the vulnerability of each tissue type.

The outcome of mitochondrial tRNA mutations is hardly predictable. Nevertheless, it could be expected that mutations in a tRNA and in its cognate aminoacyl-tRNA synthetase resulted in similar phenotypes. Indeed, tRNA^{His} mutations can cause hearing loss, as do *HARS2* mutations [23, 42]. But, for example, *AARS2* defects lead to cardiomyopathy in infants [24] whereas tRNA^{Ala} mutations have been described in adult-onset ophthalmoplegia or pure myopathy [43, 44].

From the molecular diagnostic point of view, mitochondrial synthetase diseases are difficult to identify as translation defects because they do not manifest in fibroblasts. In distinct clinical phenotypes such as Perrault syndrome, direct sequencing of *HARS2* is feasible but in, for example, mitochondrial cardiomyopathy, multiple genes need to be screened. Next-generation sequencing techniques such as exome sequencing may soon become the most cost-effective way to find the genes responsible for these mitochondrial disease patients.

Identification of additional patients with mutations in the genes described here or in new synthetase genes will increase our understanding of the range of clinical phenotypes caused by mitochondrial translation defects and of the molecular mechanisms by which the synthetase defects impair protein synthesis in mitochondria.

References

1. Schimmel P, Söll D (2005) The world of aminoacyl-tRNA synthetases. In: Ibba M, Francklyn C, Cusack S (eds) The aminoacyl-tRNA synthetases. Landes Bioscience, Georgetown, pp 1–3
2. Watanabe K (2010) Unique features of animal mitochondrial translation systems. The non-universal genetic code, unusual features of the translational apparatus and their relevance to human mitochondrial diseases. Proc Jpn Acad Ser B Phys Biol Sci 86:11–39

3. Nagao A, Suzuki T, Katoh T, Sakaguchi Y (2009) Biogenesis of glutaminyl-mt tRNAGln in human mitochondria. Proc Natl Acad Sci U S A 106:16209–16214
4. van der Knaap MS, van der Voorn P, Barkhof F et al (2003) A new leukoencephalopathy with brainstem and spinal cord involvement and high lactate. Ann Neurol 53:252–258
5. Linnankivi T, Lundbom N, Autti T et al (2004) Five new cases of a recently described leukoencephalopathy with high brain lactate. Neurology 63:688–692
6. Scheper GC, van der Klok T, van Andel RJ et al (2007) Mitochondrial aspartyl-tRNA synthetase deficiency causes leukoencephalopathy with brain stem and spinal cord involvement and lactate elevation. Nat Genet 39:534–539
7. van Berge L, Dooves S, van Berkel CG, Polder E, van der Knaap MS, Scheper GC (2012) Leukoencephalopathy with brain stem and spinal cord involvement and lactate elevation is associated with cell type-dependent splicing of mtAspRS mRNA. Biochem J 441(3):955–962
8. Lin J, Chiconelli Faria E, Da Rocha AJ et al (2010) Leukoencephalopathy with brainstem and spinal cord involvement and normal lactate: a new mutation in the DARS2 gene. J Child Neurol 25:1425–1428
9. Tzoulis C, Tran GT, Gjerde IO et al (2012) Leukoencephalopathy with brainstem and spinal cord involvement caused by a novel mutation in the DARS2 gene. J Neurol 259(2):292–296
10. Synofzik M, Schicks J, Lindig T et al (2011) Acetazolamide-responsive exercise-induced episodic ataxia associated with a novel homozygous DARS2 mutation. J Med Genet 48:713–715
11. Miyake N, Yamashita S, Kurosawa K et al (2011) A novel homozygous mutation of DARS2 may cause a severe LBSL variant. Clin Genet 80:293–296
12. Edvardson S, Shaag A, Kolesnikova O et al (2007) Deleterious mutation in the mitochondrial arginyl-transfer RNA synthetase gene is associated with pontocerebellar hypoplasia. Am J Hum Genet 81:857–862
13. Rankin J, Brown R, Dobyns WB et al (2010) Pontocerebellar hypoplasia type 6: a British case with PEHO-like features. Am J Med Genet A 152A:2079–2084
14. Namavar Y, Barth PG, Kasher PR et al (2011) Clinical, neuroradiological and genetic findings in pontocerebellar hypoplasia. Brain 134:143–156
15. Kasher PR, Namavar Y, van Tijn P et al (2011) Impairment of the tRNA-splicing endonuclease subunit 54 (tsen54) gene causes neurological abnormalities and larval death in zebrafish models of pontocerebellar hypoplasia. Hum Mol Genet 20:1574–1584
16. Yoshihisa T, Yunoki-Esaki K, Ohshima C, Tanaka N, Endo T (2003) Possibility of cytoplasmic pre-tRNA splicing: the yeast tRNA splicing endonuclease mainly localizes on the mitochondria. Mol Biol Cell 14:3266–3279
17. Bykhovskaya Y, Casas K, Mengesha E, Inbal A, Fischel-Ghodsian N (2004) Missense mutation in pseudouridine synthase 1 (PUS1) causes mitochondrial myopathy and sideroblastic anemia (MLASA). Am J Hum Genet 74:1303–1308
18. Patton JR, Bykhovskaya Y, Mengesha E, Bertolotto C, Fischel-Ghodsian N (2005) Mitochondrial myopathy and sideroblastic anemia (MLASA): missense mutation in the pseudouridine synthase 1 (PUS1) gene is associated with the loss of tRNA pseudouridylation. J Biol Chem 280:19823–19828
19. Riley LG, Cooper S, Hickey P et al (2010) Mutation of the mitochondrial tyrosyl-tRNA synthetase gene, YARS2, causes myopathy, lactic acidosis, and sideroblastic anemia—MLASA syndrome. Am J Hum Genet 87:52–59
20. Fechter P, Rudinger-Thirion J, Theobald-Dietrich A, Giege R (2000) Identity of tRNA for yeast tyrosyl-tRNA synthetase: tyrosylation is more sensitive to identity nucleotides than to structural features. Biochemistry 39:1725–1733
21. Belostotsky R, Ben-Shalom E, Rinat C et al (2011) Mutations in the mitochondrial seryl-tRNA synthetase cause hyperuricemia, pulmonary hypertension, renal failure in infancy and alkalosis, HUPRA syndrome. Am J Hum Genet 88:193–200
22. Chimnaronk S, Gravers Jeppesen M, Suzuki T, Nyborg J, Watanabe K (2005) Dual-mode recognition of noncanonical tRNAs(Ser) by seryl-tRNA synthetase in mammalian mitochondria. EMBO J 24:3369–3379

23. Pierce SB, Chisholm KM, Lynch ED et al (2011) Mutations in mitochondrial histidyl tRNA synthetase HARS2 cause ovarian dysgenesis and sensorineural hearing loss of Perrault syndrome. Proc Natl Acad Sci U S A 108:6543–6548
24. Gotz A, Tyynismaa H, Euro L et al (2011) Exome sequencing identifies mitochondrial alanyl-tRNA synthetase mutations in infantile mitochondrial cardiomyopathy. Am J Hum Genet 88:635–642
25. Guo M, Chong YE, Shapiro R, Beebe K, Yang XL, Schimmel P (2009) Paradox of mistranslation of serine for alanine caused by AlaRS recognition dilemma. Nature 462:808–812
26. t Hart LM, Hansen T, Rietveld I et al (2005) Evidence that the mitochondrial leucyl tRNA synthetase (LARS2) gene represents a novel type 2 diabetes susceptibility gene. Diabetes 54:1892–1895
27. Reiling E, Jafar-Mohammadi B, van 't Riet E et al (2010) Genetic association analysis of LARS2 with type 2 diabetes. Diabetologia 53:103–110
28. Tolkunova E, Park H, Xia J, King MP, Davidson E (2000) The human lysyl-tRNA synthetase gene encodes both the cytoplasmic and mitochondrial enzymes by means of an unusual alternative splicing of the primary transcript. J Biol Chem 275:35063–35069
29. Chang KJ, Wang CC (2004) Translation initiation from a naturally occurring non-AUG codon in Saccharomyces cerevisiae. J Biol Chem 279:13778–13785
30. Antonellis A, Ellsworth RE, Sambuughin N et al (2003) Glycyl tRNA synthetase mutations in Charcot-Marie-Tooth disease type 2D and distal spinal muscular atrophy type V. Am J Hum Genet 72:1293–1299
31. He W, Zhang HM, Chong YE, Guo M, Marshall AG, Yang XL (2011) Dispersed disease-causing neomorphic mutations on a single protein promote the same localized conformational opening. Proc Natl Acad Sci U S A 108:12307–12312
32. Nangle LA, Zhang W, Xie W, Yang XL, Schimmel P (2007) Charcot-Marie-Tooth disease-associated mutant tRNA synthetases linked to altered dimer interface and neurite distribution defect. Proc Natl Acad Sci U S A 104:11239–11244
33. Motley WW, Talbot K, Fischbeck KH (2010) GARS axonopathy: not every neuron's cup of tRNA. Trends Neurosci 33:59–66
34. McLaughlin HM, Sakaguchi R, Liu C et al (2010) Compound heterozygosity for loss-of-function lysyl-tRNA synthetase mutations in a patient with peripheral neuropathy. Am J Hum Genet 87:560–566
35. Jordanova A, Irobi J, Thomas FP et al (2006) Disrupted function and axonal distribution of mutant tyrosyl-tRNA synthetase in dominant intermediate Charcot-Marie-Tooth neuropathy. Nat Genet 38:197–202
36. Latour P, Thauvin-Robinet C, Baudelet-Mery C et al (2010) A major determinant for binding and aminoacylation of tRNA(Ala) in cytoplasmic Alanyl-tRNA synthetase is mutated in dominant axonal Charcot-Marie-Tooth disease. Am J Hum Genet 86:77–82
37. Zuchner S, Mersiyanova IV, Muglia M et al (2004) Mutations in the mitochondrial GTPase mitofusin 2 cause Charcot-Marie-Tooth neuropathy type 2A. Nat Genet 36:449–451
38. Li R, Guan MX (2010) Human mitochondrial leucyl-tRNA synthetase corrects mitochondrial dysfunctions due to the tRNALeu(UUR) A3243G mutation, associated with mitochondrial encephalomyopathy, lactic acidosis, and stroke-like symptoms and diabetes. Mol Cell Biol 30:2147–2154
39. Perli E, Giordano C, Tuppen HA et al (2012) Isoleucyl-tRNA synthetase levels modulate the penetrance of a homoplasmic m.4277T>C mitochondrial tRNAIle mutation causing hypertrophic cardiomyopathy. Hum Mol Genet 21(1):85–100
40. Rorbach J, Yusoff AA, Tuppen H et al (2008) Overexpression of human mitochondrial valyl tRNA synthetase can partially restore levels of cognate mt-tRNAVal carrying the pathogenic C25U mutation. Nucleic Acids Res 36:3065–3074
41. Guo M, Yang XL, Schimmel P (2010) New functions of aminoacyl-tRNA synthetases beyond translation. Nat Rev Mol Cell Biol 11:668–674
42. Crimi M, Galbiati S, Perini MP et al (2003) A mitochondrial tRNA(His) gene mutation causing pigmentary retinopathy and neurosensorial deafness. Neurology 60:1200–1203

43. Spagnolo M, Tomelleri G, Vattemi G, Filosto M, Rizzuto N, Tonin P (2001) A new mutation in the mitochondrial tRNA(Ala) gene in a patient with ophthalmoplegia and dysphagia. Neuromuscul Disord 11:481–484
44. Swalwell H, Deschauer M, Hartl H et al (2006) Pure myopathy associated with a novel mitochondrial tRNA gene mutation. Neurology 66:447–449

Chapter 17
Mitochondrial Protein Translation-Related Disease: Mitochondrial Ribosomal Proteins and Translation Factors

Brett H. Graham

Introduction

Mitochondria are cellular organelles that are involved in a broad variety of cellular processes within the eukaryotic cell. They are dynamic structures that can vary in their morphology and numbers to adjust to the changing cellular environment [1]. While mitochondria play a fundamental role in energy metabolism and oxidative phosphorylation, they are also involved in many other cellular functions such as the intrinsic pathway of apoptosis, intracellular Ca^{2+} buffering, intermediate metabolism, and cellular homeostasis [2–4]. Mitochondria have a characteristic structure and topology, consisting of double lipid bilayers that delineate an intermembrane space (the space between the mitochondrial outer and inner membranes) and matrix (the space within the mitochondrial inner membrane). The components of the TCA cycle and other enzymes are typically located in the matrix, while the electron transport chain (ETC) is embedded in the mitochondrial inner membrane with oxidative phosphorylation occurring at the interface of the inner membrane and the matrix [5].

The biology of mitochondria is remarkable for the organelle having its own multicopy circular genome, the mitochondrial DNA (mtDNA), which encodes 13 polypeptides (a subset of ETC complex subunits), 22 transfer RNAs, and 2 ribosomal RNAs [6]. While mtDNA encodes 13 proteins, the mitochondrion contains at least 1,300 proteins that contribute to all of its functions, >99% of which are nuclear encoded and imported into the mitochondrion [7]. These nuclear-encoded proteins include components of the ETC as well as the protein machinery, distinct from the nucleus and cytoplasm, for replication and maintenance of the mtDNA, and transcription and translation of the mtDNA [7].

Mitochondrial disease can be caused by mutations in mtDNA or in nuclear-encoded mitochondrial genes. Genes required for proper mitochondrial translation is

---

B. H. Graham (✉)
Department of Molecular and Human Genetics, Baylor College of Medicine,
One Baylor Plaza, MS:BCM225, Houston, TX 77030, USA
e-mail: bgraham@bcm.edu

Medical Genetics, Texas Children's Hospital, Houston, TX USA

one subset of these disease genes. This subset includes mtDNA-encoded genes (encoding ribosomal RNAs (rRNA) and transfer RNAs (tRNA)) and nuclear-encoded genes that encode mitochondrial ribosomal proteins (MRPs), translation factors, aminoacyl-tRNA synthetases, and tRNA-modifying enzymes [8]. While disease genes encoding mitochondrial aminoacyl-tRNA synthetases and tRNA-modifying enzymes are discussed elsewhere in this book, this chapter is devoted to mitochondrial disease caused by mutations in genes encoding MRPs and mitochondrial translation factors.

Mitochondrial Translation

Overview

The scope of this chapter allows for only a brief description of mitochondrial translation; however, the reader is referred to more detailed recent reviews [9, 10]. Protein synthesis or translation is a cyclic process that consists of four stages: initiation, elongation, termination, and ribosome recycling. Initiation involves multiple initiation factors that orchestrate the pairing of messenger RNA (mRNA) with ribosomal complexes, resulting in the alignment of the start codon with an fMet-tRNA^{Met} in the ribosomal P-site [6]. During elongation, the mRNA is advanced through the ribosome by three nucleotides (one codon) through the coordinated action of elongation factors, allowing each subsequent codon to be recognized by the proper related aminoacylated tRNA and the corresponding amino acid to be covalently bonded to the growing polypeptide [11, 12]. Termination occurs when a transacting termination (release) factor recognizes a stop codon occupying the ribosomal A-site and separates the nascent polypeptide from the tRNA in the ribosomal P-site. This action releases the polypeptide from the ribosomal complex [10]. During the final ribosomal recycling stage, the ribosomal complex is separated into individual large and small ribosomal subunits, with the release of the mRNA transcript, thereby allowing the utilization of these components in a fresh cycle of translation [9]. In addition to the described stages above, individual mitochondrial ribosomal proteins (MRP) must be assembled into the large and small ribosomal subunits with the mitochondrial rRNAs and posttranslationally modified MRPs. Also, individual tRNAs must be posttranscriptionally modified and aminoacylated with the appropriate amino acid. Each of these steps requires additional nuclear-encoded proteins to perform the specific functions [9].

Reflecting the evolutionary prokaryotic-endosymbiotic origin of mitochondria, translation in the mitochondrion is distinct from cytoplasmic translation and shares many similarities with eubacterial translation (translation factors, ribosomal sensitivity to chloramphenicol, etc.); however, there are also clear differences. The mitochondrial genetic code contains multiple differences in terms of codon usage when compared with the standard genetic code utilized by cytosolic and prokaryotic translation; and the mitochondrial genetic code itself varies among species

[6]. While eubacterial and mitochondrial ribosomes have a similar molecular mass of 2.7×10^6 Da, the mitoribosome is more buoyant with a lower sedimentation coefficient, reflecting a higher protein:nucleic acid ratio [6, 9].

Mitochondrial Ribosomes

Mammalian mitoribosomes have a sedimentation coefficient of 55 S, compared with 80 S for cytoplasmic ribosomes and 70 S for prokaryotic ribosomes. The mitochondrial ribosomal small subunit (SSU) is a 28 S particle (compared with 40 S for cytoplasmic and 30 S for bacterial SSU), while the ribosomal large subunit (LSU) is a 39 S particle (compared with 60 S for cytoplasmic and 50 S for bacterial LSU) [9]. The mammalian SSU contains a minimum of 29 proteins with 14 proteins being clear homologs of bacterial ribosomal proteins and 15 proteins being mitochondrial-specific ribosomal proteins [13–17]. The mammalian LSU contains a minimum of 48 proteins with 28 proteins being clear homologs of bacterial ribosomal proteins and 20 proteins being mitochondrial-specific ribosomal proteins [18]. The mitochondrial large and small rRNAs present in the mitoribosome are shorter than their cytoplasmic and bacterial homologs with the large rRNA being 16 S/1600 bases (compared with 23 S/2900 bases for bacteria and 28 S/4800 bases for eukaryotic cytoplasm) and the small rRNA being 12 S/950 bases (compared with 16 S/1540 bases for bacteria and 18 S/1900 bases for eukaryotic cytoplasm) [6]. It has been a long-standing contention that metazoan mitoribosomes lack the 5 S rRNA present in bacterial, cytoplasmic, and plant mitochondrial ribosomes; however, recent studies have suggested that, in mammals, the nuclear-encoded 5 S rRNA is imported into the mitochondrion from the cytoplasmic pool and associates with the mitoribosome through a mechanism dependent upon rhodanese (a 5 S rRNA import factor) and mitochondrial ribosomal protein L18 [19].

Translation Factors

Prokaryotic translation initiation utilizes three initiation factors (IF1-3), while eukaryotic cytoplasmic translation exhibits a repertoire of five initiation factors (eIF1-5). For mitochondria translation initiation, only two initiation factors have been identified (IF2$_{mt}$ and IF3$_{mt}$), with a putative mitochondrial homolog to IF1 unidentified in any species to date [9]. IF2$_{mt}$ and IF3$_{mt}$ work in concert to initiate translation, with IF3$_{mt}$ promoting the dissociation of the mitoribosome into the LSU and SSU and with IF2$_{mt}$ promoting the binding of fMet-tRNA^{Met} to the SSU, followed by assembly of the mitoribosome-fMet-tRNA^{Met}-mRNA complex and concomitant release of IF2$_{mt}$ and IF3$_{mt}$ [9].

Mitochondrial translation elongation closely resembles eubacterial elongation with both processes involving three elongation factors: EFTu, EFTs, and EFG. The mitochondrial elongation factors are termed EF-Tu$_{mt}$, EF-Ts$_{mt}$, and EF-G1$_{mt}$. First, an EF-Tu$_{mt}$-GTP complex shuttles the proper aminoacylated-tRNA to the A-site of the mitoribosome. With proper placement of the aminoacylated-tRNA in the A-site, GTP is hydrolyzed to GDP, and the EF-Tu$_{mt}$-GDP complex is released from the mitoribosome. EF-Ts$_{mt}$ functions by binding the EF-Tu$_{mt}$-GDP complex and stimulates the exchange of GDP for GTP, resulting ultimately in a renewed EF-Tu$_{mt}$-GTP complex that is free to bind another aminoacylated-tRNA. In the meantime, the LSU in the mitoribosome catalyzes peptide bond formation between the nascent amino acid and the growing polypeptide chain, which is then transferred to the tRNA occupying the A-site. Finally, a EF-G1$_{mt}$–GTP complex binds to the A-site and catalyzes the translocation of the mitoribosome with the peptidyl-tRNA moving from the A-site to the P-site, resulting in the release of the deacylated tRNA and the EF-G1$_{mt}$–GDP complex [9].

The mitochondrial translation factors involved in termination and ribosomal recycling are mtRF1a (mtRF1-L), RRF1$_{mt}$ (mtRRF), and RRF2$_{mt}$ (EF-G2$_{mt}$). When a termination codon enters the mitoribosomal A-site, a mtRF1a-GTP complex binds to the A-site, promoting GTP hydrolysis and release of the polypeptide from the peptidyl-tRNA in the P-site. After mtRF1a is released from the ribosome (mechanism is unclear since there is no identified homolog to the bacterial RF3), RRF1$_{mt}$/mtRRF and RRF2$_{mt}$/EF-G2$_{mt}$ bind to the A-site and promote recycling with the dissociation of the ribosomal subunits as well as the release of the deacylated tRNA and the mRNA [9, 10]. While mtRF1a is the sole mitochondrial termination factor that recognizes termination codons, there are three paralogs in mammals that localize to the mitochondrion: mtRF1, C12orf65, and ICT1 [9]. While the specific function during translation is unknown for mtRF1 and C12orf65, ICT1 has been demonstrated in vitro to function as a ribosome-dependent, codon-independent, peptidyl-tRNA hydrolase and may play a role in the surveillance of ribosomal stalling [20, 21].

MRPs and Disease

MRPS16

While the mitochondrial ribosome consists of at least 77 proteins, mutations in only two genes encoding SSU ribosomal proteins (*MRPS16*, *MRPS22*) have been reported to cause mitochondrial disease in patients. The first mitochondrial ribosomal protein defect was described in a female infant from a consanguineous Bedouin family who died at 3 days of age from intractable lactic acidosis in the context of normal ventricular function [22]. Prenatally, the patient presented with agenesis of the corpus callosum, cerebral ventricular dilatation, and intrauterine growth retardation. Postnatally, the patient exhibited encephalopathy, hypotonia, elevated liver transaminases, and dysmorphic features. Respiratory chain enzymatic analyses in muscle, liver, and fibroblasts demonstrated multiple deficiencies with sparing of complex II activity. Analysis of translation in fibroblasts by pulse-labeling studies demonstrated a translational defect, while analysis of RNA by Northern blot demonstrated a specific reduction in the 12 S rRNA. Sequencing of the 12 S rRNA mtDNA-encoded gene and flanking sequences was normal, but candidate sequencing of the 14 SSU ribosomal protein genes that are homologs of bacterial ribosomal proteins exhibited a homozygous nonsense mutation (R111X) in *MRPS16* [22].

MRPS22

The second family with a mitochondrial ribosome protein defect was described in 2007 with the presentation of three affected children from a consanguineous couple (first cousins) [23]. Each of the children presented with subcutaneous edema and ascites at birth with subsequent development of hypertrophic cardiomyopathy and renal tubulopathy. All of the children died between 2 and 22 days of life. Analysis of respiratory chain enzyme activities from muscle demonstrated multiple deficiencies with the sparing of complex II activity. Analysis of RNA transcripts by qRT-PCR from patient fibroblasts demonstrated significant reduction of 12 S rRNA levels. Homozygosity mapping revealed a large area of homozygosity on chromosome 3 for all three patients that contained three candidate genes for a defect in mitochondrial translation: *GFM1* (encodes EF-G1$_{mt}$), *MRPL3*, and *MRPS22*. While sequencing of *GFM1* and *MRPL3* was normal, sequencing of *MRPS22* revealed a homozygous missense mutation (R170H) that cosegregated with disease. Pathogenicity was confirmed by in vitro rescue experiments where wild-type *MRPS22* cDNA expressed in patient fibroblasts increased 12 S rRNA levels and restored cytochrome *c* oxidase (complex IV) activity [23]. More recently, a second unrelated family with an affected child with mitochondrial disease and a pathogenic *MRPS22* mutation was described [24]. The child, born from a first-degree consanguineous Pakistani family presented prenatally with microcephaly and hypertrophic cardiomyopathy by fetal ultrasound. Postnatally, the child exhibited encephalopathy, hypertrophic cardiomyopathy, lactic acidosis, dysmorphic features, and poor growth, but was alive at 5.5 years of age at the time of the report. Analysis of patient fibroblasts demonstrated multiple respiratory chain deficiencies with complex II activity spared, reduced steady state levels of complexes I and IV by blue native gel electrophoresis (BN-PAGE), and a translational defect by pulse labeling of mitochondria. Homozygosity mapping resulted in the identification of a homozygous missense mutation in *MRPS22* (L215P). A subsequent cDNA rescue experiment in the patient fibroblasts confirmed pathogenicity of the mutation [24].

Mitochondrial Translation Factors and Disease

EF-G1$_{mt}$

The first family with members affected with mitochondrial disease and defective EF-G1$_{mt}$ was described in 2004 [25]. Two affected siblings from first-degree consanguineous Lebanese parents presented with encephalopathy, hepatopathy, growth failure, and lactic acidosis. Both patients died from fulminant hepatic failure between 4 and 8 weeks of life. Analysis of fibroblasts demonstrated complex I and complex IV deficiencies and reduced steady state levels of assembled complexes I, III, IV, and V by BN-PAGE with normal levels of complex II. Pulse chase analysis of mitochondrial translation demonstrated a significant translational defect. Complementation testing by microcell-mediated chromosome transfer mapped the locus to chromosome 3 and subsequent microsatellite mapping narrowed the critical region containing the candidate disease gene to an interval that contained genes that encode EF-G1$_{mt}$ (*GFM1*) and MRPS22. Sequencing of *MRPS22* was normal, but analysis of *GFM1* demonstrated a homozygous missense mutation (N174S) that affects a conserved residue of the GTP-binding domain of EF-G1$_{mt}$. Rescue experiments by cDNA complementation confirmed the pathogenicity of the identified mutation [25]. Subsequently, mutations in *GFM1* have been identified in three other unrelated families, with affected members exhibiting lethal hepatic failure and/or encephalopathy [26–28].

EF-Ts$_{mt}$

Two unrelated patients with mitochondrial disease and mutations in *TSFM*, the gene-encoding EF-TS$_{mt}$, have been reported [29]. The first patient was a boy born from a consanguineous Turkish couple that presented with encephalomyopathy, seizures, and lactic acidosis and died at 7 weeks of age from progressive muscle weakness and respiratory failure. The second patient was a girl born from a consanguineous couple of Kurdish Jewish descent that presented with hypotonia, hypertrophic cardiomyopathy, and lactic acidosis and died at 7 weeks of age from progressive cardiomyopathy and lactic acidosis. Analysis of respiratory chain activities from fibroblasts and/or muscle from both patients demonstrated multiple deficiencies with normal complex II activity. Pulse labeling of mitochondrial translation products in fibroblasts from both patients showed generalized translational defects with normal mtDNA sequencing. In addition, reduced steady state levels of assembled respiratory complexes as well as EF-Ts$_{mt}$ and EF-Tu$_{mt}$ (but not EF-G1$_{mt}$) were demonstrated. Candidate gene sequencing in the first patient and homozygosity mapping in the second patient resulted in identification of *TSFM* as the likely disease gene with the identical homozygous missense mutation (R333W) found in both. This mutation affects a conserved residue in the domain required for dimerization of EF-Ts$_{mt}$ and EF-Tu$_{mt}$. Ectopic expression of either EF-Ts$_{mt}$ or EF-Tu$_{mt}$ rescued the translational defects [29].

EF-Tu$_{mt}$

To date, a single patient with mitochondrial disease and mutations in *TUFM*, the gene-encoding EF-Tu$_{mt}$, has been described [27]. The patient was born to nonconsanguineous, unaffected parents who originated from an isolated valley of the Italian

South Tyrol region. The patient presented with encephalopathy, lactic acidosis, and progressive leukodystrophy with microcephaly. She died at 14 months of age from progressive neurological dysfunction. Analyses of muscle biopsy and fibroblasts from the patient demonstrated multiple mtDNA-dependent deficiencies and defective mitochondrial translation. Sequencing of candidate genes resulted in the identification of a homozygous missense mutation in *TUFM* (R339Q). Pathogenicity of the mutation was confirmed by cDNA complementation in the patient's fibroblasts [27].

C12orf65

In 2010, patients from two unrelated families affected with mitochondrial disease and mutations in *C12orf65* were reported [30]. The first family had a child born to a first-degree consanguineous couple of Turkish descent who presented with a Leigh disease-like phenotype, optic atrophy, and ophthalmoplegia. She ultimately died at 8 years of age. The second family had two affected brothers born to nonconsanguineous Dutch parents who presented with progressive neurological deterioration, optic atrophy, and ophthalmoplegia. One brother died at 22 years of age after acute decompensation following a respiratory tract infection. The other brother was alive at age 20 at the time of the report. Analysis of patient fibroblasts from both families demonstrated a translational defect evidenced by deficiency of complex IV activity, reduced levels of mitochondrial translation products by pulse labeling, and reduced steady state levels of assembled respiratory complexes by BN-PAGE. However, normal levels of mitochondrial RNAs, tRNAs, and rRNAs as well as normal steady state levels of mitochondrial translational elongation factors and select ribosomal proteins were demonstrated. Homozygosity mapping led to the identification of homozygous frameshift mutations in *C12orf65*. While each family had a distinct frameshift mutation, both resulted in the same premature stop codon at position 84. Ectopic expression of wild-type *C12orf65* in patient fibroblasts rescued the translational defects. While C12orf65 is a paralog of mtRF1a, it does not exhibit peptidyl-tRNA hydrolase activity with bacterial ribosomes in vitro and its specific function in mitochondrial protein translation is unclear [30].

Conclusion

Defects in nuclear-encoded proteins important for mitochondrial protein translation are an important subset of mitochondrial disease that undoubtedly is under diagnosed. To date, only a small fraction of genes encoding the known mitochondrial translation factors and ribosomal proteins have been identified in patients with mitochondrial disease. Biochemical and molecular investigations of patient tissues and cells are required to recognize a potential translational defect. With the rapid advancement of next-generation sequencing technologies in both the research laboratory and the clinic, the promise of identification of many more disease genes causing mitochondrial translation defects and disease is ever present.

References

1. Bereiter-Hahn J, Voth M (1994) Dynamics of mitochondria in living cells: shape changes, dislocations, fusion, and fission of mitochondria. Microsc Res Tech 27(3):198–219
2. Green DR, Galluzzi L, Kroemer G (2011) Mitochondria and the autophagy-inflammation-cell death axis in organismal aging. Science 333(6046):1109–1112
3. Parone P, Priault M, James D, Nothwehr SF, Martinou JC (2003) Apoptosis: bombarding the mitochondria. Essays Biochem 39:41–51
4. Rambold AS, Lippincott-Schwartz J (2011) Mechanisms of mitochondria and autophagy crosstalk. Cell Cycle 10(23):4032–4038
5. Bereiter-Hahn J, Jendrach M (2010) Mitochondrial dynamics. Int Rev Cell Mol Biol 284:1–65
6. Scheffler IE (2008) Mitochondria, 2nd ed. Wiley, Hoboken
7. Pagliarini DJ, Calvo SE, Chang B et al (2008) A mitochondrial protein compendium elucidates complex I disease biology. Cell 134(1):112–123
8. Rotig A (2011) Human diseases with impaired mitochondrial protein synthesis. Biochim Biophys Acta 1807(9):1198–1205
9. Christian BE, Spremulli LL (2012) Mechanism of protein biosynthesis in mammalian mitochondria. Biochim Biophys Acta 1819(9-10):1035–1054
10. Chrzanowska-Lightowlers ZM, Pajak A, Lightowlers RN (2011) Termination of protein synthesis in mammalian mitochondria. J Biol Chem 286(40):34479–34485
11. Korostelev A, Ermolenko DN, Noller HF (2008) Structural dynamics of the ribosome. Curr Opin Chem Biol 12(6):674–683
12. Wen JD, Lancaster L, Hodges C et al (2008) Following translation by single ribosomes one codon at a time. Nat 452(7187):598–603
13. Koc EC, Burkhart W, Blackburn K, Koc H, Moseley A, Spremulli LL (2001) Identification of four proteins from the small subunit of the mammalian mitochondrial ribosome using a proteomics approach. Protein Sci 10(3):471–481
14. Cavdar Koc E, Burkhart W, Blackburn K, Moseley A, Spremulli LL (2001) The small subunit of the mammalian mitochondrial ribosome. Identification of the full complement of ribosomal proteins present. J Biol Chem 276(22):19363–19374
15. Koc EC, Burkhart W, Blackburn K, Moseley A, Koc H, Spremulli LL (2000) A proteomics approach to the identification of mammalian mitochondrial small subunit ribosomal proteins. J Biol Chem 275(42):32585–32591
16. Cavdar Koc E, Blackburn K, Burkhart W, Spremulli LL (1999) Identification of a mammalian mitochondrial homolog of ribosomal protein S7. Biochem Biophys Res Commun 266(1):141–146
17. Goldschmidt-Reisin S, Kitakawa M, Herfurth E, Wittmann-Liebold B, Grohmann L, Graack HR (1998) Mammalian mitochondrial ribosomal proteins. N-terminal amino acid sequencing, characterization, and identification of corresponding gene sequences. J Biol Chem 273(52):34828–34836
18. Koc EC, Burkhart W, Blackburn K et al (2001) The large subunit of the mammalian mitochondrial ribosome. Analysis of the complement of ribosomal proteins present. J Biol Chem 276(47):43958–43969
19. Smirnov A, Entelis N, Martin RP, Tarassov I (2011) Biological significance of 5 S rRNA import into human mitochondria: role of ribosomal protein MRP-L18. Genes Dev 25(12):1289–1305
20. Haque ME, Spremulli LL (2010) ICT1 comes to the rescue of mitochondrial ribosomes. Embo J 29(6):1019–1020

21. Richter R, Rorbach J, Pajak A et al (2010) A functional peptidyl-tRNA hydrolase, ICT1, has been recruited into the human mitochondrial ribosome. Embo J 29(6):1116–1125
22. Miller C, Saada A, Shaul N et al (2004) Defective mitochondrial translation caused by a ribosomal protein (MRPS16) mutation. Ann Neurol 56(5):734–738
23. Saada A, Shaag A, Arnon S et al (2007) Antenatal mitochondrial disease caused by mitochondrial ribosomal protein (MRPS22) mutation. J Med Genet 44(12):784–786
24. Smits P, Saada A, Wortmann SB et al (2011) Mutation in mitochondrial ribosomal protein MRPS22 leads to Cornelia de Lange-like phenotype, brain abnormalities and hypertrophic cardiomyopathy. Eur J Hum Genet 19(4):394–399
25. Coenen MJ, Antonicka H, Ugalde C et al (2004) Mutant mitochondrial elongation factor G1 and combined oxidative phosphorylation deficiency. N Engl J Med 351(20):2080–2086
26. Smits P, Antonicka H, van Hasselt PM et al (2011) Mutation in subdomain G' of mitochondrial elongation factor G1 is associated with combined OXPHOS deficiency in fibroblasts but not in muscle. Eur J Hum Genet 19(3):275–279
27. Valente L, Tiranti V, Marsano RM et al (2007) Infantile encephalopathy and defective mitochondrial DNA translation in patients with mutations of mitochondrial elongation factors EFG1 and EFTu. Am J Hum Genet 80(1):44–58
28. Antonicka H, Sasarman F, Kennaway NG, Shoubridge EA (2006) The molecular basis for tissue specificity of the oxidative phosphorylation deficiencies in patients with mutations in the mitochondrial translation factor EFG1. Hum Mol Genet 15(11):1835–1846
29. Smeitink JA, Elpeleg O, Antonicka H et al (2006) Distinct clinical phenotypes associated with a mutation in the mitochondrial translation elongation factor EFTs. Am J Hum Genet 79(5):869–877
30. Antonicka H, Ostergaard E, Sasarman F et al (2010) Mutations in C12orf65 in patients with encephalomyopathy and a mitochondrial translation defect. Am J Hum Genet 87(1):115–122

Chapter 18
Disorders of Mitochondrial RNA Modification

William J. Craigen

Introduction

Naturally occurring RNA molecules contain a variety of chemically modified nucleosides derived from the four standard nucleosides: cytidine, adenosine, uridine, and guanosine. Over the last 6 decades, more than 100 structurally distinct modified nucleosides have been identified in all forms of RNA, including ribosomal RNAs (rRNAs), transfer RNAs (tRNAs), messenger RNAs (mRNAs), small nuclear RNAs (snRNAs), small nucleolar RNAs (snoRNAs), and most recently in regulatory RNAs, including microRNAs (miRNAs), small interfering RNAs (siRNAs), and Piwi-interacting RNAs (piRNAs). These alterations are catalogued in web-based databases such as rna-mdb.cas.albany.edu/RNAmods [1]. Analysis of genome sequences of eukaryotes, archaea and bacteria has predicted that between 3 and 11 % of the coding capacity is devoted to enzymes that modify RNA, reflecting the essential nature of this process [2].

The Functions of RNA Modification

The functions of these modified RNAs vary with the types of RNAs. The process of RNA modification itself may contribute to refolding of the RNA [3]. In tRNAs, posttranscriptional modifications function to stabilize the tertiary structure of the RNA, enhance the accuracy of aminoacylation, direct or alter the codon:anticodon specificity, and can modulate the rate of translational frame-shifting [4]. tRNA modifications may not be nucleotide sequence specific, but rather can be directed by the structure of the tRNA.

Similarly, the modification of mRNAs has several potential functions, including altering mRNA stability, translatability, and even the codon composition. The

---

W. J. Craigen (✉)
Department of Molecular and Human Genetics, Baylor College of Medicine,
S842, One Baylor Plaza, Houston, TX 77030, USA
e-mail: wcraigen@bcm.edu

deamination of adenosine to inosine by a family of proteins known as Adenosine Deaminases Acting on RNAs (ADARs) occurs in both nuclear encoded RNAs and invasive viral RNAs [5]. For mRNAs, ADAR editing can alter secondary structure of the RNA, change the amino acid that is incorporated into a protein, or even generate a stop codon and thus truncate a protein prematurely (e.g., ApoB lipoprotein [6]). The modification of ion channel mRNAs by ADARs has been shown to both alter the biophysical properties of the channels and induce the modifications in response to physiologic inputs [7, 8]. While ADAR-mediated editing can be either highly selective for a specific nucleotide or nonselective in the case of viral RNAs and introns, the multiple factors that influence these activities remain an area of investigation. The recent recognition the ADARs act on regulatory RNAs further underlines the complexity of these RNA modifications [9].

tRNA Nucleoside Modifications

The majority of altered nucleosides have been identified in tRNAs, followed in number by rRNA, mRNA, snoRNA and snRNA, and regulatory RNA modifications. Approximately 15–20 % of the tRNA nucleotides are modified, occurring on the base and/or the ribose, and may require a series of enzyme catalyzed steps. While numerous tRNA modifications occur, with the average of about 12 modifications per tRNA [1], only a few examples will be briefly discussed. Similar to inosine formation in mammalian mRNAs, inosine in tRNA is the result of a deamination reaction catalyzed by ADATs (Adenosine Deaminases Acting on tRNA). However, unlike mRNA editing, which can potentially target a number of sites in an mRNA, adenosine-to-inosine editing in tRNA appear to be restricted to only three positions, including the anticodon, where it contributes to codon specificity and acyl-tRNA synthetase activity [10].

A less well understood but similar tRNA editing activity deaminates cytosine to uracil, but till date this modification appears to be limited to organelle tRNAs such as mitochondrial tRNAs, where it allows for a single tRNA species to be charged by more than one acyl-tRNA synthetase [11].

Modifications of uridine are common [12]. In *Escherichia coli*, the uridine at position 34 of tRNA (the wobble position) is always modified, and the modifications are either derivatives of 5-hydroxyuridine, 5-methyluridine, or 5-methyl-aminomethyl-2-thiouridine. In particular, the modified nucleoside 5-methyl-aminomethyl-2-thiouridine (mnm^5s^2U) has been found at the wobble position of the bacterial tRNA^{glu}, tRNA^{gln}, and tRNA^{lys}. The s^2U (thiolation) modification is required for many of the biochemical functions of tRNAs; the s^2U in *E. coli* tRNA^{Glu} is a recognition element for its cognate aminoacyl tRNA synthetase [13], increases the translation rate of GAA codons [14], and mutants lacking s^2U display a significant increase in ribosomal frameshifting, a feature common to many tRNA modification mutants [15]. The synthesis of mnm^5s^2U occurs in multiple enzymatic steps in bacterial tRNAs. The synthesis of s^2U34 in *E. coli* requires the 2-thiouridylase *MnmA* and the cysteine

Fig. 18.1 Thiolation of uridine

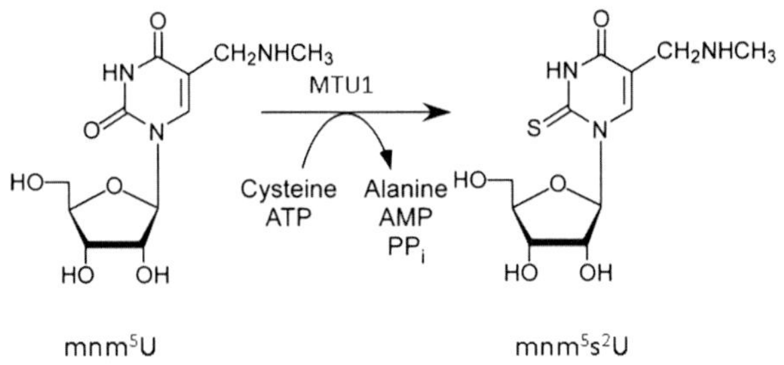

desulfurase *IscS* [16]. Additional steps in bacteria require transfer RNA methyltransferase U (*trmU*) encoding 5-methylaminomethyl-2-thiouridylate-methyltransferase. The names and identifiable activities of these bacterial genes have changed as a better understanding of their functions has occurred [17, 18]. *TrmU*, which actually encodes the thiouridylase activity, became *MnmA*, while *TrmC*, which in fact performs the final methylation step, became *MnmC* [17]. Thus, there is confusion in the nomenclature of eukaryotic genes. While the official symbol for the human gene that performs the *MnmA* function is *TRMU*, the eukaryotic orthologue responsible for the thiolation at position 2 of mnm^{5}s^{2}U34 is better termed as *MTU1* (Fig. 18.1). In yeast disruption of the yeast orthologue of *MnmA*, the *MTU1* gene that has also been termed *MTO2*, eliminates the 2-thio modification of mitochondrial tRNAs and impairs mitochondrial protein synthesis, leading to reduced respiratory activity [19, 20]. Human disorders associated with altered thiolation are discussed below.

Pseudouridine (5-ribosyluracil: Ψ (Psi)) is one of the most prevalent posttranscriptional RNA modifications [21], occurring on tRNAs, rRNAs, snoRNAs and snRNAs, and regulatory RNAs. Pseudouridine is a C-glycoside rotation isomer of uridine that is generated in either a guide RNA-independent or guide RNA-dependent process (Fig. 18.2). The isomerization of uridine to Ψ may improve nucleotide stacking interactions and therefore may contribute to the stabilization of the local RNA structure. A distinction between uridine and Ψ that likely is biologically important is the ability of Ψ to form an additional hydrogen bond in the major groove of an RNA duplex and thus increase the rigidity and stability of RNAs with Ψ. Certain Ψ residues in rRNA have been shown to play a role in ribosome biogenesis and in protein synthesis, while several Ψ sites in U2 spliceosomal snRNA contribute to pre-mRNA splicing [12, 21].

rRNA Nucleoside Modifications

rRNA modifications generally include methylation or pseudouridinylation, but a few examples of more complex modifications exist [22]. Methylation of RNA enhances base stacking by increasing the hydrophobicity of nucleosides, blocks hydrogen bonds through increased steric hindrance, and affects the strength of hydrogen bonds by introducing a positive charge to the nucleoside [3]. In eucaryotic rRNAs, targeting

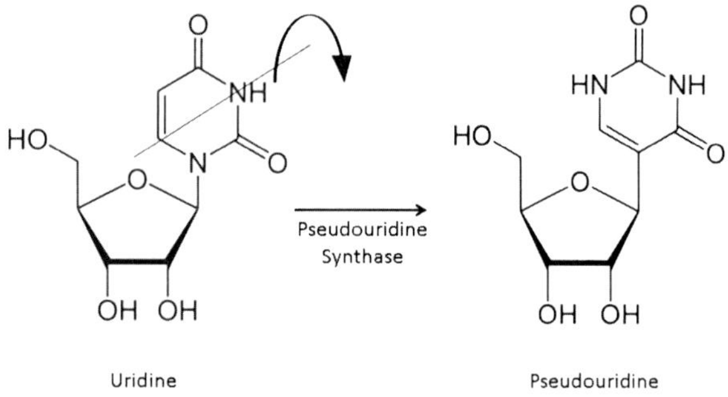

Fig. 18.2 Conversion of uridine to pseudouridine

of pseudouridine synthesis and 2'-O-ribose methylation is guided by snoRNAs that are complementary to the rRNA region flanking the nucleotide to be modified, with snoRNAs themselves undergoing prior Ψ modifications [23]. In vertebrates, most snoRNAs are intronic and cotranscribed with the host gene transcripts, then processed out of the excised introns via snRNA-dependent spliceosomal complexes, but a small number are transcribed from either independent RNA polymerase II or III units. The host genes of intronic snoRNAs are often involved in processes related to snoRNA functions [24]. SnoRNAs can be classified into three structurally and functionally defined classes. SnoRNAs carrying the conserved box C/D elements function primarily as guides in the site-specific 2'-O-methylation of rRNAs via the methyltransferase fibrillarin [25]. Box H/ACA snoRNAs direct the pseudouridylation of rRNAs via a ribonucleoprotein protein-RNA complex (box H/ACA snoRNP). A third class of snoRNAs, the small Cajal body-specific RNAs (scaRNAs) are involved in the posttranscriptional modification of snRNAs. The scaRNAs differ from the other two snoRNAs in that they contain the box C/D and H/ACA motifs and an additional CAB motif, and are thus longer than the other two snoRNAs. The Cajal bodies; small subcompartments of the nucleus, are the sites of scaRNP-mediated spliceosome snRNA modification [26].

The formation of Ψ is catalyzed by an evolutionarily ancient family of proteins, the pseudouridine synthases [12]. Pseudouridine synthase sequences from Archaea, bacteria, and eukaryotes can be classified into five families named after the *E. coli* enzymes: RluA, RsuA, TruA, TruB, and TruD. The latter three are involved primarily in tRNA modification, while the RluA and RsuA family members are directed more at rRNA sites. Despite minimal sequence similarity between the various enzymes of the different families, the only universally shared element is an aspartic acid residue in the catalytic site, structural comparisons have revealed that all Ψ synthases share a core protein motif with a common fold and a conserved active-site cleft. A subset of enzymes of the RluA and RsuA groups have N-terminal domains similar to that of the ribosomal protein S4 in keeping with their role in rRNA modification, while the enzymes of the TruB family have a C-terminal domain termed PUA that is shared amongst pseudouridine synthases and archaeosine transglycosylases (that generate a 7-deazaguanosine derivative unique to Archaea [27]).

A remarkable feature of the pseudouridine synthases is their substrate specificity. Pseudouridine synthases recognize their substrate uridine only in the context of a target RNA. This appears necessary because catalysis of free uridine would potentially deplete the cellular pool of uridine, the precursor of thymidine needed for DNA synthesis. With the exception of the TruB type pseudouridine synthase found in box H/ACA RNPs (Cbf5 in yeast, dyskerin in humans), where the RNA component directs substrate selection, all pseudouridine synthases examined to date are capable of recognizing their substrates without any accessory factors [28]. Two heritable human disorders have been identified that involve mutations in pseudouridine synthases. Mutations in dyskerin, the catalytic component of box H/ACA RNPs involved in rRNA, tRNA, and telomere processing, are associated with a highly pleiotropic X-linked syndrome dyskeratosis congenita (OMIM 305000) and the more clinically severe Hoyeraal-Hreidarsson syndrome, with features including bone marrow failure, increased cancer risk, intellectual disabilities, and ectodermal changes [29]. Mutations in the human TruA family member PUS1 leads to sideroblastic anemia, mitochondrial dysfunction, and variable intellectual disabilities, as discussed below.

Disorders of Mitochondrial RNA Modification

It has been suggested that RNA modifications provide a mechanism for decoding multiple codons with a limited number of tRNAs [30], and allow for the differences in the universal genetic code and those found in mitochondria [31]. Defects in mitochondrial RNA modification can be due to primary alterations in the RNA sequence encoded by the mitochondrial genome and thus altered substrate, or Mendelian defects in nuclear-encoded modifying enzymes. It can be imagined that other classes of disorders may be uncovered related to the perturbations in the guide function of snoRNAs, similar to that seen in the RNA component (*TERC*) of telomerase [29], or in other as yet poorly understood processes regulating post-transcriptional modifications. The difficulty in identifying when changes to the primary sequence of an RNA gene lead to alterations in the modification of nearby or distant nucleosides that have functional consequences cannot be overstated.

Two well-described examples of primary mutations in mtDNA-encoded genes are the mitochondrial myopathy, encephalopathy, lactic acidosis, and stroke-like episodes (MELAS) syndrome and the myoclonus epilepsy associated with ragged red fibers (MERFF) syndrome. These two disorders are the most frequent causes of the mitochondrial encephalomyopathies, MELAS being primarily caused by a single base replacement, an A to G transition, at nucleotide position 3243 in the tRNA^{Leu} gene (*MT-TL1*) that is responsible for the translation of the UUR (R = A or G) leucine codons (tRNA$^{Leu(UUR)}$), while MERRF syndrome is typically caused by an A to G transition at position 8344 in the tRNA$^{Lys(AAR)}$ gene (*MT-TK*) needed for decoding lysine codons. Both tRNAs have a taurinomethyl group at position 5 of the uridine (5-taurinomethyluridine), while tRNA^{Lys} has an additional thiol-group at position 2 (5-taurinomethyl-2-thiouridine). Loss of taurine modification leads to a failure of codon:anticodon interactions within the context of the mitochondrial ribosome and thus a failure of protein translation [32]. In the case of tRNA^Leu(UUR), it appears to be specific for the UUG codon, with continued decoding of the UUA codon, while for tRNA^Lys(AAR) both AAA and AAG codons fail to be accurately decoded. Depending on codon usage in specific mitochondrial proteins, differing efficiencies of protein synthesis can be anticipated, with the MERRF mutation having a global impact on translation while the MELAS mutation potentially having a more selective defect, a feature that has been observed [33].

TRMU Deficiency

As discussed previously, the nomenclature in the field is inconsistent; *TRMU* purportedly encodes a mitochondria-specific tRNA-modifying enzyme, tRNA 5-methylaminomethyl-2-thiouridylate methyltransferase (OMIM 610230). However, the term that describes the human disease more accurately is a 2-thiouridylase and corresponds to the bacterial gene *MnmA* [19, 34]. The mitochondrial tRNA-specific 2-thiouridylase, MTU1 (currently also known as TRMU or MTO2) is the orthologue of the bacterial *MnmA* gene that carries out the sulfation of uridine at the 2-position of U34. The initial report of a human disease caused by deficiency of TRMU (OMIM 613070) described eight patients in seven unrelated families of Yemenite Jewish origin who all presented in infancy with acute liver failure [35]. All affected infants had physiologic hyperbilirubinemia as newborns but were described as healthy during the early neonatal period. The infants were hospitalized at 2–4 months because of irritability, poor feeding, and vomiting, and found to be in liver failure, with jaundiced sclerae, distended abdomens, hepatomegaly, and a coagulopathy. The liver synthetic failure was characterized by a vitamin K-unresponsive coagulopathy that included low coagulation factors 5 and 11 and a low plasma albumin. Other notable metabolic abnormalities included a direct hyperbilirubinemia, a metabolic acidosis, and an increased plasma lactic acid and alpha-fetoprotein. Plasma amino acid profiles demonstrated consistent increases in phenylalanine, tyrosine, methionine, glutamine, and alanine, while urinary organic acid analysis showed elevations of lactate, tyrosine, and phenylalanine metabolites, along with increased dicarboxylic and 3-hydroxydicarboxylic acids. While one patient also had a dilated cardiomyopathy and proteinuria, the disease in the remaining patients was limited to the liver.

Liver biopsy during the acute phase revealed minimal chronic inflammation and mild focal proliferation of bile ductules, with variable degrees of portal and sinusoidal fibrosis. In the liver parenchyma, abundant acidophilic granular cytoplasm ("oncocytic") suggestive of mitochondrial proliferation was observed in hepatocytes, as well as focal macrovesicular steatosis.

During the acute phase of the disease, ETC activities were measured in liver and skeletal muscle samples. In the liver, the activities of complexes I, III, and IV were reduced, with preservation of complex II activity, suggesting a defect in expression of the mitochondrial genome either at the level of translation, transcription, or mtDNA metabolism. Citrate synthase, a matrix enzyme that reflects mitochondrial content in the cell, was increased in almost all liver samples, while mtDNA copy number was normal or increased, differentiating the liver disease from that seen in the mtDNA depletion syndromes. The ETC enzymology in skeletal muscle was normal. The investigators took advantage of the consanguinity in the patients by carrying out SNP (single nucleotide polymorphism) genotyping of two affected individuals and identifying a candidate region through homozygosity mapping. This led to the identification of a founder mutation (Y77H) in the *TRMU* gene that was homozygous in 5 of the initial patients, and screening of other patients with a similar clinical and biochemical profile identified a variety of other compound heterozygotes. Most infants showed gradual clinical improvement over several months, although another patient who recovered during infancy had extensive hepatic fibrosis and cirrhotic changes [36]. Interestingly, an older publication of transient liver failure associated with hepatic cytochrome oxidase deficiency in a Yemenite Jewish infant also described pancreatic insufficiency as part of the clinical presentation [37].

The human *TRMU* gene located at 22q13 encodes a 421 residue protein that participates in the posttranscriptional modification of mitochondrial tRNAs. The protein is responsible for the 2-thiolation of the wobble position of the mitochondrial tRNA^{Lys}, tRNA^{Gln}, and tRNA^{Glu} molecules. The authors of the initial report demonstrated that mitochondrial protein translation is globally impaired in patient fibroblasts, and that thiolation of mitochondrial, but not cytosolic, tRNAs was reduced; however, the effect on translation has recently been questioned [34]. The authors speculated that the reason for the transient nature of the disorder is that neonates have a reduced ability to synthesize cysteine via the transsulfuration pathway needed for the sulfation of U34. The isolated nature of the disorder, with only liver involvement and spontaneous recovery in most patients, makes this disorder fairly unique in the catalogue of human mitochondrial diseases. The recent report of a patient previously diagnosed with Infantile Mitochondrial Myopathy due to reversible cytochrome oxidase deficiency and persistent hypotonia who had transient liver failure and in whom mutations in *TRMU* were found [38] suggests that long term follow up studies will broaden the phenotype.

TRMU has also been implicated as a modifier of mtDNA encoded hearing loss. The homoplasmic A1555G and C1494T mutations are located in the highly conserved aminoacyl-tRNA decoding site of the mitochondrial 12S rRNA (OMIM 561000) and have been associated with both aminoglycoside-induced [OMIM 580000) and nonsyndromic deafness in many populations. There is considerable variability in penetrance within and between families. A survey of several hundred individuals from multiple Arab-Israeli, Chinese, and European families revealed a missense variant (A10S) within TRMU in numerous Arab-Israeli and European pedigrees, but not Chinese pedigrees [39]. The same authors showed that homozygosity of the variant in the absence of the mtDNA mutations has no effect on hearing per se, but in cell lines from individuals with both genotypes they observed undermodification of tRNAs, reduced steady-state levels of mitochondrial tRNALeu$^{(UUR)}$, tRNASer$^{(UCN)}$, tRNA^{Met}, and tRNA^{His} transcripts, and reduced protein synthesis of mtDNA encoded proteins.

Myopathy, Lactic Acidosis, and Sideroblastic Anemia

The sideroblastic anemias are a heterogeneous group of hematologic disorders characterized by anemia and ineffective erytheropoiesis reflected in a low reticulocyte count. In the bone marrow, the ring sideroblasts are the hallmark of these disorders, with a perinuclear ring of mitochondria with accumulated iron associated with abnormal heme biosynthesis [40]. Myopathy, lactic acidosis, and sideroblastic anemia (MLASA [OMIM 600462]) is a mitochondrial respiratory chain disorder characterized by progressive exercise intolerance and sideroblastic anemia. The condition was originally described in 1974 in two brothers found to have sideroblastic anemia, a cardiomyopathy and skeletal myopathy, and chronic lactic acidemia [41]. Electron microscopy studies of a muscle biopsy showed inclusions in mitochondria. Subsequently, a similar disorder, but with additional mild facial dysmorphism, was described in Jewish Iranian families from an isolated population [42]. The severity of symptoms varies between and within affected families and sometimes includes intellectual disabilities in some individuals. Muscle biopsy in other patients likewise revealed paracrystalline inclusions [43]. MLASA is associated with mutations in pseudouridylate synthase 1 (PUS1 [OMIM 610957]), resulting in decreased pseudouridylation of both cytoplasmic and mitochondrial tRNAs, leading to decreased protein translation. A founder mutation (R116W) was identified in families of Iranian Jewish origin [44, 45], while more recently a nonsense mutation has been identified in an Italian family [46]. The Italian cases involved two brothers with different clinical phenotypes; one was of normal intelligence, with muscle hypoplasia and ragged red fibers, growth hormone deficiency, and a severe, transfusion-dependent sideroblastic anemia, while his brother exhibited moderate intellectual disability, a less severe myopathy, and a mild anemia. Muscle and fibroblast extracts showed a significant reduction in complex 1 and complex IV ETC activities in both boys, and each was found to harbor a truncating nonsense mutation in PUS1 [46]. The basis for the intrafamilial variability in the reported cases remains unexplained.

The PUS1 locus at 12q24 encodes two distinct transcripts that direct the expression of two protein isoforms due to alternative splicing of exon 1, one containing a cleavable mitochondrial targeting polypeptide that after proteolytic processing gives rise to a 37 kDa protein (PUS1-1). The second isoform (PUS1-2) lacks the amino terminal leader peptide but has additional amino terminal residues not found in the mature PUS1-1 protein. The amino terminal 72 amino acid residues of PUS1-2 match residues 29–103 of the PUS1-1 unprocessed form but are not found in the mature mitochondrial PUS1-1 polypeptide, hence the PUS1-2 isoform has a larger mass of 44 kDa. Depending on the isoform-specific transcript, the mutation in the two boys is either E220X (PUS1-1) or E192X (PUS1-2) [46]. It was suggested that the additional protein sequence found in PUS1-2, in addition to determining the correct nuclear localization of the enzyme, may also affect its kinetic properties. The two mutations described to date would affect both isoforms, but isoform-specific mutations could be envisioned.

PUS1 also acts as a regulator of nuclear receptor activity through the modification of steroid receptor RNA activator (SRA). SRA functions as an RNA molecule to activate nuclear receptors (members of a superfamily of ligand-inducible transcription factors), although certain 5′ splice variants also encode a translated coactivator protein termed SRAP [47]. The pseudouridine synthase activity of PUS1 is required for SRA coactivation, and is part of the nuclear receptor complex that interacts with the response element on DNA [48]. Hence, some aspects of MLASA may be caused by loss of SRA-dependent activities.

Recently, through homozygosity mapping, a second MLASA locus (OMIM 613561) was identified in Australians of Lebanese ancestry. Clinically, the phenotype is very similar to that of PUS1 deficiency, with sideroblastic anemia, progressive myopathy and variable cardiomyopathy; however intellectual disability was not a feature and the onset of disease was at a somewhat younger age. Biochemically, chronic lactic acidemia was associated with reduction in complex I, III, and IV ETC activities, increased citrate synthase activity, and reduced expression of mtDNA-encoded proteins. By sequencing candidate genes, and after excluding *PUS1*, a missense substitution (F56L) in the mitochondrial tyrosyl-tRNA synthetase (*YARS2*, OMIM 610957) was identified. The *YARS2* gene encodes a 477 amino acid protein that localizes to the mitochondrial matrix, and the missense substitution leads to a ninefold reduction in catalytic efficiency. Given that it was identified in two apparently unrelated families on a common haplotype at the *YARS2* locus, it was assumed to be due to a founder mutation.

In summary, disorders of RNA modification can be due to primary changes in the RNA moiety or in enzymes directly involved in the modification process. To date, two enzymes, TRMU and PUS1, have been identified, which when mutated reduces the accurate modifications of different classes of RNAs, leading to altered protein expression and RNA stability. Undoubtedly, additional components of the elaborate modification pathways will be found to cause human mitochondrial disease.

References

1. Dunin-Horkawicz S, Czerwoniec A, Gajda MJ, Feder M, Grosjean H, Bujnicki JM (2006) MODOMICS: A database of RNA modification pathways. Nucleic Acids Res 34:D145–149
2. Anantharaman V, Koonin EV, Aravind L (2002) Comparative genomics and evolution of proteins involved in RNA metabolism. Nucleic Acids Res 30:1427–1464
3. Ishitani R, Yokoyama S, Nureki O (2008) Structure, dynamics, and function of RNA modification enzymes. Current opinion in structural biology 18:330–339
4. Waas WF, Druzina Z, Hanan M, Schimmel P (2007) Role of a tRNA base modification and its precursors in frameshifting in eukaryotes. J Biol Chem 282:26026–26034
5. Gallo A, Locatelli F (2012) ADARs: Allies or enemies? The importance of A-to-I RNA editing in human disease: from cancer to HIV-1. Biol Rev Cambridge Philos Soc 87:95–110
6. Blanc V, Davidson NO (2011) Mouse and other rodent models of C to U RNA editing. Methods Mol Biol 718:121–135
7. Orlandi C, Barbon A, Barlati S (2012) Activity regulation of adenosine deaminases acting on RNA (ADARs). Mole Neurobiol 45:61–75
8. Tan BZ, Huang H, Lam R, Soong TW (2009) Dynamic regulation of RNA editing of ion channels and receptors in the mammalian nervous system. Mol Brain 2:13
9. Nishikura K (2010) Functions and regulation of RNA editing by ADAR deaminases. Annu Rev Biochem 79:321–349

10. Paris Z, Fleming IM, Alfonzo JD (2011) Determinants of tRNA editing and modification: Avoiding conundrums, affecting function. Semin Cell Dev Biol
11. Randau L, Stanley BJ, Kohlway A, Mechta S, Xiong Y, Soll D (2009) A cytidine deaminase edits C to U in transfer RNAs in Archaea. Sci 324:657–659
12. Ofengand J (2002) Ribosomal RNA pseudouridines and pseudouridine synthases. FEBS Lett 514:17–25
13. Madore E, Florentz C, Giege R, Sekine S, Yokoyama S, Lapointe J (1999) Effect of modified nucleotides on Escherichia coli tRNAGlu structure and on its aminoacylation by glutamyl-tRNA synthetase. Predominant and distinct roles of the mnm5 and s2 modifications of U34. Eur J Biochem FEBS 266:1128–1135
14. Kruger MK, Pedersen S, Hagervall TG, Sorensen MA (1998) The modification of the wobble base of tRNAGlu modulates the translation rate of glutamic acid codons in vivo. J Mol Biol 284:621–631
15. Urbonavicius J, Stahl G, Durand JM, Ben Salem SN, Qian Q, Farabaugh PJ, Bjork GR (2003) Transfer RNA modifications that alter+1 frameshifting in general fail to affect−1 frameshifting. RNA 9:760–768
16. Kambampati R, Lauhon CT (2003) MnmA and IscS are required for in vitro 2-thiouridine biosynthesis in Escherichia coli. Biochem 42:1109–1117
17. Hagervall TG, Pomerantz SC, McCloskey JA (1998) Reduced misreading of asparagine codons by Escherichia coli tRNALys with hypomodified derivatives of 5-methylaminomethyl-2-thiouridine in the wobble position. J Mol Biol 284:33–42
18. Sullivan MA, Cannon JF, Webb FH, Bock RM (1985) Antisuppressor mutation in Escherichia coli defective in biosynthesis of 5-methylaminomethyl-2-thiouridine. J Bacteriol 161:368–376
19. Umeda N, Suzuki T, Yukawa M, Ohya Y, Shindo H, Watanabe K (2005) Mitochondria-specific RNA-modifying enzymes responsible for the biosynthesis of the wobble base in mitochondrial tRNAs. Implications for the molecular pathogenesis of human mitochondrial diseases. J Biol Chem 280:1613–1624
20. Wang X, Yan Q, Guan MX (2007) Deletion of the MTO2 gene related to tRNA modification causes a failure in mitochondrial RNA metabolism in the yeast Saccharomyces cerevisiae. FEBS Lett 581:4228–4234
21. Charette M, Gray MW (2000) Pseudouridine in RNA: What, where, how, and why. IUBMB Life 49:341–351
22. Guymon R, Pomerantz SC, Ison JN, Crain PF, McCloskey JA (2007) Post-transcriptional modifications in the small subunit ribosomal RNA from Thermotoga maritima, including presence of a novel modified cytidine. RNA 13:396–403
23. Reichow SL, Hamma T, Ferre-D'Amare AR, Varani G (2007) The structure and function of small nucleolar ribonucleoproteins. Nucleic Acids Res 35:1452–1464
24. Kiss T, Fayet-Lebaron E, Jady BE (2010) Box H/ACA small ribonucleoproteins. Mol Cell 37:597–606
25. Lin J, Lai S, Jia R, Xu A, Zhang L, Lu J, Ye K (2011) Structural basis for site-specific ribose methylation by box C/D RNA protein complexes. Nat 469:559–563
26. Hamma T, Ferre-D'Amare AR (2010) The box H/ACA ribonucleoprotein complex: interplay of RNA and protein structures in post-transcriptional RNA modification. J Biol Chem 285:805–809
27. Phillips G, de Crecy-Lagard V (2011) Biosynthesis and function of tRNA modifications in Archaea. Curr Opin Microbiol 14:335–341
28. Hamma T, Ferre-D'Amare AR (2006) Pseudouridine synthases. Chem Biol 13:1125–1135
29. Mason PJ, Bessler M (2011) The genetics of dyskeratosis congenital. Cancer Genet 204:635–645
30. Agris PF, Vendeix FA, Graham WD (2007) tRNA's wobble decoding of the genome: 40 years of modification. J Mol Biol 366:1–13
31. Watanabe K, Yokobori S (2011) tRNA Modification and genetic code variations in animal mitochondria. J Nucleic Acids 2011:623095

32. Kirino Y, Suzuki T (2005) Human mitochondrial diseases associated with tRNA wobble modification deficiency. RNA Biol 2:41–44
33. Kirino Y, Yasukawa T, Ohta S, Akira S, Ishihara K, Watanabe K, Suzuki T (2004) Codon-specific translational defect caused by a wobble modification deficiency in mutant tRNA from a human mitochondrial disease. Proc Nat Acad Sci U S A 101:15070–15075
34. Sasarman F, Antonicka H, Horvath R, Shoubridge EA (2011) The 2-thiouridylase function of the human MTU1 (TRMU) enzyme is dispensable for mitochondrial translation. Hum Mol Genet 20:4634–4643
35. Zeharia A, Shaag A, Pappo O, Mager-Heckel AM, Saada A, Beinat M, Karicheva O, Mandel H, Ofek N, Segel R, Marom D, Rotig A, Tarassov I, Elpeleg O (2009) Acute infantile liver failure due to mutations in the TRMU gene. Am J Hum Genet 85:401–407
36. Schara U, von Kleist-Retzow JC, Lainka E, Gerner P, Pyle A, Smith PM, Lochmuller H, Czermin B, Abicht A, Holinski-Feder E, Horvath R (2011) Acute liver failure with subsequent cirrhosis as the primary manifestation of TRMU mutations. J Inherited Metab Dis 34:197–201
37. Lev D, Gilad E, Leshinsky-Silver E, Houri S, Levine A, Saada A, Lerman-Sagie T (2002) Reversible fulminant lactic acidosis and liver failure in an infant with hepatic cytochrome-c oxidase deficiency. J Inherited Metab Dis 25:371–377
38. Uusimaa J, Jungbluth H, Fratter C, Crisponi G, Feng L, Zeviani M, Hughes I, Treacy EP, Birks J, Brown GK, Sewry CA, McDermott M, Muntoni F, Poulton J (2011) Reversible infantile respiratory chain deficiency is a unique, genetically heterogenous mitochondrial disease. J Med Genet 48:660–66839
39. Guan MX, Yan Q, Li X, Bykhovskaya Y, Gallo-Teran J, Hajek P, Umeda N, Zhao H, Garrido G, Mengesha E, Suzuki T, del Castillo I, Peters JL, Li R, Qian Y, Wang X, Ballana E, Shohat M, Lu J, Estivill X, Watanabe K, Fischel-Ghodsian N (2006) Mutation in TRMU related to transfer RNA modification modulates the phenotypic expression of the deafness-associated mitochondrial 12 S ribosomal RNA mutations. Am J Hum Genet 79:291–302
40. Bergmann AK, Campagna DR, McLoughlin EM, Agarwal S, Fleming MD, Bottomley SS, Neufeld EJ (2010) Systematic molecular genetic analysis of congenital sideroblastic anemia: Evidence for genetic heterogeneity and identification of novel mutations. Pediatr Blood Cancer 54:273–278
41. Rawles JM, Weller RO(1974) Familial association of metabolic myopathy, lactic acidosis and sideroblastic anemia. Am J Med 56:891–897
42. Casas KA, Fischel-Ghodsian N (2004) Mitochondrial myopathy and sideroblastic anemia. Am J Med Genet Part A 125A:201–204
43. Zeharia A, Fischel-Ghodsian N, Casas K, Bykhocskaya Y, Tamari H, Lev D, Mimouni M, Lerman-Sagie T (2005) Mitochondrial myopathy, sideroblastic anemia, and lactic acidosis: an autosomal recessive syndrome in Persian Jews caused by a mutation in the PUS1 gene. J Child Neurol 20:449–452
44. Bykhovskaya Y, Casas K, Mengesha E, Inbal A, Fischel-Ghodsian N (2004) Missense mutation in pseudouridine synthase 1 (PUS1) causes mitochondrial myopathy and sideroblastic anemia (MLASA). Am J Hum Genet 74:1303–1308
45. Casas K, Bykhovskaya Y, Mengesha E, Wang D, Yang H, Taylor K, Inbal A, Fischel-Ghodsian N (2004) Gene responsible for mitochondrial myopathy and sideroblastic anemia (MSA) maps to chromosome 12q24.33. Am J Med Genet Part A 127A:44–49
46. Fernandez-Vizarra E, Berardinelli A, Valente L, Tiranti V, Zeviani M (2007) Nonsense mutation in pseudouridylate synthase 1 (PUS1) in two brothers affected by myopathy, lactic acidosis and sideroblastic anaemia (MLASA). J Med Genet 44:173–180
47. Colley SM, Leedman PJ (2011) Steroid Receptor RNA Activator – A nuclear receptor coregulator with multiple partners: Insights and challenges. Biochimie 93:1966–1972
48. Zhao X, Patton JR, Ghosh SK, Fischel-Ghodsian N, Shen L, Spanjaard RA (2007) Pus3p- and Pus1p-dependent pseudouridylation of steroid receptor RNA activator controls a functional switch that regulates nuclear receptor signaling. Mol Endocrinol 21:686–699

Part V
Others

Chapter 19
Pyruvate Dehydrogenase Complex Deficiencies

Suzanne D. DeBrosse and Douglas S. Kerr

Pyruvate dehydrogenase complex (PDC) deficiency is an inherited disorder of energy metabolism and one of the primary lactic acidemias. The lactic acidemia due to PDC deficiency results from a metabolic block within the enzyme complex that catalyzes conversion of pyruvate to acetyl-CoA, the first step in oxidative metabolism of carbohydrate, and lactic acidemia is due to the increase in pyruvate. Elevated lactate accompanied by elevated pyruvate with a normal lactate-to-pyruvate ratio is a hallmark of PDC deficiency, shared by a few other inborn errors. Several prior clinical reviews of PDC deficiency have been published [1–3].

Biochemistry of the Pyruvate Dehydrogenase Complex

Pyruvate dehydrogenase complex is a mitochondrial multienzyme complex that, in several steps, catalyzes the production of acetyl-CoA from pyruvate produced by glycolysis. Conversion to acetyl-CoA is one of the several possible fates of pyruvate, others being reduction to lactic acid, gluconeogenesis via oxaloacetate, or transamination to alanine. Pyruvate, produced in the cytoplasm from glycolysis, is actively transported into the mitochondrion to undergo oxidative decarboxylation, forming acetyl-CoA and NADH. Acetyl-CoA enters the tricarboxylic acid cycle to produce more NADH and $FADH_2$ substrates for generation of most of the ATP derived from carbohydrate via oxidative phosphorylation by the mitochondrial electron transport chain. PDC contains 3 catalytic enzymes, 2 regulatory enzymes, and a binding protein, and requires the cofactors thiamine pyrophosphate (TPP), lipoic acid, and flavin adenine dinucleotide (FAD). The three catalytic enzymes are present in multiple copies within PDC and associate into a dodecahedron with an E2 core (see below), whereby they can efficiently act in concert [4–6].

D. S. Kerr (✉) · S. D. DeBrosse
Pediatric Endocrinology and Metabolism, Department of Pediatrics,
Case Western Reserve University, Rainbow Babies and Children's Hospital,
11100 Euclid Avenue, Cleveland, OH 44106–6004, USA
e-mail: douglas.kerr@case.edu

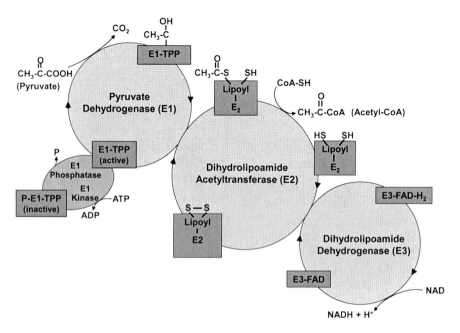

Fig. 19.1 The overall biochemical reactions of the pyruvate dehydrogenase complex, including its regulation by phosphorylation/dephosphorylation

Pyruvate Dehydrogenase and its Regulatory Enzymes

The first enzyme of the PDC complex is pyruvate dehydrogenase (E1), which contains both alpha and beta subunits, encoded by separate genes, in a heterotetramer. E1, in the presence of bound TPP, removes carbon dioxide (decarboxylation) from pyruvate to form hydroxyethylthiamine pyrophosphate (HETPP; Fig. 19.1). The active site of E1 and location of the TPP cofactor is between the alpha and beta subunits [7].

The activity of E1 is regulated by PDH phosphatases and kinases. PDH phosphatases convert E1 into the dephosphorylated active form. The most active isoform of PDH phosphatase is PDP1, a heterodimer. PDP1 includes a regulatory subunit responsive to ionized calcium and magnesium concentrations and binds to the lipoyl domain of E2 [5]. A second isoform of PDH phosphatase, PDP2, has much lower activity. PDH kinases catalyze ATP-dependent phosphorylation of E1 into an inactive form, by phosphorylating three serine residues. PDH kinases are regulated and are inhibited by pyruvate or dichloroacetate (an analogue of pyruvate). Thus, E1 is the rate-limiting step of PDC and is physiologically regulated. In the usual fed state, when pyruvate is plentiful and inhibits PDH kinase, flux through PDC is increased. In the fasted state, when carbohydrate is relatively scarce, flux through PDC is decreased, and energy metabolism shifts towards utilization of fat and amino acids. Increases in the ratios of the products of pyruvate oxidation (ATP/ADP, NADH/NAD and acetyl-CoA/CoA) downregulate PDC via activation of PDH kinases [8].

Dihydrolipoamide S-Acetyltransferase

The next step in conversion of pyruvate to acetyl-CoA is catalyzed by dihydrolipoamide S-acetyltransferase (E2). E2 transfers the hydroxyethyl group from TPP to covalently-bound oxidized lipoamide, which oxidizes the hydroxyethyl group to form an acetylthioester of reduced lipoamide (acetyldihydrolipoamide). The acetyl group is then transferred to free coenzyme A to form acetyl-CoA and reduced dihydrolipoamide-E2 (Fig. 19.1).

Dihydrolipoamide Dehydrogenase

Finally, dihydrolipoamide dehydrogenase (E3), a flavoprotein, reoxidizes the lipoyl group of dihydrolipoamide-E2 to lipoamide-E2. FAD, bound to E3, accepts the electrons, with reduction to $FADH_2$, then passes them on to NAD+, producing $NADH + H^+$ (Fig. 19.1). E3 is not exclusive to the pyruvate dehydrogenase complex, as it is common to alpha-ketoglutarate and branched-chain 2-ketoacid dehydrogenase complexes and the glycine cleavage enzyme. E3 is bound to E1 and E2 via the E3 binding protein, which is specific to PDC.

Genetics of PDC

Each catalytic and regulatory enzyme and the E3 binding protein in PDC is encoded by a separate nuclear gene (Table 19.1). Transcripts of these nuclear genes are translated in the cytoplasm as precursor proteins, which are targeted and imported into mitochondria, where the targeting sequence is removed. Mutations in all of these genes have been associated with PDC deficiency. Defects in PDH kinase have not been documented and would not be expected to result in PDC deficiency, as loss of function of this inhibitory kinase would result in unregulated overactivity of PDC. It has been postulated that mutations in PDH kinase may be responsible for some cases of hypoglycemia, but that has not been established [5].

In addition to the genes that directly encode the catalytic, structural, and regulatory proteins of PDC (Table 19.1), mutations affecting biosynthesis of required cofactors, TPP and lipoic acid, have recently been described in association with PDC and other enzyme deficiencies (see section "Mutations Specifically Associated with PDC Deficiency"). These include *TPK1* (encoding thiamine pyrophosphokinase), *LIPS* (encoding lipoate synthase), *NFU1* and *BOLA3* (encoding iron-sulfate cluster scaffolds required for lipoate biosynthesis) [9–11]. Ubiquitination and proteosome-mediated degradation of the E1-beta subunit has been described as a mechanism for PDC deficiency; the genetic basis of this observation has not been determined [12]. Correlations between the diverse genotypes of PDC deficiency and phenotypic characteristics are discussed below (see section "Genotypes and Phenotypes").

Table 19.1 PDC genes and mutations

Gene	Location[a]	Exons[a]	Amino acids (primary protein)[a]	Reported mutations[b,c]	Reported cases[c,d]
PDHA1	Xp22.12	11	390	127	195
PDHB	3p14.3	10	359	12	15
DLAT	11q23.1	14	647	4	3
PDHX	11p13	11	509	19	24
DLD	7q31.1	14	501	15	13
PDP1	8q22.1	2	595	2	3

Numbers of reported mutations and cases are approximate estimates at time of review
[a]NCBI/Entrez Gene [69]
[b]HGMD [70]
[c]Imbard et al. [16]
[d]Patel K et al. [22]

Mutations Specifically Associated with PDC Deficiency

Point mutations, small insertions and deletions, exon skipping mutations, larger deletions and contiguous gene deletion syndromes have been identified within and around the *PDHA1* gene in association with PDC deficiency. These account for more cases of PDC deficiency (60 % or more of biochemically diagnosed cases) than all other genes involved in the complex (Table 19.1) [13–16]. The location of this gene on chromosome X accounts for the initial observation that males were more likely to be affected with PDC deficiency (Dahl 1995) [17]. Subsequently, it was discovered that females are frequently affected, but that the types and locations of mutations observed in symptomatic patients differ between males and females. Of note, some mutations, particularly in *PDHA1*, have been reported with or without their mitochondrial targeting site, leading to alternate numbering conventions. Males have mostly missense mutations, usually in exons 3, 7, 8, and 11 [13, 16]. Symptomatic females have a higher incidence of deletions or insertions, usually in exons 10–11 [13, 15]. One analysis described 36 out of 55 affected females as having insertion/deletion mutations, typically small insertions or deletions [13]. Null mutations are found almost exclusively in females, whose intact second copy of *PHDA1* permits survival despite a severe defect in one copy of the gene. Several apparent exceptions in males prove this point, as the survivors with apparently null mutations invariably retain some expression of PDC activity, due to mosaicism or intermittent exon skipping [18, 19].

The mode of inheritance of *PDHA1* mutations can be classified as X-linked, and is neither "dominant" nor "recessive" [13]. This X-linked mode of inheritance also accounts for differences in the survival and neurological outcomes between the sexes (discussed below). Numbers vary between authors and types of *PDHA1* mutation, but in approximately 5–25 % of cases tested, the mother of an affected child was found to be a carrier of the mutation [13, 15, 20]. The mode of inheritance of mutations in the remainder of the genes associated with PDC deficiency is autosomal recessive. Affected individuals may be homozygotes or compound heterozygotes for these mutations, and consanguinity is not uncommon in the pedigrees of affected individuals. Although these other genes account for a much smaller proportion of cases of PDC deficiency compared to *PDHA1*, mutations in *PDHX* are the second most common reported cause of PDC deficiency [16, 21, 22].

Mutations that are Non-Specifically Associated with PDC Deficiency

This group of disorders includes pathogenic mutations of *DLD* and genes involved in the biosynthesis of the co-factors associated with PDC, TPP, and lipoic acid, which are common to the three alpha-ketoacid dehydrogenases and other enzymes. E3 (*DLD*) deficiency is the most commonly reported of this group; the other defects were not recognized until recently and their frequency has not yet been ascertained. Although E3 deficiency has been reported less commonly than defects of E1 or E3BP, it is relatively more common in certain populations, especially Ashkenazi Jews [23, 24]. In addition to its role in the three alpha-ketoacid dehydrogenases, E3 is also part of the glycine cleavage system, where it is known as the "L protein" [24, 25]. However, E3 deficiency is not usually associated with hyperglycinemia.

Thiamine pyrophosphokinase deficiency has recently been described due to mutations of the responsible gene (*TPK1*), including 3 mutations in 5 individuals from three families [9]. As expected, TPK deficiency is associated with variable evidence of loss of activity of all three of the mitochondrial alpha-ketoacid dehydrogenase complexes, similar to E3 deficiency, as well as transketolase, a cytosolic enzyme. Lipoic acid synthetase deficiency, described recently in one individual, is due to a pathogenic mutation of the *LIAS* gene [10]. Mutations of two other genes also have been identified recently that affect different iron-sulfur cluster scaffolds needed for lipoate biosynthesis [11]. These disorders were found in four cases in two families with mutations in one of two different genes: *NFU1* or *BOLA3*. The biochemical consequences are similar, affecting not only the alpha-ketoacid dehydrogenases and the glycine cleavage enzyme, but additionally electron transport chain complexes I, II, and III, whose assembly depends on the same iron-sulfur clusters.

Clinical Presentations and Manifestations

The signs and symptoms of PDC deficiency vary significantly between patients and none are pathognomonic of PDC deficiency. The manifestations, with a few exceptions, tend to be exclusively neurological. Generally speaking, there is a correlation between the age at presentation and the spectrum of clinical findings, which also is influenced by the sex of the affected individual (see section "Genotypes and Phenotypes") and contiguous gene deletion syndromes [1, 26].

Table 19.2 PDC deficiency: "Phenotypic" spectrum. (Modified from Barnerias et al. [26])

Clinical feature	Age at clinical presentation		
	Newborns	Infants	Children
Lactic acidosis	+		
Encephalopathy	+		
Hypotonia	+	+	
Collosal agenesis/dysgenesis	+	+	
Cortical malformation	+	+	
Seizures	+	+	
Developmental delay		+	
Cerebral atrophy		+	
Leigh syndrome		+[a]	
Ataxia, relapsing		+	+
Neuropathy (typically axonal)		+	+
Dystonia, paroxysmal		+	+

[a]Leigh syndrome due to PDC deficiency has been reported primarily in males

The neonatal presentation of PDC deficiency is typically characterized by severe lactic acidemia (associated with tachypnea), hypotonia, and coma (Table 19.2) [26, 27]. Brain malformations, as described below, are common. Neonatal congenital lactic acidosis, often associated with death in the neonatal period, is seen in both males and females. The phenotypes of infants presenting after the neonatal period have been classified into two different groups [26]. Severe developmental delay, often with epilepsy, predominates in females with early infantile onset. Males presenting in infancy are more likely to have Leigh syndrome or brainstem dysfunction without radiologically demonstrated lesions.

Recurrent ataxia with axonal neuropathy, in the absence of intellectual disabilities, is a presentation seen mostly in older children and adolescents, especially males [26, 24]. Motor and sensory deficits, including proprioceptive, are present, and the cerebellum is not involved. Rare cases of late-onset PDC deficiency have been described. Two brothers have been reported who presented in their late 40s with extra-pyramidal symptoms and lytic lesions of the putamina, associated with partial deficiency of PDC activity in cultured skin fibroblasts; no mutation was identified [28].

Many of the neurological findings associated with PDC deficiency are seen independent of age of onset (Table 19.3). Hypotonia and neurodevelopmental delay are the most consistently observed neurological manifestations of PDC deficiency [22, 26, 27, 29]. Hypotonia has been reported in 46–89 % of PDC deficient subjects. Developmental delay has been reported in 57–80 % of cases with PDC deficiency, although criteria for these estimates are not the same (Table 19.3). Although they have superior survival, symptomatic females as a group have worse cognitive outcomes than surviving males, with females having an average developmental quotient (DQ) in the severe range (DQ 25–39), and males having an average degree of intellectual disability in the moderate range (DQ 40–55) [27].

Table 19.3 Reported clinical features in PDC deficiencies (%)

Clinical feature	Robinson et al. [29]	Patel et al. [22]	Barnerias et al. [26]	DeBrosse et al. [27]
Hypotonia	53	46	54	89
Hypertonia	13		9[b]	47
Developmental delay	63	57	68	80[c]
Seizures	40	26	41	57
Ataxia[a]	17	19	27	22
Ventriculomegaly	50	35	36	67
Corpus callosum abnormalities	10	31	18	57
Leigh syndrome	10	27	18	37
Peripheral neuropathy[a]	3	7	41	5

[a]Uncertain ascertainment
[b]Pyramidal signs
[c]Developmental quotient < 70

Seizures, ataxia, and hypertonia, and structural brain abnormalities including microcephaly, ventriculomegaly, and agenesis or dysgenesis of the corpus callosum (see section "Neuropathology"), are also commonly seen (Table 19.3). Seizures, including infantile spasms, have been described in some studies as more common in females than males [16, 30]. This has not been consistently observed by others [26, 27]. Ataxia has previously been reported in 17–27 % of PDC deficient subjects, but among individuals observed after achieving ambulation, 13 out of 15 subjects were described as ataxic, including 9 out of 9 ambulatory males [27]. Peripheral neuropathy is not routinely objectively evaluated, but was reported in 3–41 % of subjects reviewed (Table 19.3).

In general, organ dysfunction outside the nervous system such as myopathy, cardiomyopathy, hepatic, pancreatic, ocular, or hematological dysfunction has rarely been reported in isolated PDC deficiency, which is in contrast to distal defects of the mitochondrial electron transport chain. These features may occur in the more complex disorders associated with PDC deficiency, such as E3, TPP, or lipoic acid synthetic deficiencies described above. An exception is facial dysmorphology, resembling fetal alcohol syndrome, reported in some patients [15, 29]. This facial dysmorphology, reported in about a third of cases with E1 defects, was attributed to impaired pyruvate metabolism, since acetaldehyde produced during metabolism of ethanol also impairs pyruvate metabolism.

Survival

Two recent estimates of mortality and survival in PDC deficiency have noted greater survival in females, as expected due to the prevalence of *PDHA1* mutations [22, 27]. In 60 cases of mixed ages up to 25 years with mutation-confirmed PDC deficiency, death occurred in 59 % of males (17 out of 29) compared to 26 % of females

(8 out of 31) [27]. Mortality was highest in the first year, and diminished significantly after age 4 years. It is postulated that antenatal demise also is frequent in severely affected human males [31]. Surviving males reportedly tend to have residual PDC activity between 20 and 50 % or more of normal [13]. Antenatal demise of males with severe deficiency, coupled with the presence of asymptomatic or subtly affected female carriers, would explain the approximately equal numbers of affected males and females, and the difference in observed mutation types seen in the two sexes [17]. Females with null mutations and a relatively even pattern of X-inactivation would be expected to survive longer than those with skewed X-inactivation of the normal allele [13].

Neuropathology

The brain abnormalities that may be seen with PDC deficiency are heterogeneous and include both congenital malformations and destructive lesions. Agenesis or dysgenesis of the corpus callosum, subependymal cysts, widespread injury to white matter including cystic lesions, and colpocephaly may be seen [32]. Dysgenesis of the corpus callosum is typically associated with other defects of neuronal migration, such as absence of medullary pyramids, olivary nuclei ectopia, dentate nuclei dysplasia, cerebellar Purkinje cell abnormalities, subcortical heterotopias and pachygyria [33]. Multifocal dysplastic lesions have been reported in female patients [31]. Cerebral atrophy and/or ventriculomegaly is commonly reported (Table 19.3). Frank hydrocephalus is occasionally seen [27, 29]. Microcephaly is observed in approximately half of cases [27]. In some patients, especially males, Leigh syndrome, including radiographically visualized lesions of the basal ganglia, thalamus, and brainstem, is observed. Leigh syndrome, subacute necrotizing encephalopathy, is a radiographic or postmortem diagnosis not exclusively associated with PDC deficiency. Leigh syndrome can be caused by a number of different nuclear and mitochondrial DNA mutations, including those encoding all complexes of the electron transport chain, particularly complexes I, IV, and V.

The effects of PDC deficiency on the developing brain have, in some cases, been ultrasonographically documented to begin antenatally, including the onset of ventriculomegaly and cerebral atrophy [31]. Callosal dysgenesis, identified prenatally by ultrasound and fetal MRI, also points to disease onset during the first half of gestation [32]. In the mouse brain, expression of the E1-alpha subunit becomes more prominent after the midorganogenesis stage [34]; whereas knockout of the *PDHA1* gene is not compatible with embryonic survival after 12 days post conception [35]. Brain injury may be progressive during postnatal life, whereas the normal developmental progression of myelination and the cerebellum tend to be spared [32].

Autopsy reports on a few individuals with PDC deficiency have noted a lack of cardiac or muscle or other chronic organ pathology. Significant chronic pathological findings were confined to the CNS, including focal cystic necrosis of the basal ganglia (including putamina and caudate nuclei), associated with neuronal loss and astrocytosis. Cystic lesions of the brainstem and cortex have also been reported [29]. In one subject, there was mild hypomyelination of cerebrum and central axons of optic nerves and posterior columns of spinal cord [36]. Ventriculomegaly or hydrocephalus, and agenesis of the corpus callosum and/or other structural malformations (absence of the putamina or external olives) have been reported [29], consistent with MRI findings in individuals with PDC deficiency.

Pathophysiology

Why is brain energy metabolism so dependent on pyruvate oxidation? In the brain, for the most part, acetyl-CoA is synthesized almost entirely from pyruvate derived from glucose, with little alternative energy source in utero, in the carbohydrate-fed state, or in early fasting. In addition, congenital lactic acidosis may further disrupt energy metabolism and synthesis of neurotransmitters, brain proteins, and brain lipids [37]. Since these neuropathological features are not unique to PDC deficiency and are associated with defects of the mitochondrial electron transport chain, it is presumed that they may result from common deficiency of ATP. Disorders of fatty acid oxidation do not have similar neuropathology, but fatty acids are not taken up by the brain directly as an energy source; they provide energy indirectly through production of ketone bodies and gluconeogenesis. On the other hand, lack of cardiac or muscle pathology in PDC deficiency, which is common to defects of the electron transport chain and fatty acid oxidation disorders, presumably correlates with the ability of these tissues to use fatty acids directly as an energy source, bypassing pyruvate as a source of acetyl-CoA.

Reduced energy generation in PDC deficiency may be responsible for white matter injury [32]. Necrotic lesions may also occur in areas of higher energy demand, such as the basal ganglia [29]. Impairment of glutamate uptake by astrocytes, which is normally coupled with glucose oxidation and ATP generation, may result in excitotoxicity [38]. Glutamate and free radicals may both injure immature oligodendrocytes. A murine model of brain specific E1-alpha deficiency shows defects in neuronal cytoarchitecture in grey matter and reduced size of white matter, associated with complete loss of glutamine and glutamate [39]. These pathophysiologic mechanisms are not exclusive to PDC deficiency, and are common to other mitochondrial disorders that impair ATP formation.

Genotypes and Phenotypes

Some genotype–phenotype correlation exists between severe (e.g., null) mutations, and clinical outcomes, particularly in males or with autosomal recessive mutations, but in general, genotype–phenotype correlations are limited by the relatively large number of different mutations and small number of cases with the same mutation

[22, 27]. A notable exception is phenotypic distinctions between mutations *of DLD, TPK1, LIAS, and BOLA3*, which affect all alpha-ketoacid complexes and other enzymes, *vs.* mutations that are specific to PDC deficiency.

PDHA1 mutations display a widely varied clinical presentation with limited genotype–phenotype correlation [29]. The variable degree of symptoms in females with *PDHA1* mutations can generally be attributed to the degree of unfavorable skewing of X-inactivation [17], although at least one family with a *PDHA1* mutation has been reported with a heterozygous female paradoxically more severely affected than her hemizygous brother [40]. This phenotypic variation is to be expected in females, but also has been found in males, usually unexplained. Somatic mosaicism for *PDHA1* mutations has been described in a few males and females. This can result in a milder phenotype for a male with a severe mutation, similar to heterozygous females [18]. Although most patients with *PDHA1* mutations have unique mutations, there are several reported "hotspots" with at least 15 reported cases sharing a specific mutation. The R263G mutation in *PDHA1* represents one of these hotspots, and is predominantly reported in males with PDC deficiency [13, 22]. They share similar clinical characteristics of longer survival and slowly progressive neurological disorders, with a relatively high frequency of asymptomatic carrier mothers. Some areas of *PDHA1* affected by mutations, including R263G, have been shown to be in areas that link alpha and beta subunits, namely the region extending from amino acid residues 236–290, resulting in loss of immunoreactivity of both subunits [41]. *PDHB* mutations are associated with phenotypic variation similar to that of *PDHA1* mutations [14]. Prior assertions that ataxia is less common in *PDHB* may result from the early age at examination of some patients and early death of others.

Defects in *DLAT* encoding E2 have been observed to be associated with Leigh syndrome with episodic dystonia [42, 43]. However, dystonia is not unique to *DLAT* deficiency and has been reported in patients with *PDHA1* mutations and *PDHX* mutations as well [26, 44]. Patients with E3BP deficiency, resulting from homozygous or compound heterozygous mutations in *PDHX*, have been described as having severe mutations with no expected E3BP protein product. Lactic acidosis, delayed psychomotor development, and hypotonia are the most consistently reported manifestations. However, survival in *PDHX* mutations is claimed to be better than with *PDHA1* mutations, despite an apparent lack of protein product and a mean residual PDC activity of 17 % [21].

Brothers with *PDP1* mutations have been described with a 3-bp deletion resulting in the absence of a leucine residue; both had a relatively mild presentation, but had delayed psychomotor development [45]. A more severely affected female child with a *PDP1* null mutation survived only 6 months [46]. Her survival for such a short period was attributed to the role of PDP2, the other phosphatase isoform. Under normal conditions, PDP2 is not thought to play a major role in PDC regulation, as it has tissue-specific expression, and cannot completely compensate for loss of functional PDP1 protein [46]. A hallmark of PDP1 deficiency is the "rescue" of PDC enzyme activity in the presence of DCA, which inhibits PDC kinase and allows the PDP2 isoform to dephosphorylate E1 [45].

DLD mutations result in a spectrum of biochemical phenotypes ranging from mild to severe, and may affect PDC and the alpha-ketoacid dehydrogenase complex differentially [24, 47]. In cases with relatively more severe alpha-ketoglutarate dehydrogenase complex (KDC) deficiency, severe cardiomyopathy, and hepatic dysfunction have been reported [47, 48]. A mouse E3 knockout model is embryonic lethal [49].

Cases of TPK deficiency had variable structural and functional neurological abnormalities, similar to those with only PDC deficiency, but one child also had left ventricular hypertrophy. TPP is not transported across cell membranes and supplementation with TPP was not found to be effective in restoring PDC activity in vitro. Thiamine supplementation has been tried, as is common in PDC deficiencies, with uncertain benefit [9].

Lipoic acid synthetase deficiency was discovered in an apparently normal infant, born to consanguineous parents, who developed neonatal seizures and rapid neurological deterioration associated with lactic acidosis and hyperglycinemia [10]. Biochemical testing showed severe deficiency of PDC with slight reductions in activity of electron transport chain complexes. This affected child also developed structurally abnormal muscle mitochondria and mild cardiomyopathy prior to death. Lipoate supplementation is not expected to be helpful in this disorder, as experimental models in yeast and mice have shown that eukaryotes are dependent on intramitochondrial thiolation of covalently bound lipoyl groups. The clinical consequences of the fatal disorders of iron-sulfur cluster scaffolds needed for lipoate biosynthesis include those biochemical and neurological features reported with lipoate synthase deficiency and also cardiomyopathy in one case [11, 50].

Diagnosis of PDC Deficiency

The presence of signs and symptoms consistent with PDC deficiency should prompt further investigation. Laboratory investigations for PDC deficiency typically begin with measurements of lactate and pyruvate, and the ratio of lactate to pyruvate. Typically, the lactate to pyruvate ratio is normal (between 10 and 20), reflecting an equilibrium between lactate/pyruvate and NAD/NADH without impairment of NADH oxidation. However, blood lactate may be normal in PDC deficiency, particularly in the fasting state. Separate measurements on several occasions after carbohydrate intake may be required to document lactic acidemia, and in some patients, increased blood lactate or pyruvate has not been detected [51]. Measurements of lactate in cerebrospinal fluid (CSF) or determination of brain lactate by MRS may be more sensitive, but again, increased CSF or brain lactate is not seen in every case of PDC deficiency. However, if present, lactic acidemia or CSF lactate elevation should increase suspicion for PDC deficiency. The concomitant rise in both pyruvate and lactate in PDC deficiency, while a key finding, is not unique to this disorder. Conditions resulting in a normal ratio of lactate to pyruvate include disorders of gluconeogenesis and recovery from transient hypoxemia. In more common clinical situations of insufficient tissue oxygenation or in disorders of the electron transport chain, in which NADH oxidation is impaired, a high ratio of NADH/NAD shifts equilibrium between pyruvate and lactate toward lactate and an increase in the lactate to pyruvate ratio is observed.

Plasma amino acid analysis is useful, since transamination of pyruvate to alanine typically results in increased plasma alanine. Urine organic acid analysis also is valuable, to distinguish specific lactic and pyruvic aciduria from defects of the tricarboxylic acid cycle and other organic acidurias. Hyperammonemia is not characteristic of PDC deficiency and is not known to be a direct consequence of a PDC enzyme deficiency; however, unexplained transient hyperammonemia has been observed in some young children with PDC deficiency [24, 31]. Hypoglycemia and/or ketosis are not characteristic of untreated PDC deficiency, in contrast to pyruvate carboxylase deficiency and other gluconeogenic defects, as there is no associated impairment of gluconeogenesis or oxidation of acetyl-CoA. On the other hand, ketosis may be beneficial for those on ketogenic diets for treatment of PDC deficiency (see section "Treatment of PDC Deficiency").

In deficiencies of E3, TPP, or lipoate synthesis, one would expect increases in branched chain amino acids and/or glycine in plasma, and lactate, pyruvate, alpha-ketoglutaric acid, and other alpha-keto or alpha-hydroxy branched chain organic acids in urine. However, frequently these expected associated metabolic findings are not found or only partially evident [3, 9, 10, 50].

Biochemical and genetic testing are generally needed to confirm a diagnosis of PDC deficiency and to identify the specific protein deficiency within the complex. The enzyme activity of PDC and its components may be measured by direct assay of enzyme activity. Genetic testing includes sequencing and deletion/duplication (dosage) testing of one or more of the genes involved in the pyruvate dehydrogenase complex [52]. Immunohistochemistry of PDC proteins is a potential adjunct to enzyme activity assays, but not currently available in most clinical reference laboratories [53].

PDC enzyme activity may be measured in several tissue types [54]. The least invasive assay measures enzyme activity in lymphocytes isolated from peripheral blood. Skin fibroblast PDC enzyme testing is more commonly used as a minimally invasive assay and has the advantage of an ongoing source of cells for further testing. PDC activity can be assayed in frozen muscle biopsies. Typically reserved for postmortem specimens, assay of PDC activity in other tissues is also feasible [36]. However, variation of PDC activity in cells and tissues has been observed, especially in females with *PDHA1* mutations but also in some males [55].

The clinical sequelae of PDC deficiency presumably are dependent on the extent of the reduction of PDC enzyme activity [1]. Very low activity is often but not exclusively seen in children with severe lactic acidosis and death before age 6 months, and mild cases with isolated episodic ataxia tend to have higher (>20 % of normal) residual activity [29]. However, measured activity in cultured skin fibroblasts, lymphocytes, or muscle may not reflect PDC activity in the nervous system, and therefore determination of PDC enzyme activity, or a specific mutation, is not reliable for prediction of outcome. Paradoxically, one of the oldest, most mildly affected cases of PDC deficiency had no detectable enzyme activity in his skin fibroblasts, so other unidentified factors probably contribute to favorable outcome [56].

Treatment of PDC Deficiency

The historically first and now most commonly used treatment has been to employ a low carbohydrate, high fat "ketogenic" diet [57]. The basic rationale for this approach is to provide an alternate source of acetyl-CoA from fatty acid oxidation, and to sustain ketosis in the fed state, replacing glucose as the primary energy source for the brain, as occurs naturally during extended fasting. Use of a ketogenic diet is effective in achieving these results, but may be difficult to implement. Typically, a 3:1 or 4:1 ketogenic diet (grams fat: grams protein plus carbohydrate) is employed, providing from 87–90 % of energy from fat, about 8–10 % from protein, and less than 5 % from carbohydrate. The effect of diet on degree of ketosis varies by age and individual, so it is important to monitor blood beta-hydroxybutyrate (and, if possible, acetoacetate) to sustain a level of total ketosis similar to that achieved in fasting (about 4–6 mM). This is easier to achieve in an infant than in an older child or adolescent. Although there is anecdotal clinical evidence of the benefit of ketogenic diet treatment from case reports, particularly comparison of the outcome of males with identical mutations of *PDHA1*, the efficacy of this approach has not been objectively proven by controlled clinical trials, and PDC deficient subjects may have fatal outcomes while receiving a ketogenic diet [58, 59]. However, ketogenic nutrition is beneficial in experimental models of PDC deficiency [60, 61].

The second widely used therapy for PDC deficiency is supplementation with thiamine. The rationale is that since TPP is necessary for the E1 component of PDC, the rate-limiting step of the complex, then providing more thiamine could result in more TPP and more enzymatic activity. This rationale is not supported by experimental evidence that providing supplemental thiamine to rats increases their tissue levels of TPP. However, several "thiamine-responsive" cases have been described, mostly with mutations of *PDHA1*, and thiamine supplementation is inexpensive and apparently harmless [16, 62].

The third less commonly used treatment is dichloroacetate (DCA). The rationale is that DCA is an inhibitor of pyruvate kinase, and facilitates keeping E1 in the dephosphorylated or active form [63]. After many case reports of the benefit of DCA for treatment of PDC deficiency and other mitochondrial disorders, two clinical trials were conducted, one open label and the other a controlled double-blind trial in children, neither of which showed significant benefits for the total heterogeneous subject groups or for the PDC deficient subgroup [64, 65]. A subsequent controlled trial of DCA in older subjects with MELAS had to be discontinued because of the development of neuropathy in many of the subjects receiving DCA [66]. This toxic neuropathic effect may be less in children [67]. Finally, the potential for gene therapy for *PDHA1* mutations has been investigated in mice, using a recombinant adeno-associated viral (rAAV) vector, with some positive results [68]. This approach depends on adequate expression within the central nervous system and progress in the safe use and efficacy of gene therapy for humans in general.

References

1. Robinson BH (2001) Lactic acidemia: disorders of pyruvate carboxylase and pyruvate dehydrogenase. In: Scriver CR, Sly WS, Valle D, Beaudet AL (eds) The metabolic and molecular basis of inherited disease. McGraw-Hill, New York, pp 2275–2295
2. De Meirleir L, Van Coster R, Lissens W (2006) Disorders of pyruvate metabolism and the tricarboxylic acid cycle. In: Fernandes J, Saudubray J-M, van den Berg H (eds) Inborn metabolic diseases: diagnosis and treatment, 4th edn. Springer Medzin Verlag, Heidelberg, pp 163–174
3. Kerr DS, Zinn AB (2009) Disorders of pyruvate metabolism and the tricarboxylic acid cycle. In: Sarafoglu K (ed) Essentials of pediatric endocrinology and metabolism. McGraw-Hill, New York
4. Zhou ZH, McCarthy DB, O'Connor CM, Reed LJ, Stoops JK (2001) The remarkable structural and functional organization of the eukaryotic pyruvate dehydrogenase complexes. Proc Natl Acad Sci USA 98:14802–14807
5. Maj MC, Cameron JM, Robinson BH (2006) Pyruvate dehydrogenase phosphatase deficiency: orphan disease or an under-diagnosed condition? Mol Cell Endocrinol 249:1–9
6. Yu X, Hiromasa Y, Tsen H, Stoops JK, Roche TE, Zhou ZH (2008) Structures of the human pyruvate dehydrogenase complex cores: a highly conserved catalytic center with flexible N-terminal domains. Struct 16:104–114
7. Ciszak EM, Korotchkina LG, Dominiak PM, Sidhu S, Patel MS (2003) Structural basis for flip-flop action of thiamin pyrophosphate-dependent enzymes revealed by human pyruvate dehydrogenase. J Biol Chem 278:21240–21246
8. Patel MS, Korotchkina LG (2006) Regulation of the pyruvate dehydrogenase complex. Biochem Soc Trans 34:217–222
9. Mayr JA, Freisinger P, Schlachter K et al (2011) Thiamine pyrophosphokinase deficiency in encephalopathic children with defects in the pyruvate oxidation pathway. Am J Hum Genet 89:806–812
10. Mayr JA, Zimmermann FA, Fauth C et al (2011) Lipoic acid synthetase deficiency causes neonatal-onset epilepsy, defective mitochondrial energy metabolism, and glycine elevation. Am J Hum Genet 89:792–797
11. Cameron JM, Janer A, Levandovskiy V et al (2011) Mutations in iron-sulfur cluster scaffold genes NFU1 and BOLA3 cause a fatal deficiency of multiple respiratory chain and 2-oxoacid dehydrogenase enzymes. Am J Hum Genet 89:486–495
12. Han Z, Zhong L, Srivastava A, Stacpoole PW (2008) Pyruvate dehydrogenase complex deficiency caused by ubiquitination and proteasome-mediated degradation of the E1 subunit. J Biol Chem 283:237–243
13. Lissens W, De Meirleir L, Seneca S et al (2000) Mutations in the X-linked pyruvate dehydrogenase (E1) alpha subunit gene (PDHA1) in patients with a pyruvate dehydrogenase complex deficiency. Hum Mutat 15:209–219
14. Okajima K, Korotchkina LG, Prasad C et al (2008) Mutations of the E1 beta subunit gene (PDHB) in four families with pyruvate dehydrogenase deficiency. Mol Genet Metab 93:371–380
15. Quintana E, Gort L, Busquets C et al (2010) Mutational study in the PDHA1 gene of 40 patients suspected of pyruvate dehydrogenase complex deficiency. Clin Genet 77:474–482
16. Imbard A, Boutron A, Vequaud C et al (2011) Molecular characterization of 82 patients with pyruvate dehydrogenase complex deficiency. Structural implications of novel amino acid substitutions in E1 protein. Mol Genet Metab 104:507–516
17. Dahl HHM (1995) Pyruvate dehydrogenase E1a deficiency: males and females differ yet again. Am J Hum Genet 56:553–557
18. Okajima K, Warman ML, Byrne LC, Kerr DS (2006) Somatic mosaicism in a male with an exon skipping mutation in PDHA1 of the pyruvate dehydrogenase complex results in a milder phenotype. Mol Genet Metab 87:162–168
19. Boichard A, Venet L, Naas T et al (2008) Two silent substitutions in the PDHA1 gene cause exon 5 skipping by disruption of a putative exonic splicing enhancer. Mol Genet Metab 93:323–330

20. Tulinius M, Darin N, Wiklund LM et al (2005) A family with pyruvate dehydrogenase complex deficiency due to a novel C > T substitution at nucleotide position 407 in exon 4 of the X-linked Epsilon1alpha gene. Eur J Pediatr 164:99–103
21. Brown RM, Head RA, Morris AA et al (2006) Pyruvate dehydrogenase E3 binding protein (protein X) deficiency. Dev Med Child Neurol 48:756–760
22. Patel KP, O'Brien TW, Subramony SH, Shuster J, Stacpoole PW (2012) The spectrum of pyruvate dehydrogenase complex deficiency: clinical, biochemical and genetic features in 371 patients. Mol Genet Metab 105:34–43
23. Shaag A, Saada A, Berger I et al (1999) Molecular basis of lipoamide dehydrogenase deficiency in Ashkenazi Jews. Am J Med Genet 82:177–182
24. Cameron JM, Levandovskiy V, MacKay N et al (2006) Novel mutations in dihydrolipoamide dehydrogenase deficiency in two cousins with borderline-normal PDH complex activity. Am J Med Genet A 140:1542–1552
25. Patel MS, Hong YS, Kerr DS (2000) Genetic defects in E3 component of alpha-keto acid dehydrogenase complexes. Methods Enzymol 324:453–464
26. Barnerias C, Saudubray JM, Touati G et al (2010) Pyruvate dehydrogenase complex deficiency: four neurological phenotypes with differing pathogenesis. Dev Med Child Neurol 52:e1–e9
27. DeBrosse S, Okajima K, Schmotzer C, Frohnapfel M, Kerr DS (2012) Spectrum of neurological outcomes in pyruvate dehydrogenase complex deficiencies (manuscript in preparation)
28. Mellick G, Price L, Boyle R (2004) Late-onset presentation of pyruvate dehydrogenase deficiency. Mov Disord 19:727–729
29. Robinson BH, MacMillan H, Petrova-Benedict R, Sherwood WG (1987) Variable clinical presentation in patients with defective E1 component of pyruvate dehydrogenase complex. J Pediatr 111:525–533
30. De Meirleir L (2002) Defects of pyruvate metabolism and the Krebs cycle. J Child Neurol 17 Suppl 3:3S26–3S33
31. Wada N, Matsuishi T, Nonaka M, Naito E, Yoshino M (2004) Pyruvate dehydrogenase E1 alpha subunit deficiency in a female patient: Evidence of antenatal origin of brain damage and possible etiology of infantile spasms. Brain Dev 26:57–60
32. Soares-Fernandes JP, Teixeira-Gomes R, Cruz R et al (2008) Neonatal pyruvate dehydrogenase deficiency due to a R302H mutation in the PDHA1 gene: MRI findings. Pediatr Radiol 38:559–562
33. Michotte A, De Meirleir L, Lissens W et al (1993) Neuropathological findings of a patient with pyruvate dehydrogenase E1 alpha deficiency presenting as a cerebral lactic acidosis. Acta Neuropathol (Berl) 85:674–678
34. Takakubo F, Dahl HH (1994) Analysis of pyruvate dehydrogenase expression in embryonic mouse brain: localization and developmental regulation. Brain Res Dev Brain Res 77:63–76
35. Johnson MT, Mahmood S, Hyatt SL et al (2001) Inactivation of the murine pyruvate dehydrogenase (PDHA1) gene and its effect on early embryonic development. Mol Genet Metab 74:293–302
36. Kerr DS, Ho L, Berlin CM et al (1987) Systemic deficiency of the first component of the pyruvate dehydrogenase complex. Pediatr Res 22:312–318
37. De Meirleir L, Lissens W, Denis R et al (1993) Pyruvate dehydrogenase deficiency: clinical and biochemical diagnosis. Pediatr Neurol 9:216–220
38. Yoshioka A, Bacskai B, Pleasure D (1996) Pathophysiology of oligodendroglial excitotoxicity. J Neurosci Res 46:427–437
39. Pliss L, Mazurchuk R, Spernyak JA, Patel MS (2007) Brain MR imaging and proton MR spectroscopy in female mice with pyruvate dehydrogenase complex deficiency. Neurochem Res 32:645–654
40. De Meirleir L, Specola N, Seneca S, Lissens W (1998) Pyruvate dehydrogenase E1 alpha deficiency in a family: different clinical presentation in two siblings. J Inherit Metab Dis 21:224–226
41. Wexler ID, Hemalatha SG, Liu TC, Berry SA, Kerr DS, Patel MS (1992) A mutation in the E1a subunit of pyruvate dehydrogenase associated with variable expression of pyruvate dehydrogenase complex deficiency. Pediatr Res 32:169–174

42. Head RA, Brown RM, Zolkipli Z et al (2005) Clinical and genetic spectrum of pyruvate dehydrogenase deficiency: dihydrolipoamide acetyltransferase (E2) deficiency. Ann Neurol 58:234–241
43. McWilliam CA, Ridout CK, Brown RM, McWilliam RC, Tolmie J, Brown GK (2010) Pyruvate dehydrogenase E2 deficiency: a potentially treatable cause of episodic dystonia. Eur J Paediatr Neurol 14:349–353
44. Head RA, de Goede CG, Newton RW et al (2004) Pyruvate dehydrogenase deficiency presenting as dystonia in childhood. Dev Med Child Neurol 46:710–712
45. Maj MC, MacKay N, Levandovskiy V et al (2005) Pyruvate dehydrogenase phosphatase deficiency: identification of the first mutation in two brothers and restoration of activity by protein complementation. J Clin Endocrinol Metab 90:4101–4107
46. Cameron JM, Maj M, Levandovskiy V et al (2009) Pyruvate dehydrogenase phosphatase 1 (PDP1) null mutation produces a lethal infantile phenotype. Hum Genet 125:319–326
47. Odievre MH, Chretien D, Munnich A et al (2005) A novel mutation in the dihydrolipoamide dehydrogenase E3 subunit gene (DLD) resulting in an atypical form of alpha-ketoglutarate dehydrogenase deficiency. Hum Mutat 25:323–324
48. Aptowitzer I, Saada A, Faber J, Kleid D, Elpeleg ON (1997) Liver disease in the Ashkenazi-Jewish lipoamide dehydrogenase deficiency. J Pediatr Gastroenterol Nutr 24:599–601
49. Johnson MT, Yang HS, Magnuson T, Patel MS (1997) Targeted disruption of the murine dihydrolipoamide dehydrogenase gene (Dld) results in perigastrulation lethality. Proc Natl Acad Sci USA 94:14512–14517
50. Seyda A, Newbold RF, Hudson TJ et al (2001) A novel syndrome affecting multiple mitochondrial functions, located by microcell-mediated transfer to chromosome 2p14–2p13. Am J Hum Genet 68:386–396
51. Brown GK, Haan EA, Kirby DM et al (1988) "Cerebral" lactic acidosis: defects in pyruvate metabolism with profound brain damage and minimal systemic acidosis. Eur J Pediatr 147:10–14
52. Gene Tests. http://www.ncbi.nlm.nih.gov/sites/GeneTests/. Accessed 30 Jan 2012
53. Lib MY, Brown RM, Brown GK, Marusich MF, Capaldi RA (2002) Detection of pyruvate dehydrogenase E1 alpha-subunit deficiencies in females by immunohistochemical demonstration of mosaicism in cultured fibroblasts. J Histochem Cytochem 50:877–884
54. Kerr DS, Grahame G, Nakousi G (2012) Assays of the pyruvate dehydrogenase complex and pyruvate carboxylase. In: Wong LJ (ed) Biochemical and molecular analysis of mitochondrial disorders. Springer-Humana Press (in press)
55. Kerr DS, Berry SA, Lusk MM, Ho L, Patel MS (1988) A deficiency of both subunits of pyruvate dehydrogenase which is not expressed in fibroblasts. Pediatr Res 24:95–100
56. Sheu KFR, Hu CWC, Utter MF (1981) Pyruvate dehydrogenase complex activity in normal and deficient fibroblasts. J Clin Invest 67:1463–1471
57. Falk RE, Cederbaum SD, Blass JP, Gibson GE, Pieter Kark RA, Carrel RE (1976) Ketogenic diet in the management of pyruvate dehydrogenase deficiency. Pediatr 58:713–721
58. Wexler ID, Hemalatha SG, McConnell J et al (1997) Outcome of pyruvate dehydrogenase deficiency treated with ketogenic diets. Studies in patients with identical mutations. Neurol 49:1655–1661
59. Saenz MS, Pickler L, Elias E, Fenton L, Kerr DS, Van Hove J (2010) Resolution of lesions on MRI with ketogenic diet in pyruvate dehydrogenase d0eficiency. (abstract) Saenz MS, Pickler L, Elias E, Fenton L, Kerr DS, Van Hove J (eds) Mol Genet Metab 99:231–232
60. Taylor MR, Hurley JB, Van Epps HA, Brockerhoff SE (2004) A zebrafish model for pyruvate dehydrogenase deficiency: rescue of neurological dysfunction and embryonic lethality using a ketogenic diet. Proc Natl Acad Sci USA 101:4584–4589
61. Sidhu S, Gangasani A, Korotchkina LG et al (2008) Tissue-specific pyruvate dehydrogenase complex deficiency causes cardiac hypertrophy and sudden death of weaned male mice. Am J Physiol Heart Circ Physiol 295:H946–H952
62. Naito E, Ito M, Yokota I et al (2002) Thiamine-responsive pyruvate dehydrogenase deficiency in two patients caused by a point mutation (F205L and L216F) within the thiamine pyrophosphate binding region. Biochim Biophys Acta 1588:79–84

63. Stacpoole PW, Barnes CL, Hurbanis MD, Cannon SL, Kerr DS (1997) Treatment of congenital lactic acidosis with dichloroacetate: a review. Arch Pediatr Adolesc Med 77:535–541
64. Barshop BA, Naviaux RK, McGowan KA et al (2004) Chronic treatment of mitochondrial disease patients with dichloroacetate. Mol Genet Metab 83:138–149
65. Stacpoole PW, Kerr DS, Barnes C et al (2006) Controlled clinical trial of dichloroacetate for treatment of congenital lactic acidosis in children. Pediatr 117:1519–1531
66. Kaufmann P, Engelstad K, Wei Y et al (2006) Dichloroacetate causes toxic neuropathy in MELAS: a randomized, controlled clinical trial. Neurol 66:324–330
67. Stacpoole PW (2011) The dichloroacetate dilemma: environmental hazard versus therapeutic goldmine–both or neither? Environ Health Perspect 119:155–158
68. Owen R IV, Lewin AP, Peel A et al (2000) Recombinant adeno-associated virus vector-based gene transfer for defects in oxidative metabolism. Mol Ther 11:2067–2078
69. NCBI/Entrez Gene. http://www.ncbi.nlm.nih.gov/gene. NCBI; Accessed 30 Jan 2012
70. Human Gene Mutation Database. http://www.biobase-international.com/product/hgmd. BioBase; Accessed 30 Jan 2012

Chapter 20
Nuclear Genes Causing Mitochondrial Cardiomyopathy

Stephanie M. Ware and Jeffrey A. Towbin

Mitochondrial Function and the Cardiomyocyte

The heart functions to pump blood throughout the body and is capable of adjusting the frequency and intensity of its repetitive contractions to meet energetic demands. Cardiac myocytes are connected in series and, unlike skeletal muscle fibers, do not assemble in parallel arrays but bifurcate and join to form a three-dimensional network [1–4]. Mitochondria occupy greater than 30 % of cardiomyocyte volume and are organized in densely packed rows under the sarcolemma and between myofilaments. This ordered arrangement ensures that a constant diffusion distance exists between mitochondria and the core of the myofilament. The myofibrils are formed by repeating sarcomere units. The sarcomeres are the basic contractile units and are composed of thin filaments consisting of cardiac actin, alpha tropomyosin, and troponins T, I, and C and thick filaments composed of myosin heavy chain, myosin light chains, and myosin-binding proteins C, H, and X. Myosin heavy chain contains the ATPase activity that drives cardiac contraction.

The heart is one of the major energy-consuming organs. Energy is primarily stored in the form of ATP or phosphocreatine, which is formed by creatine kinase-mediated phosphorylation of creatine by ATP. Overall, the heart uses approximately 1 mM ATP per second and requires complete energy substrate renewal about every 20 seconds. [5]. The heart is promiscuous in its utilization of energy substrates and can use fatty acids, carbohydrates, lactate, ketone bodies, or amino acids. Prenatally, energy is primarily derived from glucose as the substrate undergoing glycolysis. At the time of birth, the myocardium switches from anaerobic glycolysis to fatty acid oxidation and oxidative phosphorylation as a means of ATP production [6, 7]. Fatty acids remain the preferred substrate throughout postnatal life. Mouse models have demonstrated that there is a strong activation of mitochondrial biogenesis in

---

S. M. Ware (✉) · J. A. Towbin
Cincinnati Children's Hospital Medical Center and the University of Cincinnati
College of Medicine, 240 Albert Sabin Way,
MLC 7020 USA, Cincinnati, OH 45229, USA
e-mail: stephanie.ware@cchmc.org

the myocardium during the perinatal period. Ablation of transcription factors critical for this metabolic switch, such as TFAM or PGC-1 isoforms, results in upregulation of glycolytic genes and rapidly fatal neonatal heart failure [8, 9]. A recent human study demonstrated that there are important differences in the expression of nuclear-encoded mitochondrial genes during fetal versus postnatal development, suggesting that developmental stage-specific control of mitochondrial biogenesis exists [10]. In pathological states, hypertrophied or dilated hearts switch to a fetal pattern that favors utilization of glucose over fatty acids [11, 12].

Cardiomyopathy and Heart Failure

Heart failure is defined as a progressive deterioration of cardiac pump function leading to an inability of the heart to meet the body requirements. Heart failure therefore represents an economic problem—a mismatch in supply and demand. In fact, it has been proposed that energy starvation is a unifying mechanism underlying cardiac contractile failure [5, 13, 14]. This is frequently the result of a downward spiral of events in which decreases in oxygen and substrate availability trigger adaptive mechanisms including neuroendocrine overdrive, activation of signaling pathways, extracellular remodeling, and alterations in mechanical load among others. While this stabilizes contractile function in the short term, over the long term it may result in further compromise due to increased mismatch of the supply to demand ratio [15]. As metabolic remodeling progresses from adaptation to maladaptation, the failing heart loses the ability to switch to the most energetically efficient fuel [11].

Symptoms of heart failure include decreased exercise capacity, increased fatigability, dyspnea, and fluid retention and swelling. In infants and children, symptoms may be more subtle. Irritability or lethargy can be accompanied by additional nonspecific complaints of failure to thrive, nausea, vomiting, or abdominal pain. Respiratory symptoms (tachypnea, wheezing, cough, or dyspnea on exertion) are often present. The diagnosis of a mitochondrial disorder can be challenging in a patient with heart failure, especially in infants for whom little medical history is available. Because any cause of hypoperfusion can lead to secondary mitochondrial dysfunction, heart failure can be either a cause or a result of mitochondrial dysfunction [16].

In 2006, an American Heart Association Scientific Statement delineated a contemporary classification of five subtypes of cardiomyopathy: dilated, hypertrophic, restrictive, left ventricular noncompaction (LVNC), and arrhythmogenic right ventricular cardiomyopathy (ARVC) [17]. Dilated cardiomyopathy, the most common form of cardiomyopathy, is characterized predominantly by left ventricular dilation and decreased left ventricular systolic function. Hypertrophic cardiomyopathy demonstrates increased ventricular myocardial wall thickness, normal or increased systolic function, and often, diastolic (relaxation) abnormalities. Restrictive cardiomyopathy is characterized by nearly normal ventricular chamber size and wall thickness with preserved systolic function, but dramatically impaired diastolic function (relaxation) leading to elevated filling pressures and atrial enlargement.

Arrhythmogenic right ventricular cardiomyopathy and LVNC are characterized by specific morphologic abnormalities and heterogeneous functional disturbances [17–19].

Cardiomyopathy as a Symptom of Mitochondrial Disorders

Since the first clinical recognition of mitochondrial disorders in 1962, a vast clinical spectrum of presentations has emerged characterized by a progressive course and frequent involvement of unrelated organs or tissues [20]. Consensus diagnostic criteria have been proposed for mitochondrial disorders in adults, and have been modified to better capture their occurrence in the pediatric population, where mtDNA mutations are rare and mutations in nuclear genes predominate [21–23]. Because of the common association with neuromuscular or encephalopathic presentations, cardiomyopathy is often an overlooked feature of mitochondrial disorders. Cardiomyopathy may be the only presenting sign or symptom of a mitochondrial disorder, especially in infants, but a high level of suspicion is required to make the diagnosis lacking additional organ involvement.

A large study examining 113 pediatric patients with definite mitochondrial disorders based on modified Walker criteria identified nonspecific encephalomyopathy as the presenting symptom in 44/113 patients and cardiomyopathy and myopathy as the presenting symptom in 45/113, indicating that cardiac involvement is a common feature in these disorders [24]. Importantly, patients with cardiomyopathy had significantly increased mortality in comparison to patients with noncardiac features, with 82 % mortality at the age of 16 versus 5 % [24]. In a second study of 101 patients with CNS and neuromuscular disease caused by an underlying mitochondrial disorder, 17 % had cardiomyopathy. Mortality in this subgroup was higher (71 %) than those without cardiomyopathy (26 %) indicating that this finding is generalizable [25].

Hypertrophic cardiomyopathy, dilated cardiomyopathy, LVNC, or mixed phenotypes have all been identified in patients with mitochondrial disorders. In the study by Scaglia et al. [24], 58 % exhibited hypertrophic cardiomyopathy, 29 % dilated cardiomyopathy, and 13 % LVNC. Idiopathic or familial hypertrophic cardiomyopathy is characterized by hyperdynamic contractility and normal to increased shortening fractions and ejection fractions. In contrast, in hypertrophic cardiomyopathy resulting from mitochondrial disorders, there is often a paradoxical decrease in systolic function and a characteristic hypocontractile-hypertrophic phenotype [26–28]. The mechanistic basis for this finding is unclear but speculation exists about diminished ATP needed to drive contraction as well as calcium dysregulation. Because the clinical management and natural history of cardiomyopathy caused by mitochondrial disorders may differ from other causes of cardiomyopathy, early recognition and diagnosis are important [26, 27, 29, 30]. Arrhythmias are also common in patients with mitochondrial disorders but little is known about their onset or natural history with the exception of the relatively well-described findings of progressive heart block in Kearns–Sayre syndrome.

In the following sections, five nuclear-encoded genes that cause mitochondrial disease characterized by cardiomyopathy are described. Interestingly, the proteins encoded by these four genes are important for disparate mitochondrial functions including phospholipid remodeling (taffazin), inner mitochondrial membrane phosphate transport (SLC25A3), Complex IV assembly (SURF1 and SCO2), and Complex V function (TMEM70).

Barth Syndrome and Mutations in TAZ

Barth syndrome, also referred to as 3-methylglutaconic aciduria type II, is a disorder affecting boys classically characterized by cardiomyopathy, neutropenia, growth retardation, and 3-methyglutaconic aciduria. Exercise intolerance, low muscle mass, and skeletal muscle myopathy are additional features. In 1983, Barth et al. [31] identified a large pedigree with X-linked recessive inheritance in which boys died in infancy from septicemia or heart failure. Ultrastructural abnormalities in mitochondria of the cardiomyocytes, neutrophils, and to a lesser extent skeletal muscle cells, were identified. After the locus was mapped to a gene-dense region on Xq28, Bione et al. [32] identified mutations in *G4.5/TAZ* in patients with Barth syndrome in 1996. Shortly thereafter, *TAZ* mutations were also identified as the cause for other forms of infantile cardiomyopathies including X-linked LVNC and some forms of X-linked dilated cardiomyopathy [33, 34]. The majority of patients with Barth syndrome manifest all four principal diagnostic criteria at some time during childhood. However, there are patients with clinical diagnostic features without mutations in *TAZ*, leading to speculation about the possibility of a second locus. In other cases, patients have been shown to have other forms of mitochondrial disease. Barth syndrome is rare and the prevalence is unknown but the disease is thought to be underdiagnosed. Barth syndrome is inherited in an X-linked recessive pattern. Males affected with Barth syndrome generally inherit the *TAZ* mutation from their mother although de novo mutations have been reported. No female carriers with normal 46, XX chromosome analysis have been affected, most likely due to skewed X inactivation [35]. Females with abnormal X chromosomes, including ring chromosomes with Xq28 deletions, and Barth syndrome have been identified.

Laboratory abnormalities in patients with Barth syndrome include abnormal urine organic acid analysis and complete blood counts. Quantitative urine organic acid analysis classically reveals increased levels of 3-methylglutaconic acid. Although this is a common finding in many primary mitochondrial disorders, levels tend to be much higher in Barth syndrome, typically in the range of 5–20-fold elevated over normal. In addition, elevations in citric acid cycle intermediates are frequently identified including aconitate, fumarate, and 2-ketoglutarate as well as 2-ethylhydacrylic acid [36]. Dicarboxylic aciduria, suggesting a long chain fatty acid oxidation defect, may be present and can frequently be seen in infants with cardiomyopathy, complicating the differential. The complete blood count classically shows neutropenia, which may be chronic or cyclical.

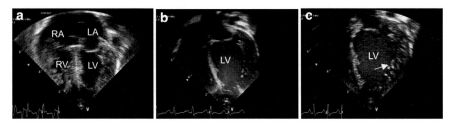

Fig. 20.1 Undulating cardiac phenotypes in a patient with Barth syndrome. Apical four-chamber serial echocardiograms are shown. **a** Echocardiogram on day of life one showing moderate biventricular hypertrophy with increased thickness of both the ventricular walls and the interventricular septum. There was severe biventricular systolic dysfunction with an endocardial shortening fraction of 16 % (z = −9.76). **b** Echocardiogram at 3 months. The heart has transitioned from a hypertrophic to a dilated phenotype. There is a dilated, globular left ventricle with prominent trabeculations in the left ventricular cavity. There is mild to moderate right ventricular hypertrophy. Globally depressed systolic function persists. **c** Echocardiogram at 1 year. There is left ventricular hypertrophy with left ventricular noncompaction (*arrow*). The left ventricular end diastolic dimension is increased. The function was more severely depressed with global dyskinesis. *LA* left atrium; *LV* left ventricle; *RA* right atrium; *RV* right ventricle

Echocardiographically, patients with Barth syndrome typically have LVNC with left ventricular dilation, endocardial fibroelastosis, or a dilated hypertrophic left ventricle. The cardiomyopathy often undergoes and undulating course and may transition between hypertrophic and dilated forms and back again (Fig. 20.1). Dilated cardiomyopathy has developed in utero in a few documented pregnancies [33, 37]. Recently, investigation of kindreds with Barth syndrome identified 32 % of families with fetal loss or neonatal illness or death including nine stillbirths and 14 neonatal or infant deaths in six kindreds. These included pregnancies with unexplained male hydrops, dilated cardiomyopathy, LVNC, or pregnancy loss as well as neonates with hypoglycemia or lactic acidosis [38]. More commonly, Barth patients present in the first 6 months of life with congestive heart failure. In some cases, death results from congestive heart failure, ventricular thrombi, ventricular fibrillation, or sepsis related to neutropenia. However, most children survive past infancy although dilated cardiomyopathy persists. Cardiac transplantation has been performed in patients with Barth syndrome [39, 40]. There is also thought to be an increased incidence of structural heart defects in patients with Barth syndrome including ventricular septal defects.

Other clinical features of Barth syndrome include growth retardation and developmental delay. Muscle hypoplasia is common in patients with Barth syndrome, leading to decreased weight for length. The growth velocity is slow in the first 2 years of life and the z-score is typically - 2 to - 4. Patients tend to follow the growth curve below the third centile until puberty. Many children undergo a prolonged growth spurt at puberty leading to a normal final adult height. Development is often described as delayed due to motor problems. The majority of children have normal intelligence, but a cognitive phenotype characterized by difficulties with math and visual-spatial activities has emerged [41].

The *TAZ* gene is an 11-exon gene located on the X chromosome. It is alternatively spliced and has at least four main splice variants including full length, deleted for exon 5, deleted for exon 7, and deleted for both exon 5 and 7. The majority of mutations in *TAZ* are point mutations resulting in loss of function. However, full and partial gene deletions have also been identified. *TAZ* encodes a highly conserved acyltransferase, tafazzin, whose function is not completely understood. One major result of TAZ deficiency is defective remodeling of phospholipid side chains, particularly cardiolipin [42].

Cardiolipin is a mitochondrial phospholipid that contributes to the function of many proteins in the inner mitochondrial membrane. It is important for the stability and assembly of inner membrane protein complexes, mitochondrial protein import, apoptotic signaling, and the proper function of electron transport and mitochondrial oxidative phosphorylation [43–45]. Cardiolipin is predominantly localized to the inner mitochondrial membrane, although it can also be found at contact sites of the inner and outer mitochondrial membranes. In addition, it has recently been found to play important roles in outer membrane protein biogenesis [46]. In the adult human heart, 79 % of the four fatty acyl side chains of cardiolipin are linoleic acid (C18:2) [45]. Tafazzin is at least one of the mitochondrial proteins that act as an acyltransferase to remodel cardiolipin fatty acyl side chains. Mutations in *TAZ* result in a decrease in total cardiolipin as well as decreases in the subclass tetra-linoleoyl cardiolipin with accumulation of monolysocardiolipin [43, 47, 48]. The symmetry, composition, and amount of side chains of cardiolipin are all critical for proper function. Although the mechanism of enrichment of linoleic acid side chains is not completely understood, failure of remodeling is known to lead to mitochondrial dysfunction, increased apoptosis, and heart failure [43, 45, 49–52]. In fact, a recent study in patients with heart failure without Barth syndrome showed a 95 % decrease in the mRNA levels of *TAZ* in the heart versus controls. Similar findings were shown in rats [53].

There is significant interest in developing treatments for heart failure in general, as well as Barth syndrome specifically, based on correcting the underlying abnormality in cardiolipin. The side-chain composition of cardiolipin can be manipulated by fatty acids in the diet. Therefore, attempting to restore levels via increasing linoleic acid has been attempted. Epidemiologic studies have demonstrated that increased levels of linoleic acid in the diet correlate with decreased risk for cardiovascular disease [54]. Preliminary studies in rats indicate decreased rates of heart failure when fed a linoleic-acid rich diet [55]. L-carnitine has also been shown to increase the total content of cardiolipin and therefore may have potential to influence cardiolipin remodeling. Several animal models with tafazzin deficiency have been developed and will allow more rigorous investigation of potential therapeutics [56–59].

Mutations in SLC25A3 and Hypertrophic Cardiomyopathy

ADP enters the mitochondrial matrix on adenine nucleotide translocase (ANT1). The phosphorylation of ADP is catalyzed by the ATP synthase complex. The phosphate carrier encoded by *SLC25A3*, also termed PHC, catalyzes the uptake of phosphate across the inner mitochondrial membrane for this terminal step of oxidative phosphorylation. *SLC25A3* is a 9-exon gene that encodes two isoforms, containing 42 or 41 amino acids and resulting from alternative splicing and utilization of either exon 3A or exon 3B [60]. Transcripts of isoform 3A are expressed in heart, skeletal muscle, and pancreas at high levels. Isoform 3B is poorly expressed in all tissues [61]. Mutations in the alternatively spliced exon 3A of *SLC25A3* were identified in two siblings with lactic acidosis, hypertrophic cardiomyopathy, and hypotonia who died in infancy [62]. Both siblings presented within the first day of life with cyanosis, respiratory distress, and metabolic acidosis with elevated lactate. Echocardiography showed hypertrophic cardiomyopathy with systolic dysfunction and low cardiac output. The degree of hypertrophy was progressive. Skeletal muscle biopsy demonstrated lipid accumulation and prominent type I fibers. Functional investigation of the mitochondria showed a deficiency of ATP synthase in muscle but not in fibroblasts, consistent with a mutation in the alternatively spliced exon 3A. Both siblings died of heart failure.

Cytochrome C Oxidase (COX) Deficiency and Hypertrophic Cardiomyopathy

Cytochrome C oxidase (COX) deficiency results from deficiency of Complex IV of the electron transport chain and can result in variable clinical symptoms including failure to thrive, encephalopathy, hypotonia, hepatic or cardiac involvement, or Leigh syndrome. The molecular basis of COX deficiency is identified in less than 50–60 % of cases. A retrospective multicenter study of 180 children with COX deficiency showed that 66 % died. Isolated COX deficiency was identified in 101/180 patients whereas the remainder had evidence of abnormalities in multiple respiratory chain complexes. Pathogenic mutations in nuclear DNA were established in 56/101 patients with isolated COX deficiency, including 47 patients with mutations in surfeit locus protein 1 gene (*SURF1*) and nine with mutations in *SCO2*, a copper binding protein [63]. All 47 patients with *SURF1* mutations presented with Leigh syndrome. However, 4/47 (9 %) had mild hypertrophic cardiomyopathy, a finding not previously reported in conjunction with Leigh syndrome. All four children were more than the age of 10 years, indicating that cardiomyopathy may develop later in children with Leigh syndrome that survive infancy. All nine children with mutations in *SCO2* died before 2 years of age, and hypertrophic cardiomyopathy was found in 4/9 (44 %). Altogether, hypertrophic cardiomyopathy was identified in 24 % of patients with COX deficiency, regardless of the molecular basis. These findings are similar to those reported elsewhere [25, 64]. In all cases, patients with cardiomyopathy are noted to have a worse outcome.

A large number of proteins are involved in the assembly and maintenance of the COX complex, an enzyme complex composed of 13 structural subunits. SURF1 and SCO2 are Complex IV assembly proteins [65]. SCO1 and SCO2 are thought to be important for copper delivery to COX. Mutations in the SCO2 gene have been associated with rapidly progressing hypertrophic cardiomyopathy, hypotonia, respiratory distress, and lactic acidosis. Other varied features may include seizures, strabismus, ptosis, and failure to thrive. Patients with *SCO2* mutations generally have decreased or absent COX levels. Mutations in the *SCO2* gene are inherited in an autosomal-recessive manner. *SCO2* is a 2-exon gene located at chromosome 22q13.33. Of the nine children identified with SCO2 mutations, all of them were homozygotes or heterozgyotes for a common E140K mutation [66].

The *SURF1* gene also encodes a protein that is important for the proper assembly and function of COX as part of the electron transport chain. As above, mutations in the *SURF1* gene are the most common cause of Leigh syndrome with COX deficiency. Patients with *SURF1* mutations commonly present in infancy. Leigh disease is a heterogeneous condition with progressive neurological symptoms, hypotonia, lactic acidosis, and abnormal findings on brain imaging. Seizures, strabismus, or ptosis may be present along with failure to thrive. COX activity is generally decreased or absent by histochemical staining or electron transport chain analysis. In patients with Leigh disease and COX deficiency, mutations in SURF1 account for approximately 26 % of cases. Mutations in the *SURF1* gene are inherited in an autosomal recessive manner. *SURF1* contains nine exons and is located at chromosome 9q34.

Mutations in TMEM70 and Complex V Deficiency

In 2008, mutations in *TMEM70* were first identified as a cause of isolated ATP synthase deficiency and neonatal mitochondrial encephalocardiomyopathy in patients descended from Roma Gypsies [67]. The typical presentation in patients with ATP synthase or complex V deficiency includes lactic acidosis, 3-methylglutaconic aciduria, hypertrophic cardiomyopathy, and mitochondrial myopathy. In general, patients with inherited complex V deficiency have significant morbidity and mortality regardless of underlying etiology. Sperl et al. [68] presented data from 14 patients with isolated complex V deficiency, many of whom were later found to have TMEM70 mutations [67]. All patients had neonatal onset and elevated lactate. Twelve of 14 had 3-methylglutaconic aciduria, and seven died within the first weeks of life whereas the remainder showed evidence of developmental delay or mental retardation. Hypertrophic cardiomyopathy was found in 11 of 14. Other findings included dysmorphic features, hypotonia, microcephaly, and hepatomegaly.

Recently, patients with *TMEM70* mutations have been identified in a subset of patients originally classified within Type IV 3-methylglutaconic aciduria. This is a genetically and phenotypically heterogenous group of patients in which encephalomyopathic, myopathic, hepatocerebral, or cardiomyopathic presentations may predominate [69]. Among the cardiomyopathic subgroup, severe complex V deficiency was identified (0.4–4.9 % of lowest controls), similar to the findings of Sperl et al. [68]. Several patients also showed mild complex I deficiency and two patients had a Sengers-like presentation. A *TMEM70* mutation was confirmed in three patients with the cardiomyopathic presentation [69]. More recently, new cases have been identified that highlight the phenotypic variability that may occur with *TMEM70* mutations. One individual with Type IV 3-methylglutaconic aciduria, had mild symptoms, but persistent hypertrophic cardiomyopathy was a feature [70]. In a second study, six new patients were identified with infantile onset cataracts, early gastrointestinal dysfunction, and congenital hypertonia with contractures resembling distal arthrogryposis [71].

The ATP synthase enzyme complex is composed of 16 different subunits arranged in a globular F1 catalytic region connected by two stalks to the membrane-embedded F0 complex [68]. ATP synthase works as a molecular motor to facilitate the synthesis of ATP and is also important for ATP hydrolysis. As the final step of the electron transport chain, ATP synthase provides over 95 % of cellular ATP and thus is the key enzyme for energy production. The protein complex is encoded by two mitochondrial genes, encoding ATPase6 and ATPase8, and 14 nuclear genes. Defects in the mitochondrial-encoded ATP6 and ATP8 result in failure of ATP synthesis with preservation of the rate of ATP hydrolysis and concentration of the enzyme complex. Cardiomyopathy was a universal feature of patients described with mtDNA mutations in ATP6/8 [72]. However, several mutations in the mtDNA *ATP6* gene have been described that present with NARP (neuropathy, ataxia, and retinitis pigmentosa) or the more severe Leigh syndrome, dependent in part on the level of heteroplasmy [73]. A nuclear origin of ATP synthase deficiency was first described in 1999. Similarly, in the initial description of six families with *TMEM70* mutations, 22/25 individuals had hypertrophic cardiomyopathy as one of their clinical features. However, in contrast to the selective loss of ATP synthesis seen in mutations in the mitochondrial genome, ATP synthase defects of nuclear origin show substantial loss of both synthetic and hydrolytic activities [74, 75].

TMEM70 is a 3-exon gene localized to chromosome 8q21.11. In the Roma Gypsy population, a homozygous single-founder splice site mutation has been identified. The exact role of TMEM70 in ATP synthase assembly is not known. Recent evidence indicates that TMEM localizes to the inner mitochondrial membrane and does not appear to have a direct physical interaction with ATP synthase, but rather exists in a dimeric form [76]. TMEM70 is estimated to constitute up to 30 % of inner membrane-bound proteins and mutations lead to loss of integrity of the inner mitochondrial membrane and ultrastructural abnormalities. Using electron microscopy and immunogold labeling, Cameron et al. [75] demonstrate abnormal cristae formation with loss of invaginations and formation of concentric whorled cristae rings that disrupt the integrity of mitochondrial nucleoids in a patient with *TMEM70* mutations.

Additional Nuclear Genes Causing Mitochondrial Cardiomyopathy

Table 20.1 lists nuclear genes important for mitochondrial function which have been shown to have phenotypic presentations that include cardiomyopathy. Many of these are rare and have been identified in a single patient or family. Additional descriptions of many of these genes are provided in detail in other chapters of this book. Cardiomyopathy can be a feature of many nuclear mitochondrial disorders. The reasons underlying the phenotypic variability with regard to cardiac involvement are unclear.

Table 20.1 Nuclear genes associated with cardiomyopathy phenotypes in humans

Gene	OMIM#	Chromosome	Cardiomyopathy Phenotype	Number of cases	Reference
TAZ	300394	Xq28	LVNC, DCM	>100	[32, 88]
SLC25A3	600370	12q23.1	HCM	2	[62]
FRDA	606829	9q21.11	HCM	>200	[89–92]
Mitochondrial DNA biogenesis/integrity					
DGUOK	601465	2p13.1	DCM	1	[93]
ANT1	103220	4q35.1	HCM	2	[94, 95]
Complex subunits and assembly genes					
SCO2	604272	22q13.33	HCM	>25	[66]
TMEM70	612418	8q21.11	HCM	>25	[67, 68, 75, 96–98]
COX10	602125	17p12	HCM	1	[99]
COX15	603646	10q24	HCM	1	[100]
NDUFV2	600532	18p11.2	HCM	1	[101]
NDUFA11	612638	19p13.3	HCM	2	[102]
NDUFS2	602985	1q23.2	HCM	1	[103]
NDUFS8	602141	11q13.2	HCM	1	[104]
SDHA	600857	5p15.33	DCM	15	[105]
Mitochondrial protein translation					
TSFM	604273	12q14.1	HCM	1	[106]
MRPS22	605810	3q23	HCM/DCM	4	[107, 108]

Furthermore, only a small percent of the cases of nuclear mitochondrial disorders that have cardiomyopathy as a feature have known molecular causes. Figure 20.2 shows an example of cardiac involvement in a patient with a nuclear mitochondrial disorder. The presenting features were hypertrophic cardiomyopathy, ptosis, and lactic acidosis and the molecular cause remains to be determined.

Cardiomyopathy, Mitochondrial Dysfunction, and Treatment

Altered mitochondrial function is a common finding in cardiomyopathic disease. However, it is important to recognize that many forms of cardiac muscle disease, both genetic and acquired, are characterized by mitochondrial dysfunction even though primary mitochondrial disease caused by nuclear or mitochondrial DNA mutations is not the underlying etiology. End-stage heart failure, the aging heart, ischemia, and diabetes are all conditions in which mitochondrial dysfunction has been reported [13, 15, 45, 53, 77, 78]. Because of the substantial morbidity, mortality, and economic impact of these common diseases, understanding the pathogenesis and potential treatment of mitochondrial dysfunction is critical. Furthermore, the potential impact of mitochondrial dysfunction in patients with mutations in cardiac structural proteins such as sarcomeric or cytoskeletal proteins is important to delineate. Patients with *MYH7* mutations, the gene encoding the thick filament beta-myosin heavy chain containing the ATPase activity, have been shown to have substantial decreases in

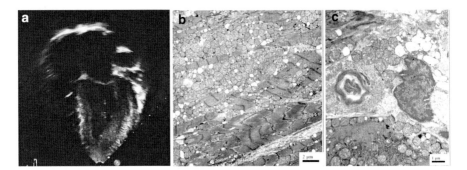

Fig. 20.2 Abnormal cardiac biopsy in hypertrophic cardiomyopathy. **a** Apical four-chamber view of a patient with hypertrophic cardiomyopathy with concentric thickening. **b** Electron micrograph from cardiac biopsy showing mitochondrial proliferation, disruption of the sarcomeres, and loss of myofilaments. The interstitium shows evidence of fibrosis. **c** Electron micrograph showing variably enlarged mitochondria, pooled granular glycogen, and lamellar lysosomal inclusions

mitochondrial respiratory function [77, 79]. In addition, the intracellular distribution of mitochondria can be abnormal in patients with mutations in desmin, a cytoskeletal protein [80], a fact that has implications for the bioenergetic efficiency of the cardiomyocyte. Thus, mitochondrial dysfunction is frequently a final common pathway resulting from a wide variety of insults that perturb cardiac structure and/or function [81, 82].

Dietary modulation has received particular attention with regard to possible treatment strategies and controversy regarding efficacy is common. Azevedo et al. [83] examined the role of L-carnitine nutritional status in children with idiopathic dilated cardiomyopathy and found that the group that received supplementation showed significant improvement in echocardiographic measures of cardiac size and function including an improvement in the ejection fraction. In addition, there was improvement in nutritional status as measured by weight and prevention or reversion of heart failure-related cachexia. Coenzyme Q10 has been reported to have beneficial effects in the treatment of fatigability and metabolic abnormalities [84]. At least one study showed dramatic improvement in mitochondrial cardiomyopathy after idebenone treatment [85, 86]. In addition, in mice, pretreatment with Coenzyme Q10 appeared to reduce the effects of oxidative injury and DNA damage in an acute myocarditis mouse model. However, a 3-month treatment of 30 patients with ischemic or idiopathic-dilated cardiomyopathy with oral Coenzyme Q10 or placebo showed no significant changes in left ventricular systolic function or quality of life [87].

Summary

Cardiomyopathy is a common but underrecognized feature of mitochondrial disease. Mutations in mtDNA or nuclear DNA have been associated with a cardiomyopathic presentation, with the latter predominating. Patients with cardiac involvement have higher morbidity and mortality associated with their mitochondrial disease. Early recognition and prompt treatment of cardiac symptoms can significantly impact disease progression. *TAZ, SLC25A3, SURF1, SCO2*, and *TMEM70* are five nuclear genes known to cause mitochondrial cardiomyopathy. In addition, secondary mitochondrial dysfunction is a common pathway in patients with end-stage heart failure, the aging heart, ischemia, and diabetes. Consequently, there is substantial interest in identifying dietary and medical therapeutics for both primary and secondary cardiac mitochondrial dysfunction.

References

1. Clark KA, McElhinny AS, Beckerle MC, Gregorio CC (2002) Striated muscle cytoarchitecture: an intricate web of form and function. Annu Rev Cell Dev Biol 18:637–706
2. Luther PK (2009) The vertebrate muscle Z-disc: sarcomere anchor for structure and signalling. J Muscle Res Cell Motil 30:171–185
3. Gregorio CC, Antin PB (2000) To the heart of myofibril assembly. Trends Cell Biol 10:355–362
4. Pyle WG, Solaro RJ (2004) At the crossroads of myocardial signaling: the role of Z-discs in intracellular signaling and cardiac function. Circ Res 94:296–305
5. Ventura-Clapier R, Garnier A, Veksler V, Joubert F (2011) Bioenergetics of the failing heart. Biochim Biophys Acta 1813:1360–1372
6. Lehman JJ, Kelly DP (2002) Transcriptional activation of energy metabolic switches in the developing and hypertrophied heart. Clin Exp Pharmacol Physiol 29:339–345
7. Fisher DJ, Heymann MA, Rudolph AM (1981) Myocardial consumption of oxygen and carbohydrates in newborn sheep. Pediatr Res 15:843–846
8. Lai L, Leone TC, Zechner C, Schaeffer PJ, Kelly SM, Flanagan DP, Medeiros DM, Kovacs A, Kelly DP (2008) Transcriptional coactivators PGC-1alpha and PGC-1beta control overlapping programs required for perinatal maturation of the heart. Genes Dev 22:1948–1961
9. Hansson A, Hance N, Dufour E, Rantanen A, Hultenby K, Clayton DA, Wibom R, Larsson NG (2004) A switch in metabolism precedes increased mitochondrial biogenesis in respiratory chain-deficient mouse hearts. Proc Natl Acad Sci U S A 101:3136–3141
10. Pejznochova M, Tesarova M, Hansikova H, Magner M, Honzik T, Vinsova K, Hajkova Z, Havlickova V, Zeman J (2010) Mitochondrial DNA content and expression of genes involved in mtDNA transcription, regulation and maintenance during human fetal development. Mitochondrion 10:321–329
11. Taegtmeyer H (2002) Switching metabolic genes to build a better heart. Circulation 106:2043–2045
12. Taegtmeyer H, Razeghi P, Young ME (2002) Mitochondrial proteins in hypertrophy and atrophy: a transcript analysis in rat heart. Clin Exp Pharmacol Physiol 29:346–350
13. Ingwall JS, Weiss RG (2004) Is the failing heart energy starved? On using chemical energy to support cardiac function. Circ Res 95:135–145
14. Neubauer S (2007) The failing heart—an engine out of fuel. N Engl J Med 356:1140–1151
15. Stanley WC, Recchia FA, Lopaschuk GD (2005) Myocardial substrate metabolism in the normal and failing heart. Physiol Rev 85:1093–1129
16. Crouser ED (2004) Mitochondrial dysfunction in septic shock and multiple organ dysfunction syndrome. Mitochondrion 4:729–741
17. Maron BJ, Towbin JA, Thiene G, Antzelevitch C, Corrado D, Arnett D, Moss AJ, Seidman CE, Young JB (2006) Contemporary definitions and classification of the cardiomyopathies: an American Heart Association Scientific Statement from the Council on Clinical Cardiology, Heart Failure and Transplantation Committee; Quality of Care and Outcomes Research

and Functional Genomics and Translational Biology Interdisciplinary Working Groups; and Council on Epidemiology and Prevention. Circulation 113:1807–1816
18. Elliott P, Andersson B, Arbustini E, Bilinska Z, Cecchi F, Charron P, Dubourg O, Kuhl U, Maisch B, McKenna WJ et al (2008) Classification of the cardiomyopathies: a position statement from the European Society Of Cardiology Working Group on Myocardial and Pericardial Diseases. Eur Heart J 29:270–276
19. Towbin JA (2010) Left ventricular noncompaction: a new form of heart failure. Heart Fail Clin 6:453–469, viii
20. Munnich A, Rustin P (2001) Clinical spectrum and diagnosis of mitochondrial disorders. Am J Med Genet 106:4–17
21. Wolf NI, Smeitink JA (2002) Mitochondrial disorders: a proposal for consensus diagnostic criteria in infants and children. Neurology 59:1402–1405
22. Bernier FP, Boneh A, Dennett X, Chow CW, Cleary MA, Thorburn DR (2002) Diagnostic criteria for respiratory chain disorders in adults and children. Neurology 59:1406–1411
23. Morava E, Van den Heuvel L, Hol F, De Vries MC, Hogeveen M, Rodenburg RJ, Smeitink JA (2006) Mitochondrial disease criteria: diagnostic applications in children. Neurology 67:1823–1826
24. Scaglia F, Towbin JA, Craigen WJ, Belmont JW, Smith EO, Neish SR, Ware SM, Hunter JV, Fernbach SD, Vladutiu GD et al (2004) Clinical spectrum, morbidity, and mortality in 113 pediatric patients with mitochondrial disease. Pediatrics 114:925–931
25. Holmgren D, Wahlander H, Eriksson BO, Oldfors A, Holme E, Tulinius M (2003) Cardiomyopathy in children with mitochondrial disease; clinical course and cardiological findings. Eur Heart J 24:280–288
26. Colan SD, Lipshultz SE, Lowe AM, Sleeper LA, Messere J, Cox GF, Lurie PR, Orav EJ, Towbin JA (2007) Epidemiology and cause-specific outcome of hypertrophic cardiomyopathy in children: findings from the Pediatric Cardiomyopathy Registry. Circulation 115:773–781
27. Cox GF, Sleeper LA, Lowe AM, Towbin JA, Colan SD, Orav EJ, Lurie PR, Messere JE, Wilkinson JD, Lipshultz SE (2006) Factors associated with establishing a causal diagnosis for children with cardiomyopathy. Pediatr 118:1519–1531
28. Kindel SJ, Miller EM, Gupta R, Cripe LH, Hinton RB, Spicer RL, Towbin JA, Ware SM (2012) Pediatric cardiomyopathy: importance of genetic and metabolic evaluation. J Card Fail 18(5):396–403
29. Towbin JA, Lowe AM, Colan SD, Sleeper LA, Orav EJ, Clunie S, Messere J, Cox GF, Lurie PR, Hsu D et al (2006) Incidence, causes, and outcomes of dilated cardiomyopathy in children. JAMA 296:1867–1876
30. Daubeney PE, Nugent AW, Chondros P, Carlin JB, Colan SD, Cheung M, Davis AM, Chow CW, Weintraub RG (2006) Clinical features and outcomes of childhood dilated cardiomyopathy: results from a national population-based study. Circulation 114:2671–2678
31. Barth PG, Scholte HR, Berden JA, Van der Klei-Van Moorsel JM, Luyt-Houwen IE, Van 't Veer-Korthof ET, Van der Harten JJ, Sobotka-Plojhar MA (1983) An X-linked mitochondrial disease affecting cardiac muscle, skeletal muscle and neutrophil leucocytes. J Neurol Sci 62:327–355
32. Bione S, D'Adamo P, Maestrini E, Gedeon AK, Bolhuis PA, Toniolo D (1996) A novel X-linked gene, G4.5. is responsible for Barth syndrome. Nat Genet 12:385–389
33. Bleyl SB, Mumford BR, Thompson V, Carey JC, Pysher TJ, Chin TK, Ward K (1997) Neonatal, lethal noncompaction of the left ventricular myocardium is allelic with Barth syndrome. Am J Hum Genet 61:868–872
34. D'Adamo P, Fassone L, Gedeon A, Janssen EA, Bione S, Bolhuis PA, Barth PG, Wilson M, Haan E, Orstavik KH et al (1997) The X-linked gene G4.5 is responsible for different infantile dilated cardiomyopathies. Am J Hum Genet 61:862–867
35. Orstavik KH, Orstavik RE, Naumova AK, D'Adamo P, Gedeon A, Bolhuis PA, Barth PG, Toniolo D (1998) X chromosome inactivation in carriers of Barth syndrome. Am J Hum Genet 63:1457–1463

36. Kelley RI, Cheatham JP, Clark BJ, Nigro MA, Powell BR, Sherwood GW, Sladky JT, Swisher WP (1991) X-linked dilated cardiomyopathy with neutropenia, growth retardation, and 3-methylglutaconic aciduria. J Pediatr 119:738–747
37. Cardonick EH, Kuhlman K, Ganz E, Pagotto LT (1997) Prenatal clinical expression of 3-methylglutaconic aciduria: Barth syndrome. Prenat Diagn 17:983–988
38. Steward CG, Newbury-Ecob RA, Hastings R, Smithson SF, Tsai-Goodman B, Quarrell OW, Kulik W, Wanders R, Pennock M, Williams M et al (2010) Barth syndrome: an X-linked cause of fetal cardiomyopathy and stillbirth. Prenat Diagn 30:970–976
39. Adwani SS, Whitehead BF, Rees PG, Morris A, Turnball DM, Elliott MJ, De Leval MR (1997) Heart transplantation for Barth syndrome. Pediatr Cardiol 18:143–145
40. Mangat J, Lunnon-Wood T, Rees P, Elliott M, Burch M (2007) Successful cardiac transplantation in Barth syndrome—single-centre experience of four patients. Pediatr Transplant 11:327–331
41. Mazzocco MM, Kelley RI (2001) Preliminary evidence for a cognitive phenotype in Barth syndrome. Am J Med Genet 102:372–378
42. Schlame M, Kelley RI, Feigenbaum A, Towbin JA, Heerdt PM, Schieble T, Wanders RJ, DiMauro S, Blanck TJ (2003) Phospholipid abnormalities in children with Barth syndrome. J Am Coll Cardiol 42:1994–1999
43. Hauff KD, Hatch GM (2006) Cardiolipin metabolism and Barth syndrome. Prog Lipid Res 45:91–101
44. Gohil VM, Hayes P, Matsuyama S, Schagger H, Schlame M, Greenberg ML (2004) Cardiolipin biosynthesis and mitochondrial respiratory chain function are interdependent. J Biol Chem 279:42612–42618
45. Sparagna GC, Lesnefsky EJ (2009) Cardiolipin remodeling in the heart. J Cardiovasc Pharmacol 53:290–301
46. Gebert N, Joshi AS, Kutik S, Becker T, McKenzie M, Guan XL, Mooga VP, Stroud DA, Kulkarni G, Wenk MR et al (2009) Mitochondrial cardiolipin involved in outer-membrane protein biogenesis: implications for Barth syndrome. Curr Biol 19:2133–2139
47. Valianpour F, Mitsakos V, Schlemmer D, Towbin JA, Taylor JM, Ekert PG, Thorburn DR, Munnich A, Wanders RJ, Barth PG et al (2005) Monolysocardiolipins accumulate in Barth syndrome but do not lead to enhanced apoptosis. J Lipid Res 46:1182–1195
48. Valianpour F, Wanders RJ, Barth PG, Overmars H, Van Gennip AH (2002) Quantitative and compositional study of cardiolipin in platelets by electrospray ionization mass spectrometry: application for the identification of Barth syndrome patients. Clin Chem 48:1390–1397
49. Chicco AJ, Sparagna GC (2007) Role of cardiolipin alterations in mitochondrial dysfunction and disease. Am J Physiol Cell Physiol 292:C33–C44
50. Houtkooper RH, Turkenburg M, Poll-The BT, Karall D, Perez-Cerda C, Morrone A, Malvagia S, Wanders RJ, Kulik W, Vaz FM (2009) The enigmatic role of tafazzin in cardiolipin metabolism. Biochim Biophys Acta 1788:2003–2014
51. Sparagna GC, Chicco AJ, Murphy RC, Bristow MR, Johnson CA, Rees ML, Maxey ML, McCune SA, Moore RL (2007) Loss of cardiac tetralinoleoyl cardiolipin in human and experimental heart failure. J Lipid Res 48:1559–1570
52. Russell LK, Finck BN, Kelly DP (2005) Mouse models of mitochondrial dysfunction and heart failure. J Mol Cell Cardiol 38:81–91
53. Saini-Chohan HK, Holmes MG, Chicco AJ, Taylor WA, Moore RL, McCune SA, Hickson-Bick DL, Hatch GM, Sparagna GC (2009) Cardiolipin biosynthesis and remodeling enzymes are altered during development of heart failure. J Lipid Res 50:1600–1608
54. Erkkila A, De Mello VD, Riserus U, Laaksonen DE (2008) Dietary fatty acids and cardiovascular disease: an epidemiological approach. Prog Lipid Res 47:172–187
55. Chicco AJ, Sparagna GC, McCune SA, Johnson CA, Murphy RC, Bolden DA, Rees ML, Gardner RT, Moore RL (2008) Linoleate-rich high-fat diet decreases mortality in hypertensive heart failure rats compared with lard and low-fat diets. Hypertension 52:549–555
56. Acehan D, Vaz F, Houtkooper RH, James J, Moore V, Tokunaga C, Kulik W, Wansapura J, Toth MJ, Strauss A et al (2011) Cardiac and skeletal muscle defects in a mouse model of human Barth syndrome. J Biol Chem 286:899–908

57. Khuchua Z, Yue Z, Batts L, Strauss AW (2006) A zebrafish model of human Barth syndrome reveals the essential role of tafazzin in cardiac development and function. Circ Res 99:201–208
58. Xu Y, Condell M, Plesken H, Edelman-Novemsky I, Ma J, Ren M, Schlame M (2006) A Drosophila model of Barth syndrome. Proc Natl Acad Sci U S A 103:11584–11588
59. Malhotra A, Edelman-Novemsky I, Xu Y, Plesken H, Ma J, Schlame M, Ren M (2009) Role of calcium-independent phospholipase A2 in the pathogenesis of Barth syndrome. Proc Natl Acad Sci U S A 106 2337–2341
60. Dolce V, Iacobazzi V, Palmieri F, Walker JE (1994) The sequences of human and bovine genes of the phosphate carrier from mitochondria contain evidence of alternatively spliced forms. J Biol Chem 269:10451–10460
61. Huizing M, Ruitenbeek W, Van den Heuvel LP, Dolce V, Iacobazzi V, Smeitink JA, Palmieri F, Trijbels JM (1998) Human mitochondrial transmembrane metabolite carriers: tissue distribution and its implication for mitochondrial disorders. J Bioenerg Biomembr 30:277–284
62. Mayr JA, Merkel O, Kohlwein SD, Gebhardt BR, Bohles H, Fotschl U, Koch J, Jaksch M, Lochmuller H, Horvath R et al (2007) Mitochondrial phosphate-carrier deficiency: a novel disorder of oxidative phosphorylation. Am J Hum Genet 80:478–484
63. Bohm M, Pronicka E, Karczmarewicz E, Pronicki M, Piekutowska-Abramczuk D, Sykut-Cegielska J, Mierzewska H, Hansikova H, Vesela K, Tesarova M et al (2006) Retrospective, multicentric study of 180 children with cytochrome C oxidase deficiency. Pediatr Res 59:21–26
64. Sacconi S, Salviati L, Sue CM, Shanske S, Davidson MM, Bonilla E, Naini AB, De Vivo DC, DiMauro S (2003) Mutation screening in patients with isolated cytochrome c oxidase deficiency. Pediatr Res 53:224–230
65. Moslemi AR, Darin N (2007) Molecular genetic and clinical aspects of mitochondrial disorders in childhood. Mitochondrion 7:241–252
66. Papadopoulou LC, Sue CM, Davidson MM, Tanji K, Nishino I, Sadlock JE, Krishna S, Walker W, Selby J, Glerum, DM et al (1999) Fatal infantile cardioencephalomyopathy with COX deficiency and mutations in SCO2, a COX assembly gene. Nat Genet 23:333–337
67. Cizkova A, Stranecky V, Mayr JA, Tesarova M, Havlickova V, Paul J, Ivanek R, Kuss AW, Hansikova H, Kaplanova V et al (2008) TMEM70 mutations cause isolated ATP synthase deficiency and neonatal mitochondrial encephalocardiomyopathy. Nat Genet 40:1288–1290
68. Sperl W, Jesina P, Zeman J, Mayr JA, Demeirleir L, VanCoster R, Pickova A, Hansikova H, Houst'kova H, Krejcik Z et al (2006) Deficiency of mitochondrial ATP synthase of nuclear genetic origin. Neuromuscul Disord 16:821–829
69. Wortmann SB, Rodenburg RJ, Jonckheere A, De Vries MC, Huizing M, Heldt K, Van den Heuvel LP, Wendel U, Kluijtmans LA, Engelke UF et al (2009) Biochemical and genetic analysis of 3-methylglutaconic aciduria type IV: a diagnostic strategy. Brain 132:136–146
70. Shchelochkov OA, Li FY, Wang J, Zhan H, Towbin JA, Jefferies JL, Wong LJ, Scaglia F (2010) Milder clinical course of type IV 3-methylglutaconic aciduria due to a novel mutation in TMEM70. Mol Genet Metab 101:282–285
71. Spiegel R, Khayat M, Shalev SA, Horovitz Y, Mandel H, Hershkovitz E, Barghuti F, Shaag A, Saada A, Korman SH et al (2011) TMEM70 mutations are a common cause of nuclear encoded ATP synthase assembly defect: further delineation of a new syndrome. J Med Genet 48:177–182
72. Ware SM, El-Hassan N, Kahler SG, Zhang Q, Ma YW, Miller E, Wong B, Spicer RL, Craigen WJ, Kozel BA et al (2009) Infantile cardiomyopathy caused by a mutation in the overlapping region of mitochondrial ATPase 6 and 8 genes. J Med Genet 46:308–314
73. Schon EA, Santra S, Pallotti F, Girvin ME (2001) Pathogenesis of primary defects in mitochondrial ATP synthesis. Semin Cell Dev Biol 12:441–448
74. Houstek J, Pickova A, Vojtiskova A, Mracek T, Pecina P, Jesina P (2006) Mitochondrial diseases and genetic defects of ATP synthase. Biochim Biophys Acta 1757:1400–1405
75. Cameron JM, Levandovskiy V, Mackay N, Ackerley C, Chitayat D, Raiman J, Halliday WH, Schulze A, Robinson BH (2011) Complex V TMEM70 deficiency results in mitochondrial nucleoid disorganization. Mitochondrion 11:191–199

76. Hejzlarova K, Tesarova M, Vrbacka-Cizkova A, Vrbacky M, Hartmannova H, Kaplanova V, Noskova L, Kratochvilova H, Buzkova J, Havlickova V et al (2011) Expression and processing of the TMEM70 protein. Biochim Biophys Acta 1807:144–149
77. Marin-Garcia J, Goldenthal MJ (2002) Understanding the impact of mitochondrial defects in cardiovascular disease: a review. J Card Fail 8:347–361
78. Taegtmeyer H, McNulty P, Young ME (2002) Adaptation and maladaptation of the heart in diabetes: part I: general concepts. Circulation 105:1727–1733
79. Thompson CH, Kemp GJ, Taylor DJ, Conway M, Rajagopalan B, O'Donoghue A, Styles P, McKenna WJ, Radda GK (1997) Abnormal skeletal muscle bioenergetics in familial hypertrophic cardiomyopathy. Heart 78:177–181
80. Milner DJ, Mavroidis M, Weisleder N, Capetanaki Y (2000) Desmin cytoskeleton linked to muscle mitochondrial distribution and respiratory function. J Cell Biol 150:1283–1298
81. Bowles NE, Bowles KR, Towbin JA (2000) The "final common pathway" hypothesis and inherited cardiovascular disease. The role of cytoskeletal proteins in dilated cardiomyopathy. Herz 25:168–175
82. Vatta M, Stetson SJ, Perez-Verdia A, Entman ML, Noon GP, Torre-Amione G, Bowles NE, Towbin JA (2002) Molecular remodelling of dystrophin in patients with end-stage cardiomyopathies and reversal in patients on assistance-device therapy. Lancet 359936–359941
83. Azevedo VM, Albanesi Filho FM, Santos MA, Castier MB, Cunha MO (2005) The role of L-carnitine in nutritional status and echocardiographic parameters in idiopathic dilated cardiomyopathy in children. J Pediatr (Rio J) 81:368–372
84. Langsjoen PH, Vadhanavikit S, Folkers K (1985) Response of patients in classes III and IV of cardiomyopathy to therapy in a blind and crossover trial with coenzyme Q10. Proc Natl Acad Sci U S A 82:4240–4244
85. Lerman-Sagie T, Rustin P, Lev D, Yanoov M, Leshinsky-Silver E, Sagie A, Ben-Gal T, Munnich A (2001) Dramatic improvement in mitochondrial cardiomyopathy following treatment with idebenone. J Inherit Metab Dis 24:28–34
86. Bhagavan HN, Chopra RK (2005) Potential role of ubiquinone (coenzyme Q10) in pediatric cardiomyopathy. Clin Nutr 24:331–338
87. Watson PS, Scalia GM, Galbraith A, Burstow DJ, Bett N, Aroney CN (1999) Lack of effect of coenzyme Q on left ventricular function in patients with congestive heart failure. J Am Coll Cardiol 33:1549–1552
88. Bolhuis PA, Hensels GW, Hulsebos TJ, Baas F, Barth PG (1991) Mapping of the locus for X-linked cardioskeletal myopathy with neutropenia and abnormal mitochondria (Barth syndrome) to Xq28. Am J Hum Genet 48:481–485
89. Durr A, Cossee M, Agid Y, Campuzano V, Mignard C, Penet C, Mandel JL, Brice A, Koenig M (1996) Clinical and genetic abnormalities in patients with Friedreich's ataxia. N Engl J Med 335:1169–1175
90. Dutka DP, Donnelly JE, Nihoyannopoulos P, Oakley CM, Nunez DJ (1999) Marked variation in the cardiomyopathy associated with Friedreich's ataxia. Heart 81:141–147
91. Dutka DP, Donnelly JE, Palka P, Lange A, Nunez DJ, Nihoyannopoulos P (2000) Echocardiographic characterization of cardiomyopathy in Friedreich's ataxia with tissue Doppler echocardiographically derived myocardial velocity gradients. Circulation 102:1276–1282
92. Isnard R, Kalotka H, Durr A, Cossee M, Schmitt M, Pousset F, Thomas D, Brice A, Koenig M, Komajda M (1997) Correlation between left ventricular hypertrophy and GAA trinucleotide repeat length in Friedreich's ataxia. Circulation 95:2247–2249
93. Buchaklian AH, Helbling D, Ware SM, Dimmock DP (2012) Recessive deoxyguanosine kinase deficiency causes juvenile onset mitochondrial myopathy. Mol Genet Metab (in press) (PMID: 22715278)
94. Palmieri L, Alberio S, Pisano I, Lodi T, Meznaric-Petrusa M, Zidar J, Santoro A, Scarcia P, Fontanesi F, Lamantea E et al (2005) Complete loss-of-function of the heart/muscle-specific adenine nucleotide translocator is associated with mitochondrial myopathy and cardiomyopathy. Hum Mol Genet 14:3079–3088

95. Echaniz-Laguna A, Chassagne M, Ceresuela J, Rouvet I, Padet S, Acquaviva C, Nataf S, Vinzio S, Bozon D, Mousson de Camaret B (2012) Complete loss of expression of the ANT1 gene causing cardiomyopathy and myopathy. J Med Genet 49:146–150
96. Honzik T, Tesarova M, Mayr JA, Hansikova H, Jesina P, Bodamer O, Koch J, Magner M, Freisinger P, Huemer M et al (2010) Mitochondrial encephalocardio-myopathy with early neonatal onset due to TMEM70 mutation. Arch Dis Child 95:296–301
97. Holme E, Greter J, Jacobson CE, Larsson NG, Lindstedt S, Nilsson KO, Oldfors A, Tulinius M (1992) Mitochondrial ATP-synthase deficiency in a child with 3-methylglutaconic aciduria. Pediatr Res 32:731–735
98. Mayr JA, Paul J, Pecina P, Kurnik P, Forster H, Fotschl U, Sperl W, Houstek J (2004) Reduced respiratory control with ADP and changed pattern of respiratory chain enzymes as a result of selective deficiency of the mitochondrial ATP synthase. Pediatr Res 55:988–994
99. Antonicka H, Leary SC, Guercin GH, Agar JN, Horvath R, Kennaway NG, Harding CO, Jaksch M, Shoubridge EA (2003) Mutations in COX10 result in a defect in mitochondrial heme A biosynthesis and account for multiple, early-onset clinical phenotypes associated with isolated COX deficiency. Hum Mol Genet 12:2693–2702
100. Antonicka H, Mattman A, Carlson CG, Glerum DM, Hoffbuhr KC, Leary SC, Kennaway NG, Shoubridge EA (2003) Mutations in COX15 produce a defect in the mitochondrial heme biosynthetic pathway, causing early-onset fatal hypertrophic cardiomyopathy. Am J Hum Genet 72:101–114
101. Benit P, Beugnot R, Chretien D, Giurgea I, De Lonlay-Debeney P, Issartel JP, Corral-Debrinski M, Kerscher S, Rustin P, Rotig A et al (2003) Mutant NDUFV2 subunit of mitochondrial complex I causes early onset hypertrophic cardiomyopathy and encephalopathy. Hum Mutat 21:582–586
102. Berger I, Hershkovitz E, Shaag A, Edvardson S, Saada A, Elpeleg O (2008) Mitochondrial complex I deficiency caused by a deleterious NDUFA11 mutation. Ann Neurol 63:405–408
103. Loeffen J, Elpeleg O, Smeitink J, Smeets R, Stockler-Ipsiroglu S, Mandel H, Sengers R, Trijbels F, Van den Heuvel L (2001) Mutations in the complex I NDUFS2 gene of patients with cardiomyopathy and encephalomyopathy. Ann Neurol 49:195–201
104. Loeffen J, Smeitink J, Triepels R, Smeets R, Schuelke M, Sengers R, Trijbels F, Hamel B, Mullaart R, Van den Heuvel L (1998) The first nuclear-encoded complex I mutation in a patient with Leigh syndrome. Am J Hum Genet 63:1598–1608
105. Levitas A, Muhammad E, Harel G, Saada A, Caspi VC, Manor E, Beck JC, Sheffield V, Parvari R (2010) Familial neonatal isolated cardiomyopathy caused by a mutation in the flavoprotein subunit of succinate dehydrogenase. Eur J Hum Genet 18:1160–1165
106. Smeitink JA, Elpeleg O, Antonicka H, Diepstra H, Saada A, Smits P, Sasarman F, Vriend G, Jacob-Hirsch J, Shaag A et al (2006) Distinct clinical phenotypes associated with a mutation in the mitochondrial translation elongation factor EFTs. Am J Hum Genet 79:869–877
107. Saada A, Shaag A, Arnon S, Dolfin T, Miller C, Fuchs-Telem D, Lombes A, Elpeleg O (2007) Antenatal mitochondrial disease caused by mitochondrial ribosomal protein (MRPS22) mutation. J Med Genet 44:784–786
108. Smits P, Saada A, Wortmann SB, Heister AJ, Brink M, Pfundt R, Miller C, Haas D, Hantschmann R, Rodenburg RJ et al (2011) Mutation in mitochondrial ribosomal protein MRPS22 leads to Cornelia de Lange-like phenotype, brain abnormalities and hypertrophic cardiomyopathy. Eur J Hum Genet 19:394–399

Chapter 21
Mitochondrial Diseases Caused by Mutations in Inner Membrane Chaperone Proteins

Lisbeth Tranebjærg

Background

The human mitochondrial DNA (mtDNA) genome encodes 13 proteins that are all components of the respiratory chain complexes embedded into the mitochondrial inner membrane [1]. Approximately 100 more nuclear-encoded proteins are involved in their translation. All proteins required for mtDNA replication, RNA transcription, modification, protein translation, and the assembly of the respiratory chain complexes are also encoded by nuclear genes. Mitochondrial protein synthesis deficiency can be caused by mutations in any part of the translation pathways, including tRNA, rRNA, mRNA, mitochondrial ribosomal protein subunits, aminoacyl tRNA synthetases (ARS2's), and translation factors. The first genetically described mitochondrial disorders had mutations in mitochondrial tRNA and rRNA genes (i.e., mutations that compromised translation). It has since become clear that maternally inherited mitochondrial disorders (i.e., mutations in the mitochondrial genome) represent only about 15–20 % of all inherited human mitochondrial disorders [2]. This estimate has emerged along with the identification of mutations in a large number of nuclear genes leading to dysfunction of tRNA-modifying enzymes, mitochondrial ribosomal proteins, ARS2's, elongation and termination factors and translational activators. Recent years have thereby revealed a huge clinical and genetic heterogeneity of mitochondrial diseases, making early diagnosis and genotype–phenotype correlation extremely challenging [3].

The mitochondrial proteome contains approximately 1,000 proteins, and there are various strategies for defining a protein as mitochondrial [4]. There are quantitative differences regarding the number of mitochondria in different tissues, with over twofold and tenfold more in heart tissue compared with brain and blood. Moreover,

---

L. Tranebjærg (✉)
Department of Audiology, Bispebjerg Hospital, 2400 Copenhagen NV, Denmark

Wilhelm Johannsen Centre for Functional Genome Research,
Department of Cellular and Molecular Medicine (ICMM),
The Panum Institute, University of Copenhagen, 2200 Copenhagen NV, Denmark
e-mail: Tranebjaerg@sund.ku.dk

mitochondria from distinct organs shared approximately 75 % of their proteins and tissues of embryonically related organs tend to have higher sharing [4].

The mitochondrial proteins encoded by nuclear genes are synthesized in the cytosol and transported into one of four mitochondrial compartments: outer membrane (OM), intermembrane space (IMS), inner membrane (IM), or matrix. The TOM (translocase of outer membrane) complex is used by all mitochondrial proteins analyzed so far for transport across the outer mitochondrial membrane. Depending on each protein's subcellular-targeting signal, the TOM complex cooperates with other mitochondrial translocases to direct the protein into one of the four compartments [5].

Import of the nuclear-encoded precursor proteins into and across the mitochondrial inner membrane is mediated by two distinct translocases, the TIM23 complex and the TIM22 complex. Both cooperate with the TOM complex. Precursors with an N-terminal-targeting signal are imported via the TIM23 complex into the matrix [6]. Integral inner membrane proteins not carrying a matrix-targeting signal, require the TIM22 complex for insertion into the inner membrane (i.e., mitochondrial carrier proteins such as the ADP/ATP carrier (AAC), the phosphate carrier (PiC) and other hydrophobic inner membrane proteins such as Tim23, Tim17, and Tim22 itself) [7].

The intermembrane space harbors two homologous small Tim (translocases of the inner membrane) complexes, namely the Tim9–Tim10 and the Tim8–Tim 13 complexes. In yeast, the Tim9–Tim10 complex is crucial for the viability [8]. In contrast, the Tim8–Tim13 complex was only strictly needed for import of Tim23 under conditions of low membrane potential in mitochondria, and not under normal growth conditions. The biogenesis of Tim23 in humans may be more dependent on the Tim8–Tim13 complex, especially under conditions with low membrane potential. In a yeast study, Beverly et al. [9] describe the crystal structure of small Tim complexes, which are hexameric, and contain three copies of each subunit. The crystal structure of both small complexes has shown that they appear as long helices extending from a central body, similar to tentacles from a jelly fish. The Tim8–Tim13 hexamer binds to the substrate Tim23 to chaperone the hydrophobic Tim23 across the aqueous intermembrane space into the inner membrane. The regions where the tentacles attach to the body of the Tim8–Tim13 complex contain six hydrophobic pockets, thought to interact with specific sequences of Tim23 and possibly other substrates. The presence of two characteristic twin CX_{3C} motifs forming two intramolecular disulfide bonds was also observed [8] and will be discussed below in terms of the location of disease mutations in the TIMM8A gene (see Table 21.1, Family 7) [10]).

The TIM chaperone complexes cooperate with the membrane translocase SAM (sorting and assembly machinery) complex of the outer membrane. Precursors of β-barrel proteins translocated via the TOM40 channel, bind to Tim9–Tim10 or Tim8–Tim13 and are transferred to SAM. By this shuttling the precursors are kept in a translocation-competent conformation [5].

The characterization of subunits of the complex TIM23 has increased during recent years. TIM23 complex contains ten components: Tim50, TIM23, Tim17, Tim44, Tim14 (Pam18), Tim16 (Pam 16), Tim21, Pam17, mtHsp70, and Mge1. Apart from Pam17 they are all highly conserved from yeast to human. For Tim14, two genes encoding

21 Mitochondrial Diseases Caused by Mutations in Inner Membrane Chaperone Proteins

Table 21.1 The clinical and genetic spectrum of reported patients with deafness-dystonia-optic neuronopathy syndrome

Family number, ethnic origin	Reference	Patients	Age at onset of hearing impairment (years)	Visual abnormalities	Age at onset of visual impairment (years)	Neurology	Age at onset of neurologic abnormalities (years)	Behavioral abnormalities	Other findings (years)	Mutation[a]/ predicted effect
1, American-French	[56]	Proband	2	NR		Dystonia	7	Emotionally labile		c.70G>T/ p.Glu24X
		Maternal uncle	6	NR		Dystonia	>6		Died in his 20s	c.70G>T/ p.Glu24X
2, Danish	[32, 33, 34, 35, 36]	Proband	2	VEP abn	20	Ataxia, no dystonia, dysphagia	19		Normal biochemical muscle enzymes	c.70G>T/ p.Glu24X
		2 maternal uncles		V.A 0.1 bilateral	14–18	Dementia, no dystonia	30	Aggression, lack of impulse control	Postmortem optic neuropathy [55]	c.70G>T/ p.Glu24X
3, Norwegian	[26, 29]	16 cases, 4 alive (K8160)	2–7	VEP abnormal	Teen age years	Dystonia, spasticity, neuropathy	7–50	Aggression, lack of impulse control	Seven males died (18, 49, 60, 61, 63, 66, and 68), postmortem studies of 6 cases, ([36, 40])	c.116delT/ p.Met39Argfs X26
4, American-White	[29]	5 cases, two generations (K8190)	NR	NR	NR	Cognitive decline, dystonia	10-early adulthood	Psychiatric abnormalities	Three died (10, 34, 40), two alive 50s	c.148_157del/ p.Lys50Glnfs X12

Table 21.1 (continued)

Family number, ethnic origin	Reference	Patients	Age at onset of hearing impairment (years)	Visual abnormalities	Age at onset of visual impairment (years)	Neurology	Age at onset of neurologic abnormalities (years)	Behavioral abnormalities	Other findings (years)	Mutation[a]/ predicted effect
5. Spanish	[28, 29]	Pt PDSM, familial case (+3)	2	NR		Dystonia	5	Three obligate female carriers affected	+XLA: recurrent infections; death at age 2, 5, and 8	21kb del entire TIMM8A+BTK, exons 11–19
6. Australian	[68, (fam C) 49]	Proband C1	<2	Visual abn	19	Pyramidal signs, mild MR, dystonia	6–19	Aggression, lack of impulse control		c.52C>T/ p.Gln18X
		Half brother C2	<2.6	N	23	Pyramidal signs, mild MR, dystonia	16–23	Aggression, lack of impulse control		c.52C>T/ p.Gln18X
		Maternal uncle C3	Deaf: NR	NR		Pyramidal signs, dystonia	35	NR		Not tested
		Maternal uncle C4	<2	NR		NR		NR		Not tested
7. Dutch	[10]	Proband	2.5	Abnormal VEP	11	dystonia	10	NR		c.198C>G/ p.Cys66Trp (de novo)
8. American-White	[63, 64]	Proband	Congenital	NR		Dystonia	28	NR		c.73delG/ p.Val25X
		Mother	–	NR		Dystonia	25	NR		c.73delG/ p.Val25X
		Sister	–	NR		Dystonia	Teens	NR		c.73delG/ p.Val25X

21 Mitochondrial Diseases Caused by Mutations in Inner Membrane Chaperone Proteins 341

Table 21.1 (continued)

Family number, ethnic origin	Reference	Patients	Age at onset of hearing impairment (years)	Visual abnormalities	Age at onset of visual impairment (years)	Neurology	Age at onset of neurologic abnormalities (years)	Behavioral abnormalities	Other findings (years)	Mutation[a]/ predicted effect
		Maternal uncle	Congenital	Blindness	Early 50s	Dementia, dystonia	20's–50's	NR	Died at age 63	Not tested
		Maternal uncle	Congenital	NR		No dystonia		NR	Died at age 22	Not tested
		Maternal great grand-mother	NR	NR		Dystonia	NR	NR		Not tested
		Two maternal male second cousins	Early onset	NR		Dystonia	Adult age	NR		Not tested
9, Japanese	[62]	Proband	6 months	N, ERG:N		Mild cognitive impairment, dystonia	30	Uneven temperament (38)	Mild cortical atrophy frontally	c.238C>T/ p.Arg80X
		Brother, case 2	4	N		Mild MR, dystonia	16	No personality changes		c.238C>T/ p.Arg80X
		Maternal grand-father, case 3	+ (age unknown)	NR		Dystonia	NR		Died at age 89	Not tested
		Maternal male cousin, case 4	3–4	NR		Dystonia	NR			c.238C>T/ p.Arg80X
		Maternal nephew, case 5	8–9	NR		NR				c.238C>T/ p.Arg80X

Table 21.1 (continued)

Family number, ethnic origin	Reference	Patients	Age at onset of hearing impairment (years)	Visual abnormalities	Age at onset of visual impairment (years)	Neurology	Age at onset of neurologic abnormalities (years)	Behavioral abnormalities	Other findings (years)	Mutation[a]/ predicted effect
10, Croatia	[42, 73] (2430)	Proband (pt 1)	2–3, progressive, deaf at age 6	N		NR		NR	+XLA, Sepsis (10 mths), Immunodeficiency	Del *BTK* exon 5–19+del distal to *TIMM8A*
11, Croatia	[42, 73] (2433)	Proband (pt 2)	3–4, progressive, deaf at age 9	N		NR		ADHD	+XLA, Pneumonia (8 mths)	Del *BTK* exon 19+*TIMM8A*
12, Lebanon	[42, 73] (0703)	Proband (pt 3)	6, mild, progressive-deaf at age 14	N		NR			+XLA	Del *BTK* exon 19+*TIMM8A*
13, American-White	[47]	Proband (1)	18 mths–moderate, stable	N		NR		NR		c.100C>T/ p.Q34X
		Maternal uncle (4)	5,6, -profound	RP, Abn ERG	30s	Ataxia	Early 50s	Paranoid ideation	Died at age 52, postmortem eye and temporal bone histology (Prause, 2012, personal communication; Merchant, 2012, personal communication)	c.100C>T; p.Q34X
		Male cousin	2, 5- profound	N		N		NR		

21 Mitochondrial Diseases Caused by Mutations in Inner Membrane Chaperone Proteins 343

Table 21.1 (continued)

Family number, ethnic origin	Reference	Patients	Age at onset of hearing impairment (years)	Visual abnormalities	Age at onset of visual impairment (years)	Neurology	Age at onset of neurologic abnormalities (years)	Behavioral abnormalities	Other findings (years)	Mutation[a]/ predicted effect
14, German	[51, 67]	Proband	3	V.A 0.2/0.25, optic atrophy	37	Dystonia	28	NR	Abn VEP; MRI: marked cortical atrophy of occipital and parietal lobes, Oximetry: upper normal level, no RRF, histology of muscle: neurogenic atrophy	c.3G>C/ p.Met1Ile, (p.Met1?) (de novo)
15, Italian	[61]	Proband	2	V.A. 0.2 bi-laterally	15	Mild cognitive impairment, dystonia	19	N	+XLA, recurrent infections; no multiple mitDNA deletions in muscle tissue	Del TIMM8A-BTK exon 19 deleted-exon 7 of BTK intact
16, Spanish	[69]	Proband	4, deaf	Abn VEP	24	Dystonia	11	NR	MRI: N (31)	IVS1–23A>C/ altered splicing
		Male cousin	11, Mild	NR		Frontosub-cortical dysfunction, dystonia	20–29	Impulse control disorder	MRI: N (29)	IVS1–23A>C/ altered splicing

Table 21.1 (continued)

Family number, ethnic origin	Reference	Patients	Age at onset of hearing impairment (years)	Visual abnormalities	Age at onset of visual impairment (years)	Neurology	Age at onset of neurologic abnormalities (years)	Behavioral abnormalities	Other findings (years)	Mutation[a]/predicted effect
17, Spanish	[45]	Proband	3.5 (90 dB HL)	Ocular fundi N		Progressive dystonia	8	Disruptive behaviour		c.127delT/ p.Cys43ValfsX22
		Brother	7 (50 dB HL)	NR		No dystonia		Aggression, lack of impulse control		c.127delT/ p.Cys43ValfsX22
18, British	[74]	Proband	8 months	V.A 6/24	42	dystonia	25	NR	MRI: N (34)	IVS1+1G>A/altered splicing
19, Spanish	[50]	Proband	3	Progressive visual loss, Abn VEP	15	EMG axonal neuropathy, dystonia	30	Irritability	MRI occipital atrophy, mild deficiency of complex IV Respiratory chain enzyme.	c.112C>T/ p.Gln38Ter
20, Czech	[43]	Proband (pt 1)	5, progressive	NR		NR		Aggressive	+XLA Recurrent infections from age 2	30-kb del BTK, intron 18-exon 19 and TIMM8A
		Brother (pt 2)	Preschool, progressive	NR		NR		Aggressive	+XLA Recurrent respiratory infections	30-kb del BTK, intron 18-exon19 and TIMM8A

Table 21.1 (continued)

Family number, ethnic origin	Reference	Patients	Age at onset of hearing impairment (years)	Visual abnormalities	Age at onset of visual impairment (years)	Neurology	Age at onset of neurologic abnormalities (years)	Behavioral abnormalities	Other findings (years)	Mutation[a]/predicted effect
21, Czech	[43]	Proband (pt 3)	4	NR		Dystonia	5	NR	+XLA (2); Died (6)-Oximetry N; *TIMM8A* missing in fibroblast mitochondria	20kb del, *BTK* intron 18-exon 19, and *TIMM8A*
22, Estonia	[43]	Proband (pt 4)	4	NR		Psychomotor retardation		NR	+XLA (7 mths)	22kb del, *BTK* exon 6–19 and *TIMM8A*
		Brother (pt 5)	2.5	NR		NR		NR	+XLA (2 mths), maternal uncle died from infection	22kb del, *BTK* exon 6–19 and *TIMM8A*
23, Afro-American	[43, 44] (pt 3)	Proband (pt 6)	30–80 dB HL bilateral, progressive	NR		Dystonia	15	Severe outbursts of aggression	+XLA (8 mths), two sisters carriers	196kb del, *BTK* exon 2–19, exons 1–2 of *TIMM8A*, *TAF7L* and *DRP2*
24, American-White	[41]	Proband (4)	3.5 Auditory neuropathy, CI	NR		Normal neuropsychology		NR	+XLA (5), Cerebral CT N	6kb del, *BTK* exons 17–19 and exon 1 of *TIMM8A* – (de novo)

Table 21.1 (continued)

Family number, ethnic origin	Reference	Patients	Age at onset of hearing impairment (years)	Visual abnormalities	Age at onset of visual impairment (years)	Neurology	Age at onset of neurologic abnormalities (years)	Behavioral abnormalities	Other findings (years)	Mutation[a]/ predicted effect
25, Ukraine	[75]	Probands, MZ twins	2–3 profound	NR		NR		NR	+XLA (2)	155kb del, *BTK* exons 3–19, *TIMM8A*, *TAF7L* and *DRP2*
26, Spanish	[45]	Proband	4, progressive	Optic atrophy	14	Dementia, dystonia	23–30	NR	MRI: N (14)	c.132+1G>T; IVS1+1G>T/ altered splicing (*de novo*)
27, Japanese	[44]	Proband (pt 1)	1, severe and progressive	NR		No dystonia		NR	+XLA (7)	63kb del, *BTK* intron 15-exon 19, *TIMM8A*, and *TAF7L*
28	[44]	Proband (pt 2)	18 months-severe and progressive	NR		No dystonia		Autism	+XLA (8)	149.7kb del, including *BTK*, *TIMM8A*, *TAF7L* and *DRP2*

21 Mitochondrial Diseases Caused by Mutations in Inner Membrane Chaperone Proteins 347

Table 21.1 (continued)

Family number, ethnic origin	Reference	Patients	Age at onset of hearing impairment (years)	Visual abnormalities	Age at onset of visual impairment (years)	Neurology	Age at onset of neurologic abnormalities (years)	Behavioral abnormalities	Other findings (years)	Mutation[a]/ predicted effect
29, Australian	[49]	Proband fam A1	<2	Optic atrophy	40	Dementia, dystonia	30-Early 50s		MRI: caudate head atrophy; no immunodeficiency	Del *TIMM8A* exon 2
		Brother A2	<2	Optic atrophy	43	Dystonia	40			Del *TIMM8A* exon 2
30	[49]	Proband (female) fam B1	Early 40s, unilateral	V.A. 6/9	59	Dystonia, cognitive decline	Mid 40s		Proband = female, abnormal VEP	c.127delT/ p.Cys43Valfs X22
		Son B2		V.A 6/36	30	No dystonia			MRI: atrophy of occipital cortex	c.127delT/ p.Cys43Valfs X22
31, American-White	Conley, 2012, unpublished	Proband	Normal hearing	NR		NR			+XLA	Del *BTK* exon 12–19 and *TIMM8A*
		Maternal uncle and a grand maternal uncle		NR		NR			+XLA, Both died at age 16 months- 2 years from infections	No genetic testing
		Maternal Deaf male cousin		NR		No dystonia			+XLA (4)	No genetic testing

Table 21.1 (continued)

Family number, ethnic origin	Reference	Patients	Age at onset of hearing impairment (years)	Visual abnormalities	Age at onset of visual impairment (years)	Neurology	Age at onset of neurologic abnormalities (years)	Behavioral abnormalities	Other findings (years)	Mutation[a]/ predicted effect
32, Dutch	Stolte-Dijkstra, 2012, personal communication	Proband	2, deaf	V.A abn VEP		Dementia, dystonia	12–17		FDG-PET scan: reduced striatal uptake; MRI N	IVS1+5G>A/ altered splicing (de novo)
33, Australian	Gardner-Berry et al, 2012, personal communication	Proband	18 months-progressive-bilateral CI	N		NR		Severe behavior abnormalities	+XLA	Del BTK exon 3-exon 19+entire TIMM8A
34, British	Tolmie and McWilliams, 2012, unpublished	Proband	Deaf (2), CI N (9) (7.5)					Some behavior problems	+XLA (3), recurrent infections from age 18 mths	17kb del of BTK from exon 6 and TIMM8A (De-novo)
35, Dutch	Kunst, 2012,unpublished	Proband	3.5, deaf	NR		Mild MR	12	Aggressive behavior from childhood	Cerebral CT scan: N Del (3), febrile convulsions (9 mths). No immunodeficiency	TIMM8A exon 2
		Maternal uncle	Congenital-deaf	V.A 0.4/ 0.3, abnormal VEP	NR	Ataxia, no dystonia	30		Cryptorchidism, progressive neurological dysfunction, vestibular hyporeflexia	Del TIMM8A exon 2

21 Mitochondrial Diseases Caused by Mutations in Inner Membrane Chaperone Proteins   349

Table 21.1 (continued)

Family number, ethnic origin	Reference	Patients	Age at onset of hearing impairment (years)	Visual abnormalities	Age at onset of visual impairment (years)	Neurology	Age at onset of neurologic abnormalities (years)	Behavioral abnormalities	Other findings (years)	Mutation[a]/ predicted effect
36, British	Humberstones, 2012, unpublished	Proband	Age?	Severe optic atrophy	NR	Severe dystonia	NR	Severe behavioral abnormalities		c.132+5G>A; IVS1+5G>A/ altered splicing
37, French-Caribbean	Verloes et al, 2012, unpublished	Proband, IV-7	2, deaf	NR		Dystonia	8	Behavioral problems	Died at approximately age 12	c.132G>T/ p.Trp44Cys
		Maternal uncle III-3	6, deaf	NR		Ataxia, MR		Behavioral problems		c.132G>T/ p.Trp44Cys
		Maternal uncle III-4		NR		Ataxia, no dystonia, MR		NR		c.132G>T/ p.Trp44Cys

NR not reported, *pt* patient, *Del* deletion, *MR* mental retardation, *abn* abnormal, *N* normal, *V.A.* visual acuity, *RRF* ragged red fibres
[a]According to recommended mutation nomenclature. Nucleotide numbering refers to the *TIMM8A* cDNA sequence NM_004085.3 in GeneBank with nucleotide number +1 being A of the start codon ATG

protein homologs of Tim14 are present. Inactivation of either leads to distinct human diseases. In yeast genome a homolog of *Tim14*, called *Mdj2*, is present, and found in TIM23 complex. Upon overexpression it can take over the function of Tim14, and deletion of *Mdj2* has no phenotype in contrast to deletion of *Tim14*, which is lethal to yeast. With the exception of *Pam17* and *Tim21*, deletion of all other components is lethal to yeast [6].

Expanding the Definition of Mitochondrial Diseases

Traditionally, mitochondrial disease has referred primarily to disorders of oxidative phosphorylation (ATP production). As summarized by Calvo and Mootha [4], 92 coding genes were identified underlying respiratory chain diseases [4, 11]. If the definition is broadened to cover diseases caused by mutations in genes encoding mitochondrial localizing proteins, the list increases to > 150 conditions. Moreover, in a large fraction of identified diseases, the pathomechanism has not been established yet and the number is steadily increasing. A serious effort to compile information about genes, phenotypes, and biological pathways led to the creation of the mitophenome database (http://www.mitophenome.org) whereby similar features/pathways can be compared in a comprehensive fashion [12]. Such a compilation of relevant genes is extremely valuable in distinct clinical situations such as selecting exons for high throughput next-generation sequencing, as already exemplified in several cases to be the way to identify a disease mutation [13]. The increasing notion of dysregulation of mitochondrial fission and fusion involved in neurodegeneration further expands the universe of mitochondrial dysfunction. Dysregulation of mitochondrial fission and fusion can now be regarded as playing important pathogenic roles in neurodegeneration by the examples of Charcot-Marie-Tooth neuropathy type 2A (CMT2A) linked to mutations in Mitofusin 2 (*MFN2*) [14–16] and dominant optic atrophy linked to mutations in *OPA1* [17, 18].

Animal Models

An interesting study of the first mouse model of abnormal mitochondrial protein import showed neurological abnormalities and reduced lifespan in heterozygous *Tim23* knock out (KO) mice [19]. It was impossible to recover homozygous *Tim23*-/- mutants, strongly indicating that *Tim23* is essential even before implantation. Heterozygous KO mice had significantly reduced strength at grip span, reduced ability of motor coordination and reduced life span. No hearing or vision impairment was noted, but formal testing of the sensory abilities was not performed. The mice also showed alopecia and kyphosis possibly reflecting premature aging. Interestingly, during the subsequent outcrossing experiments, the heterozygous *Tim23*± lost their abnormal phenotype, which remained essentially unexplained but speculated to be due to selective factors in favor of distinct genetic backgrounds or fertility in favor of those with highest residual *Tim23* expression. The possibility of redundancy due to other proteins was not explored. A mouse model for Tim8–Tim13 complex has not been established.

Human Diseases Caused by Mutations in TIMM's: Mohr–Tranebjaerg (MTS) and Other Diseases

Human disease associated with the dysfunction of mitochondrial protein import has recently been reviewed [20]. The TIMM components of the intermembrane space belong to an evolutionarily conserved protein family with > 50 gene members, of which, six (*TIMM9, TIMM13, TIMM8A, TIMM8B, TIMM10A, and TIMM10B*) are expressed in humans [21]. The human homolog of *Tim8a* is encoded by *DDP1* (or *TIMM8A*) on Xq22 (MIM 304700).

Diseases Caused by Mutations in Non-TIMM8A

Mackenzie and Payne [20] present examples of human diseases illustrating the various steps in how nuclear-encoded proteins reach their final destination in mitochondria. Defects in mitochondrial-targeting signals can cause disease in humans. Examples include pyruvate dehydrogenase deficiency (PDH E1α, MIM 312170) and primary hyperoxaluria type 1 (AGT, MIM 259900), which, respectively, is associated with microcephaly and cortical atrophy, and a renal phenotype due to deposits of calcium oxalate in the kidneys and almost every other organ in the body.

Diseases may also occur as a consequence of mutations in nuclear genes involved in the mitochondrial import and processing machinery, as illustrated by Mohr–Tranebjaerg syndrome and dilated cardiomyopathy associated with ataxia (DCMA), caused by mutations in *TIMM8A*, and *DNAJC19* (the human homolog of *Tim14*), respectively. The syndrome of dilated cardiomyopathy with ataxia (DCMA, MIM 610198) is caused by a splice mutation in the *DNAJC19* [22]. The clinical features are characterized by autosomal recessive early and severe dilated cardiomyopathy, growth retardation, ataxia, and optic atrophy.

Tim14 (Pam18) is one of several components of the TIM23 complex, initially identified in fungi, but turning out to be highly conserved from yeast to human [6]. For some TIM23 components, multiple genes have been found. Two genes encode protein homologs of *Tim14* in humans, and inactivation of either leads to quite distinct human disorders. Loss of expression of members of this protein family (called DNAJD— homologs of yeast *Tim14*) is associated with two distinct phenotypes resistance to chemotherapeutic agents used in the treatment of ovarian cancer [23]: and a pediatric brain tumor [24]. In yeast, deletion of *Tim14* is found to be lethal, whereas deletion of *Mdj2*, a homolog of *Tim14*, has no phenotype in yeast. Currently, the full spectrum of homologs of *Tim14* and all members of the protein family of DNAJ's and their functions are far from fully understood.

Translocase of inner mitochondrial membrane 8 homolog B (*TIMM8B*) was mapped to 11q23.1 [21], and in reverse orientation to *SDHD* (MIM 602690), which is mutated in familial carotid body paragangliomas (PGL) (MIM168000). Members of the two out of four families with PGL described by Badenkop et al. (2001) also had cosegregating deafness and tinnitus, but no mutations were identified in *TIMM8B*, despite the authors' speculation of altered transcription of *TIMM8B* due to *SDHD* mutations [25]. This issue remains unresolved.

Mohr–Tranebjærg syndrome is caused by mutations in *TIMM8A* (MTS, MIM304750), the human homolog of yeast *Tim8* [20], and will be the focus of the remaining chapter.

Mohr–Tranebjaerg Syndrome (MTS)

The Mohr–Tranebjaerg syndrome (MTS) was originally named as deafness-dystonia-optic neuronopathy (DDON) syndrome, recognizing the main features of this X-linked disorder [26]. The initial description of the Norwegian family [27] was followed-up by Tranebjaerg et al. in 1995 [26]. The reinvestigation revealed that the family did not have X-linked nonsyndromic hearing impairment (called DFN-1 at the time), but rather a severe progressive neurodegenerative syndrome with hearing and vision impairment as well as neurological and psychiatric features [26]. The disease gene was mapped to Xq22. Subsequently, a patient (patient G) from a family with X-linked agammaglobulinemia (XLA) and associated deafness and dystonia was identified to have a large deletion that included the Bruton agammaglobulinemia tyrosine kinase gene (*BTK*) as well as an uncharacterized gene (*DXS1274E*) located at the centromeric side of *BTK* [28]. MTS (DFN1 or deafness/dystonia) was found in two additional families with the X-linked deafness syndrome without the immunodeficiency [29]. Mutations were identified in the Norwegian family with a 1 bp deletion (c.116delT; p.Met39ArgfsX26) and in *DXS1274E* an American family with a 10 bp deletion in exon 2 c.148_157del10 (p.Lys50GlnfsX12) both segregating with the disease [21]. The *DDP* protein showed ubiquitous expression with high homology to a predicted peptide in *S. pombe*, suggesting a conserved intrinsic role in all cell types [21].

In 1999, Koehler et al. showed that human deafness-dystonia is a mitochondrial protein import disease, the first demonstration of this type of mitochondrial disease caused by mutations in a nuclear gene [30]. By comparing their yeast data and the neurodegenerative disorder described in 1995 [26] and mutations identified in *DDP* (*TIMM8A*) on Xq22 [29] they were able to demonstrate that the MTS was the first example of a human mitochondrial dysfunction disease caused by mutations in a nuclear-encoded mitochondrial transporter gene. Since 1999, several patients with MTS and *TIMM8A* mutations have been described worldwide (Table 21.1, Fig. 21.1)

21 Mitochondrial Diseases Caused by Mutations in Inner Membrane Chaperone Proteins   353

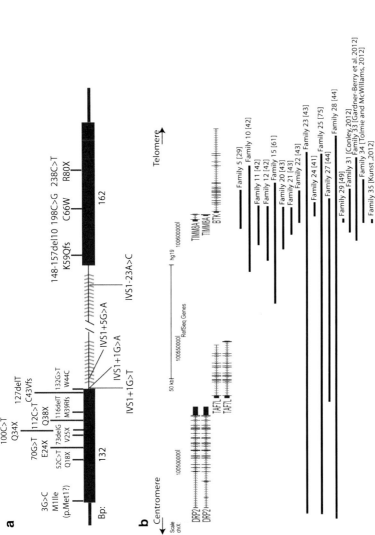

Fig. 21.1 *TIMM8A* structure and mutations. **a** Scematic representation of *TIMM8A* and identified small mutations. The *TIMM8A* genomic structure composes 2 exons. **b** Approximate extent of deletions involving *TIMM8A*. Family numbers are in accordance with the numbers used in Table 21.1

[31]. We have shown earlier that another X-linked phenotype, Jensen syndrome (MIM311150) was allelic to MTS, since the only (Danish) family reported [32, 33] to have Jensen syndrome also had a *TIMM8A* mutation segregating with the disease [34–36].

Pleiotropy of Mitochondrial Disorders

Disorders associated with mitochondrial dysfunction share a tendency to be tissue specific, and therefore they may present clinically as distinct diseases. Freidreich's ataxia (FA) [37] with severe and often lethal cardiomyopathy resembles MTS which has no cardiac phenotype but similar neurodegeneration and effect on both vision and hearing [36]. Cardiomyopathy, on the other hand is a shared phenotype between FA and DCAM. The extensive pleiotropy of mitochondrial diseases complicates the possibility to suspect mutations in specific TIMM and TOMM genes. Moreover, for some of them more than one gene encodes the peptide and the role of functional redundancy to "protect" humans from developing disease is not known. Recent next-generation sequencing of several candidate genes in patients suspected of mitochondrial dysfunction may enhance our understanding and shed light on suspect genes with a given clinical presentation [38]. A recent case report by Lieber et al. [13] showed an unexpected result in a patient with mitochondrial dysfunction by identifying mutations in the Wolframin1 gene (*WFS1*), not proven to be among mitochondrial genes.

Clinical Features of MTS

Currently, there are 91 MTS patients from 37 families, of which seven families are unpublished (Table 21.1, Families 31–37). Among them, 19 cases are familial and 18 are apparently sporadic. The *de novo* origin of six cases was confirmed (Table 21.1, Families 7, 14, 24, 26, 32, and 34).

In 16 families, MTS occurred as part of a contiguous gene deletion syndrome including the XLA gene, *BTK*- (Table 21.1, Families 5, 10, 11, 12, 15, 20, 21, 22, 23, 24, 25, 27, 28, 31, 33, and 34). As expected, 11 of 16 families were of sporadic occurrence (Table 21.1, Families 10, 11, 12, 15, 21, 24, 25, 27, 28, 33, and 34), but familial occurrence was observed in five families (Table 21.1, Families 5, 20, 22, 25 (twins), and 31). Among the 21 families with MTS alone (Table 21.1, Families 1, 2, 3, 4, 6, 7, 8, 9, 13, 14, 16, 17, 18, 19, 26, 29, 30, 32, 35, 36, and 37), 14 were familial (Table 21.1, Families 1, 2, 3, 4, 6, 8, 9, 13, 16, 17, 29, 30, 35, and 37) and seven were sporadic (Table 21.1, Families 7, 14, 18, 19, 26, 32, and 36).

Audiological Features of MTS

Typically, hearing impairment (HI) is the first recognized clinical symptom. It may be congenital or mild hearing impairment, most often symmetrical, and always sensorineural in origin. Very often the hearing impairment is rapidly progressive if not presenting as deafness initially (Norwegian family). The neurological, ophthalmological, psychiatric, and other abnormalities are more variable in penetrance and the age of onset. Long-term follow-up and large families are rarely published, and the original Norwegian family where the syndrome was characterized is still the largest family reported [26]. This family has been investigated thoroughly regarding ophthalmological, neuropathological, and genetic characteristics over several years. The results clearly demonstrated a neurodegenerative disorder and constituted one of the most well-documented examples of hereditary auditory neuropathy [39–42]. The family was originally described as an X-linked nonsyndromic form of hearing impairment (DFN-1) in 1968. However, our reinvestigation was able to reclassify this family to be MTS [26]. There does not seem to be allelic heterogeneity. Regardless of the types of mutations, point mutations, or large deletions of *TIMM8A*, clinical features appear to be characteristic.

MTS is Associated with an Auditory Neuropathy

Hearing impairment is the most prominent symptom in MTS. In most cases the hearing impairment is mild or even absent in very early infancy, but progresses rapidly to profound hearing impairment in the majority of cases [26, 42–44]. There was one example of documented normal hearing by audiograms, until age of 2.5 years and progressed to profound hearing impairment by age of 6.5 years [26]. The audiological features are often superficially described as "deafness" even in cases of only moderate hearing impairment (listed in Table 21.1), and successive audiograms have only been reported in few cases [26, 45, 46].

Treatment of Hearing Impairment in MTS

Rehabilitation with hearing aids has been useful in several instances and in one (published) patient a cochlear implant (CI) was given at the age of 4 [41], but with limited beneficial outcome.

Auditory brainstem response (ABR) recordings are abnormal, whereas there are variable reports on preserved otoacoustic emission (OAE) at young age (patient 1 in [47]), but absent in another young male patient (10-year-old patient in [46]). At later ages OAE is most often absent (as exemplified by the older affected maternal uncle in Varga et al. [47]. The only published patient with CI [41] passed neonatal hearing screening by distortion product otoacoustic emission (DPOAE), and had preserved OAE's at the age of 3.5. At that age the ABR did not show any response at 80 dB HL and cochlear microphonics (CM) was normal, indicating an auditory neuropathy. Hearing aids were instituted without improvement of speech discrimination ability or language development. CI was performed unilaterally at the age of 4. His communication skills remained substantially below age-appropriate values at the age of 6. Subsequently, a 6 kb deletion of exon 1 of *TIMM8A* and exons 17–19 of neighboring *BTK* was identified, supporting the diagnosis of contiguous gene deletion syndrome, XLA + MTS. Another male patient with XLA + MTS had bilateral CI with beneficial outcome in terms of improved speech discrimination ability and language development [42].

Otopathology in MTS

The otopathology of temporal bones from four male patients from the Norwegian family with *TIMM8A* mutation (died at the age of 66, 63, 49, and 60, respectively) uniformly showed near-total loss of cochlear neurons and severe loss of vestibular neurons, and the syndrome is thereby a true auditory neuropathy [39, 40]. Auditory neuropathy can be linked to mitochondrial dysfunction in MTS, Friedreich's ataxia, Charcot-Marie-Tooth neuropathy and in a nongenetic toxic neonatal hyperbilirubinemia [48].

Vestibular dysfunction, although not typical, has been demonstrated occasionally (Table 21.1, Family 35). The postmortem notion of severe loss of vestibular neurons has not been associated with reported vestibular dysfunction [39, 40].

Furthermore, similar pathology with near-total loss of cochlear neuronal cells and vestibular neurons and overall intact vestibular and auditory-end organs was demonstrated in a male who died at the age of 53 (patient 4 of Family 13, Table 21.1) [47, Merchant SN, 2012, personal communication].

Natural History of the Hearing Impairment in MTS

Intra and interfamilial variability in terms of age of onset and degree of HI is striking (Table 21.1, Families 3, 16, and 17). An example of the intra-familial variation is the family with contiguous gene deletion syndrome; XLA + MTS (Table 21.1, Family 31). The proband had normal hearing at the age of 10, while his maternal male cousin had HI. Other organs affected are also variable in severity and age of onset. The multigeneration Norwegian family (Table 21.1, Family 3) with a frameshift mutation in *TIMM8A* contained males dying in their youth due to rapid progression of very severe dystonia and males who lived into their sixties with more modest pace of progression and less severe neurological phenotype. A diagnostically puzzling case was the sporadic male patient who had late onset of movement disorder at the age of 30 but early childhood onset of profound hearing impairment since 8 months of age (Table 21.1, Family 18). The proband of family 9 had a similar time lapse between the childhood onset of profound hearing impairment and dystonia (Table 21.1, Family 9). In contrast, in an American family, an adult female who did not have hearing impairment was reported to have dystonia (Table 21.1, Family 8).

It is important to keep in mind that the variable cognitive dysfunction presents either as early developmental delay or as dementia later in adult life, and significant behavioral and psychiatric abnormalities may influence the audiological evaluations and their interpretations, especially if the audiological testing requires patient cooperation.

Neurological and Psychiatric Features in MTS

Dystonia is the most common neurological abnormality, and is recognized as the second most common clinical feature after hearing impairment. The age and site of onset, and degree of disease severity and progression vary tremendously (Table 21.1, Families 6, 29, and 30). Medical treatment and deep brain stimulation (DBS), in general have had little benefit. There are however, unusual cases with dystonia without hearing impairment (Table 21.1, Family 8), and cases of several years' lapse between childhood hearing impairment and development of dystonia (Table 21.1, Families 9 and 18). A recent study focusing on dystonia in a large subset of MTS families has concluded that there seems to be more inter than intrafamilial variation in the onset and progressive course of this often disabling neurological abnormality [49]. The attempts to treat dystonia are discussed below.

Cognitive impairment is a frequent, but not an invariable symptom. The study of the large five generation Norwegian family suggested a late onset cognitive decline, demonstrable by CT scan as cortical atrophy after the age of 40, but more subtle brain dysfunction and aggressive behavior have clearly been recognized in infancy in several cases (Table 21.1, Family 3). In some instances, MRI or SPECT scanning have indicated abnormalities in occipital frontal cortex [36 and unpublished observation]. The lack of impulse control, aggressive outbursts, and paranoid ideation, combined with declining cognitive function are the most disabling and antisocial components of the MTS in life time perspective, and has led to court sentences and penalties after sexually abusive behavior toward close female family members in several instances (personal communications). Paranoid ideation was prominent in several members of the Norwegian family, for example, fear of poisoned food, compulsive hand washing because of fear of microbes, and imaginary pulling hair out of one's eyes. These were judged not to stem from dual sensory deprivation due to deafblindness [26 and unpublished observations]. The behavioral and psychiatric abnormalities are a strong hindrance for an independent social life for those surviving into adolescence and adulthood, preventing advanced examinations due to the massive aggression, and complicating the regular life saving γ-globulin infusion treatment of those patients with XLA.

Several patients have undergone various neurological examinations showing some evidence of peripheral neuropathy and nonspecific histological abnormalities in muscle tissue, suggesting axonal neuropathy. Variable biochemical evidence for minor respiratory chain deficiency but not multiple mitochondrial DNA deletions, have been found in a few cases analyzed (Table 21.1, Families 14, 15, and 19; Tranebjærg et al., unpublished data). Electron microscopy (EM) studies of skeletal muscle tissue have not revealed a pattern of abnormal mitochondrial morphology, and histological studies have shown no convincing evidence for ragged red fibres (RRF) [26]. OXPHOS studies of muscle and fibroblast tissues showed inconsistent mild deficiency of respiratory chain complex IV (Table 21.1, Family 19, [50]). RT-PCR demonstrated the presence of *TIMM8A* and *TIMM13* mRNA in cultured fibroblasts from the patient with a *de novo* stop codon mutation (c.112C>T; p.Gln38Ter), and with an increased amount of *TIMM8A* mRNA, but normal amount of *TIMM13* mRNA [50]. Another well-investigated patient with a *de novo* missense c.3G>C; p.Met1Ile (p.Met1?) *TIMM8A* mutation showed normal complex IV activity, no RRF, mild neurogenic atrophy by muscle light microscopy, normal metabolic parameters related to the OXPHOS system, and complete absence of the *TIMM8A* protein and reduced reduced amounts of *TIMM13* protein in muscle tissue [51]. Furthermore, neurophysiological examinations showed markedly prolonged somatosensory evoked potentials (SEPs), prolonged central motor conduction time to the lower extremities, but normal EMG. FDG-PET scan showed marked hypometabolism in both occipital lobes, including the visual cortex and in the rostral parietal lobes, and a pattern compatible with the affected subcorticocortical circuits, underlying many types of dystonia [51].

Ophthalmological Abnormalities in MTS

MTS includes degeneration of the visual pathways, but is not primarily diagnosed because of optic atrophy. Full field electroretinogram (ERG) examinations of MTS patients from the large Norwegian family showed essentially no signs of retinal degeneration, indicating that the outer layers of the retina, including photoreceptors, and horizontal and bipolar cells, were not affected [52]. Subsequent studies of several patients by neuroimaging, neurophysiology and postmortem histology have shown convincingly prolonged or absent VEPs, thus, confirming the degeneration of the central visual pathways and the affected visual cortex (Table 21.1, Families 1, 2, 3, 13 (patient 4), and 14). Clinically, the visual impairment develops very slowly and is often preceded by years of difficulty with visual acuity and almost normally looking fundus appearance by ophthalmoscopy (Table 21.1, Families 3 and 14). Examination of ERGs and visual field is usually unremarkable, but visual acuity decreases rapidly over time (see Table 21.1). Eye histopathology has been performed in five deceased males at 49, 60, 61, 63, and 66 years of age, respectively, from the Norwegian family and in one patient (at the age of 53 from Family 13) [47, 55]. All cases showed significant loss of ganglion cells in the ganglion cell layer and severely atrophic optic nerve, whereas the other structures were fairly intact without autolytic changes [55], in agreement with previous report [36]. The histopathological abnormalities are strikingly similar in MTS patients and in patients with *OPA1* mutations [53, 54].

Neuropathology of MTS

Despite the rare occurrence of MTS, a number of neuropathological studies has been conducted in several patients (Table 21.1, Families 1, 2, 3, and 13 (patient 4)). In addition to the visual and auditory aspects mentioned above, generalized cortical atrophy, dispersed micro calcifications, and neuronal cell loss, especially in the occipital lobes, were found. Loss of Purkinje cells in cerebellum, gliosis of the spinal cord, loss of motor nerve cells, and degeneration of the posterior columns have been observed. Skeletal muscle also showed neurogenic atrophy (Table 21.1, Families 2 and 3) [36].

These abnormalities were found to be consistent in two unrelated families (Table 21.1, Families 1 and 2), before the identification of *TIMM8A* mutations in both families (Table 21.1) [55, 56]. There are very striking neuropathological similarities between patients with MTS, FA, and *OPA1*-related disease [37, 53, 54].

Other Features of MTS

XLA is the most severe accompanying clinical feature in the contiguous gene deletion syndrome involving the neighboring gene, *BTK*. It is estimated that 3.5–8 % of XLA families have gross deletions of *BTK* [44, 57, 58]. About 50 % (16/37 families) of published families with MTS have XLA (Table 21.1). As we reported in 1995, fractures seem to be a coincidental finding secondary to postural instability, ataxia, and increasing walking inability, rather than a primary component of the syndrome [26].

There is no evidence of reduced fertility in affected males [26]. Daughters of affected males are obligate carriers since they inherit their father's X chromosome with the *TIMM8A* mutation. The life span of affected patients is extremely variable, even within the same family, and highly depends on the pace of progression of this neurodegenerative disorder. Dysphagia, which is common after several years of progression, is a strong risk factor for pneumonia and subsequent death. In the Norwegian family, one male died at the age of 16 after rapid progression of dystonia starting at the age of 7. In contrast, several other affected males lived into their fifties and sixties [26]. The lifespan of individuals with XLA is much shorter, and highly dependent on the age when the correct diagnosis is established and an adequate treatment of agammaglobulinemia is instituted.

Differential Diagnoses

De novo mutations in *TIMM8A* in some families may mimic autosomal-recessive inheritance, which challenges the possibility to distinguish between X-linked and autosomal cases of dystonia. Differential diagnoses are based on clinical abnormalities, family history, and the age of onset. Male patients with profound childhood hearing impairment and XLA and/or dystonia have the diagnosis of MTS until *TIMM8A* mutations/deletions have been excluded. Other differential diagnoses including MELAS, mitochondrial encephalomyopathy with *SUCLA2* mutations, McLeod neuroacanthocytosis syndrome, Usher syndrome, and *WFS1*-related diseases [59], have been discussed in details in [31]. An Algerian family with autosomal-recessive optic atrophy and auditory neuropathy caused by mutations *TMEM126A* has been reported [60]. This family had congenital poor vision, and much better audiological ability than that of MTS patients. Therefore, visual failure probably would have been first recognized clinically. Fortunately, *TIMM8A* has only two exons. It is easy to perform sequence analysis if MTS is clinically suspected.

Spectrum of *TIMM8A* Mutations in MTS and Genotype–Phenotype Correlations

Among the 21 families with MTS alone, there were 18 different mutations (5 nonsense mutations, 4 deletion/frameshift mutations, 3 missense mutations, 4 splice mutations, and 2 different deletions involving exon 2 in families 29 and 35 (Table 21.1, Fig. 21.1).

It may seem logical to distinguish between contiguous gene deletion syndrome of XLA/MTS cases and MTS cases. The clinical features due to molecular defects in *TIMM8A* and those caused by larger deletions including *BTK* are, however, indistinguishable. However, patients with large deletions involving *BTK* will have associated agammaglobulinemia. Two families, who had intragenic deletions involving the entire exon 2, had isolated MTS (Table 21.1, Families 29 and 35), whereas cases with intragenic deletion of exon 1 and part of *BTK*, expressed MTS with associated XLA (Fig. 21.1). Recent studies of deletion breakpoints have shown preferential involvement of Alu elements in smaller deletions (< 10 kb) both in isolated XLA patients with deletions and in XLA-MTS patients [44].

It has been postulated [61] that deletion mutations were less detrimental than point mutations; however, overall published studies do not support this assumption. The nonsense mutation of family 9, and frameshift mutations of families 3 and 17 [29, 46, 63] a showed a less severe phenotype than the large deletion case reported by Pizzuti et al. [61]. Moreover, the considerable intrafamilial variation in the Norwegian multigeneration family also does not support such an assumption. Since mutations occur throughout the entire gene, it appears that the whole protein is needed for its function. It remains to be explained which modifying factors play a role in determining the course and severity of the disease in the individual person.

Affected Females in MTS Families

Due to skewed X-inactivation, female carriers may develop hearing impairment and/or dystonia [26, 29, 46, 49, 63, 64]. In Ha et al. [49], (Family B), a female patient was the proband, and her son was reported to be more severely affected. Hearing impairment and/or focal dystonia have been reported in females from families with either frameshift or deletion mutations of *TIMM8A* resulting from skewed X-chromosome inactivation of the normal *TIMM8A* allele [65, 66]. In two families (Table 21.1, Families 8 and 30), female patients were the affected probands [49, 63], while in another family, family 5, with contiguous gene deletion syndrome of XLA + MTS (Table 21.1, Family 5; Jin et al. [29], patient PDSM), three obligate carrier females had hearing impairment progressing to deafness.

Treatment of Dystonia and Antisocial Behavior in MTS

Efforts to alleviate the disabling dystonia have briefly been addressed in several reports [46, 50, 51, 56, 67, 68; Stolte-Dijkstra I, 2012, personal communication]. In one instance, there was impressive effect of treatment with GABAergic substances such as alcohol, clonazepam, and gamma-hydroxybutyric acid (GHB), over a period of several months in a patient with a novel *TIMM8A* missense mutation c.3G>C; p.Met1Ile/p.Met1? [51]. In contrast, repeated botulinum toxin injections and pimozide was associated with considerable side effects without any improvement of clinical symptoms [50]. Subsequently, treatment with GABA β agonist, baclofen, resulted in moderate improvement of dystonia more than a 2-year period [50]. Aguirre et al. [46] tried several drugs, including levodopa, clonazepam, and trihexyphenidyl, in one patient but discontinued the medication because the patient developed confusion and hallucinations. In another report, multiple treatments with intrathecal baclofen, local botulinum toxin, and left pallidotomy and right pallidal stimulation, were all unsuccessful [69]. Similarly, ineffective treatment with levodopa, carbamazepine, and trihexyphenidyl hydrochloride, has been reported [62]. However, improvement of dystonia has been reported in an Australian patient treated with carbamazepine [68]. Some improvement of the dystonia has also been noted in a patient of family 36 (Table 21.1). Dramatic improvement of muscle tone, posture, and handwriting has been reported in a patient with 2.5 months of L-DOPA treatment before the *TIMM8A* mutation was identified [56]. However, the patient regressed after a 3-month-treatment period when the treatment was discontinued due to suspicion of adverse effects. Apparently, deep brain stimulation (DBS) may have been applied in cases of severe dystonia, but no reports have been published. So far, medical alleviation of the serious outbursts of aggression and other behavioral abnormalities have had poor outcome, and environmental adjustment by providing protected housing and well-informed staff are the most successful remedies to minimize serious antisocial behavior. Worldwide, these antisocial manifestations are the most disabling behavior in young and adult patients with MTS.

Mechanisms of Mitochondrial Dysfunction in MTS

As discussed above, the mitochondrial dysfunction cannot be definitively categorized as an OXPHOS dysfunction. Several studies have tried to pinpoint the underlying pathomechanisms. Studies in yeast indicate that the yeast homolog of TIMM8A,

Tim8, is located in the intermembrane space (IMS) of mitochondria and forms a hetero-oligomeric 70 kDa complex with Tim13. Tim8–Tim13 complex was suggested to stabilize Tim23 precursor protein when this is transported into the inner mitochondrial membrane (IMM). *TIMM8A* mutations were suggested to result in impaired human Tim23 generation in affected tissues [70]. Moreover, coexpression of TIMM8A and TIMM13 with the calcium-binding mitochondrial aspartate-glutamate carrier protein (Aralar1) and prominent signals in large peripheral and CNS neurons indicate that insufficient malate-aspartate NADH shuttling combined with changes in Ca^{2+} concentration in specific cell types might explain the pathological process in cases of *TIMM8A* mutations [70, 71].

Recent studies have shown that the level of TIMM8A influences the morphology of mitochondria [72]. Live cell microscopy of fibroblasts with the *TIMM8A* mutation c.116delT; p.Met39ArgfsX26, showed long continuous mitochondrial extensions and after overexpression of *TIMM8A*, fragmentation of mitochondria was observed [72]. Further extensive studies, including animal studies are needed to clarify the significance of these novel findings, which seem to be more morphological than OXPHOS related.

Acknowledgments Some of the work took place at the Wilhelm Johannsen Center for Functional Genome Research, established by the Danish National Research Foundation. We acknowledge the financial support to the project by the Lundbeck Foundation, grant R9-A918. The data compiled here are the result of dedicated collaborative work and inspiring discussions since the 1990s involving many people: Marijke van Ghelue, Mona Nystad, Department of medical Genetics, University Hospital of Northern Norway, N-Tromsø, Norway; David Scheie, Department of Pathology, University Hospital, Oslo, Norway; Marianne Lodahl and Nanna Dahl Rendtorff, Wilhelm Johannsen Centre for Functional Genome Research, University of Copenhagen, Denmark. We would like to acknowledge and thank the colleagues allowing us to briefly mention unpublished MTS patients: Mary Ellen Conley, MD, West Research Tower, LeBonHeur Children's Hospital, Memphis TN; Dirk Kunst, Department of Otorhinolaryngology, Head and Neck Surgery, Radboud University Nijmegen Medical Centre, Nijmegen, the Netherlands and Donders Institute for Brain, Cognition and Behaviour, Radboud University Nijmegen, Nijmegen, the Netherlands; Irene Stolte-Dijkstra, Department of Genetics, University Medical Center Groningen, University of Groningen, Groningen, The Netherlands; Alain Verloes, M.D., Ph.D, Department of Genetics, Robert Debre Hospital, Referring Centre for developmental abnormalities and syndromic malformations, Paris, France; John Tolmie, and Katherine McWilliams, Clinical Genetics, Ferguson-Smith Centre, Glasgow, Scotland; Miles Humberstones, MD, Department of Neurology, Nottingham University Hospital, Nottingham, UK; Jan Ulrik Prause, MD, PhD, Department of Eye Pathology, University of Copenhagen, Denmark; Henning Laursen, MD,PhD; Department of Neuropathology, University Hospital of Copenhagen, Denmark; Saumil Merchant, MD, Harvard Medical School, Massachusetts Eye and Ear Infirmary, Boston, MA, USA; Kirsty Gardner-Berry, BSc, Department of Audiology, SCIC, Gladesville Hospital, Gladesville, NSW, Australia; Graham Reynolds, MD, Department of Paediatrics and Child Health, ANU Medical School, Canberra Hospital, Australia; Melanie Wong, Department of Allergy and Immunology, Children's Hospital, Westmead, NSW, Australia.

References

1. Christian BE, Spremulli LL (2011) Mechanism of protein biosynthesis in mammalian mitochondria. Biochim Biophys Acta. Epub 7 Dec 2011
2. DiMauro S, Davidzon G (2005) Mitochondrial DNA and disease. Ann Med 37:222–232

3. Rötig A (2011) Human diseases with impaired mitochondrial protein synthesis. Biochim Biophys Acta 1807(9):1198–1205
4. Calvo SE, Mootha VK (2010) The mitochondrial proteome and human disease. Annu Rev Genomics Genet 11:25–44
5. Kutik S, Guiard B, Meyer HE, Wiedemann N, Pfanner N (2007) Cooperation of translocase complexes in mitochondrial protein import. J Cell Biol 179(4):585–591
6. Mokranjac D, Neupert W (2010) The many faces of the mitochondrial TIN23 complex. Biochim et Biophys Acta 1797:1045–1054
7. Bauer MF, Neupert W (2001) Import of proteins into mitochondria: a novel pathomechanism for progressive neurodegeneration. J Inher Metab Dis 24:166–180
8. Mokranjac D, Neupert W (2009) Thirty years of protein translocation into mitochondria: Unexpectedly complex and still puzzling. Biochim et Biophys Acta 1793:33–41
9. Beverly KN, Sawaya MR, Schmid E, Koehler CM (2008) The Tim8-Tim13 complex has multiple substrate binding sites and binds cooperatively to Tim23. J Mol Biol 382(5):1144–1156
10. Tranebjaerg L, Hamel BC, Gabreels FJ, Renier WO, Van Ghelue M (2000a) A de novo missense mutation in a critical domain of the X-linked DDP gene causes the typical deafness-dystonia-optic atrophy syndrome. Eur J Hum Genet 8:464–467
11. Kirby DM, Thorburn DR (2008) Approaches to finding the molecular basis of mitochondrial oxidative phosphorylation disorders. Twin Res Hum Genet 11:395–411
12. Scharfe C, Lu HH, Neuenburg JK, Allen EA, Li GC, Klopstock T, Cowan TM, Enns GM, Davis RWl (2009) Mapping gene associations in human mitochondria using clinical disease phenotypes. PLoS Comput Biol 5(4):e1000374
13. Lieber DS, Vafai SB, Horton LC, Slate NG, Liu S, Borowsky ML, Calvo SE, Schmahmann JD, Mootha VK (2012) Atypical case of Wolfram syndrome revealed through targeted exome sequencing in a patient with suspected mitochondrial disease. BMC Med Genet 13:3
14. Polke JM, Laurá M, Pareyson D, Taroni F, Milani M, Bergamin G, Gibbons VS, Houlden H, Chamley SC, Blake J, DeVile C, Sandford R, Sweeney MG, Davis MB, Reilly MM (2011) Recessive axonal Charcot-Marie-Tooth disease due to compound heterozygous mitofusin 2 mutations. Neurology 77:168–173
15. Rouzier C, Bannwarth S, Chaussenot A, Chevrollier A, verschueren A, Bonello-Palot N, Fragaki K, Cano A, Pouget J, Pellisier JF, Procaccio V, Chabrol B, Paguis-Flucklinger V (2012) The MFN2 gene is responsible for mitochondrial DNA instability and optic atrophy 'plus' phenotype. Brain 135(Pt 1):23–34
16. Frank S (2006) Dysregulation of mitochondrial fusion and fission: an emerging concept in neurodegeneration. Arch neuropathol 111(2):93–100
17. Chen H, Chan DC (2005) Emerging functions of mammalian mitochondrial fusion and fission. Hum Mol Genet 14(2):R283–R289
18. Elachouri G, Vidoni S, Zanna C, Pattyn A, Boukhaddaoui H, Gaget K, Yu-Wai-Man P, Gasparre G, Sarzi E, Delettre C, Olichon A, Loiseau D, Reynier P, Chinnery PF, Rotig A, Carelli V, Hamel CP, Rugolo M, Lenaers G (2011) OPA1 links human mitochondrial genome to mtDNA replication and distribution. Genome Res 21(1):12–20
19. Ahting U, Floss T, Uez N, Schneider-Lohmar I, Becker L, Kling E, Iuso A, bender A, de Angelis MH, Gailus-Durner V, Fuchs H, Meitinger T, Wurst W, Prokisch H, Klopstock T (2009) Neurological phenotype and reduced lifespan in heterozygous Tim23 knockout mice, the first mouse model of defective mitochondrial import. Biochiim et Biophys Acta 1787:371–376
20. Mackenzie JA, Payne RM (2007) Mitochondrial protein import and human health and disease. Biochim et Biophysica Acta 1772:509–523
21. Jin H, Kendall E, Freeman TC, Roberts RG, Vetrie DLP (1999) The human family of deafness/dystonia peptide (DDP) related mitochondrial import proteins. Genomics 61:259–267
22. Davey KM, Parboosingh JS, McLeod DR, Chan R, Casey R, Ferreira P, Snyder FF, Bridge PJ, bernier FP (2006) Mutation of DNAJC19, a human homologue of yeast inner mitochondrial membrane co-chaperones, causes DCMA syndrome, a novel autosomal recessive Barth syndrome-like condition. J Med Genet 43:385–393

23. Shridhar V, Bible J, Staub R, Avula R, Lee YK, Kalli K, Huang H, Hartmann LC, Kaufmann SH, Smith DI (2001) Loss of a new member of the DNAJ protein family confers resistance to chemotherapeutic agents used in the treatment of ovarian cancer. Cancer Res 61:4258–4265
24. Lindsey JC, Lusher ME, Strathdee G, Brown R, Gilbertson RJ, Bailey S, Ellison DW, Clifford SC (2006) Epigenetic inactivation of MCJ (DNAJD1) in malignant paediatric brain tumours. Int J Cancer 118:346–352
25. Badenkop RF, Cherian S, Lord RSA, Baysal BE, Taschner PEM, Schofield PR (2001) Novel mutations in the SDHD gene in pedigrees with familial carotid body paraganglioma and sensorineural hearing loss. Genes Chromoso Cancer 31:255–263
26. Tranebjaerg L, Schwartz C, Eriksen H, Andreasson S, Ponjavic V, Dahl A, Stevenson RE, May M, Arena F, Barker D, Elverland HH, Lubs H (1995) A new X linked recessive deafness syndrome with blindness, dystonia, fractures, and mental deficiency is linked to Xq22. J Med Genet 32:257–263
27. Mohr J, Mageroy K (1960) Sex-linked deafness of a possibly new type. Acta Genet Stat Med 10:54–62
28. Vetrie D et al (1993) The gene involved in X-linked agammaglobulinemia is a member of the src family of protein-tyrosine kinases. Nature 361:226–233
29. Jin H, May M, Tranebjaerg L, Kendall E, Fontan G, Jackson J, Subramony SH, Arena F, Lubs H, Smith S, Stevenson R, Schwartz C, Vetrie D (1996) A novel X-linked gene, DDP, shows mutations in families with deafness (DFN-1), dystonia, mental deficiency and blindness. Nat Genet 14:177–180
30. Koehler CM, Leuenberger D, Merchant S, Renold A, Junne T, Schatz G (1999) Human deafness dystonia syndrome is a mitochondrial disease. Proc Natl Acad Sci U S A 96:2141–2146
31. Tranebjærg L (2009) Deafness-dystonia-optic neuronopathy syndrome. In: Pagon RA, Bird TD, Dolan CR et al (eds) Gene reviews. www.genetests.org. Accessed 29 July 2012
32. Jensen PK (1981) Nerve deafness: optic nerve atrophy, and dementia: a new X-linked recessive syndrome? Am J Med Genet 9:55–60
33. Jensen PK, Reske-Nielsen E, Hein-Sorensen O, Warburg M (1987) The syndrome of opticoacoustic nerve atrophy with dementia. Am J Med Genet 28:517–518
34. Tranebjaerg L, Van Ghelue M, Nilssen O, Hodes ME, Dlouhy SR, Farlow MR, Hamel B, Arts WFM, Jankovic J, Beach J, Jensen PKA (1997) Jensen syndrome is allelic to Mohr-Tranebjaerg syndrome and both are caused by mutations in the DDP gene. Am J Hum Genet 61S:A349
35. Tranebjaerg L, Jensen PK, van Ghelue M (2000b) X-linked recessive deafness-dystonia syndrome (Mohr-Tranebjaerg syndrome). Adv Otorhinolaryngol 56:176–180
36. Tranebjaerg L, Jensen PK, Van Ghelue M, Vnencak-Jones CL, Sund S, Elgjo K, Jakobsen J, Lindal S, Warburg M, Fuglsang-Frederiksen A, Skullerud K (2001) Neuronal cell death in the visual cortex is a prominent feature of the X-linked recessive mitochondrial deafness-dystonia syndrome caused by mutations in the TIMM8a gene. Ophthalmic Genet 22:207–223
37. Koeppen AH (2011) Friedreiach's ataxia: pathology, pathogenesis, and molecular genetics. J Neurol Sci 303(1–2):1–12
38. Calvo SE, Compton AG, Hershman SG, Lim SC, Lieber DS, Tucker EJ, Laskowski A, Garone C, Liu S, Jaffe DB, Christodoulou J, Fletcher JM, Bruno DL, Goldblatt J, DiMauro S, Thorburn DR, Mootha VK (2012) Molecular diagnosis of infantile mitochondrial disease with targeted next-generation sequencing. Sci Trans Med 4(118):118ra10
39. Merchant SN, McKenna MJ, Nadol JB, Kristiansen AG, Tropitzsch A, Lindal S, Tranebjaerg L (2001) Temporal bone histopathologic and genetic studies in Mohr-Tranebjaerg syndrome (DFN-1). Otol Neurotol 22:506–511
40. Bahmad F Jr, Merchant SN, Nadol JB Jr, Tranebjaerg L (2007) Otopathology in Mohr-Tranebjaerg syndrome. Laryngoscope 117:1202–1208
41. Brookes JT, Kanis AB, Tan LY, Tranebjaerg L, Vore A, Smith RJ (2007) Cochlear implantation in deafness-dystonia-optic neuronopathy (DDON) syndrome. Int J Pediatr Otorhinolaryngol 72:121–126
42. Richter D, Conley ME, Rohrer J, Myers LA, Zahradka K, Kelecic J, Sertic J, Stavljenic-Rukavina A (2001) A contiguous deletion syndrome of X-linked agammaglobulinemia and sensorineural deafness. Pediatr Allergy Immunol 12:107–111

43. Sedivá A, Smith CI, Asplund AC, Hadac J, Janda A, Zeman J, Hansíková H, Dvoráková L, Mrázová L, Velbri S, Koehler C, Roesch K, Sullivan KE, Futatani T, Ochs HD (2007) Contiguous X-chromosome deletion syndrome encompassing the BTK, TIMM8A, TAF7L, and DRP2 genes. J Clin Immunol 27:640–646
44. Arai T, Zhao M, Kanegane H, van Zelm MC, Futatani T, Yamada M, Ariga T, Ochs HD, Miyawaki T, Oh-ishi T (2011) Genetic analysis of contiguous X-chromosome deletion syndrome encompassing the BTK and TIMM8A genes. J Hum Genet 56(8): 577–582
45. Aguirre LA, Pérez-Bas M, Villamar M, Barcena JE, López-Ariztegui MA, Moreno-Pelayo MA, Moreno F, del Castillo I (2008) A Spanish sporadic case of deafness-dystonia (Mohr-Tranebjaerg) syndrome with a novel mutation in the gene encoding TIMM8a, a component of the mitochondrial protein translocase complexes. Neuromuscular Disorders 16:979–981
46. Aguirre LA, del Castillo I, Macaya A, Meda C, Villamar M, Moreno-Pelayo MA, Moreno F (2006) A novel mutation in the gene encoding TIMM8a, a component of the mitochondrial protein translocase complexes, in a Spanish familial case of deafness-dystonia (Mohr-Tranebjaerg) syndrome. Am J Med Genet A 140:392–397
47. Varga RJ, Lewis R, Kimberling WJ (2002) Auditory neuropathy in a family with Mohr-Tranebjaerg syndrome due to a nonsense mutation in TIMM8a. Am J Hum Genet 71:A2006
48. Cacase AT, Pinheiro JM (2011) The mitochondrial connection in auditory neuropathy. Audiol Neurotol 16(6):398–413
49. Ha AD, Parratt KL, Rendtorff ND, Lodahl M, Ng K, Rowe DB, Sue CM, Morris JG, Hayes MW, Tranebjaerg L, Fung VSC (2012) The phenotypic spectrum of dystonia in Mohr-Tranebjaerg syndrome. Mov Disord (in press)
50. Blesa JR, Solano A, Briones P, Prieto-Ruiz JA, Hernandez-Yago J, Coria F (2007) Molecular genetics of a patient with Mohr-Tranebjaerg syndrome due to a new mutation in the DDP1 gene. Neuromol Med 9:285–291
51. Binder J, Hofmann S, Kreisel S, Wohrle JC, Bazner H, Krauss JK, Hennerici MG, Bauer MF (2003) Clinical and molecular findings in a patient with a novel mutation in the deafnessdystonia peptide (DDP1) gene. Brain 126:1814–1820
52. Ponjavic V, Andreasson S, Tranebjaerg L, Lubs HA (1996) Full-field electroretinograms in a family with Mohr-Tranebjaerg syndrome. Acta Ophthalmol Scand 74:632–635
53. Williams PA, Morgan JE, Votruba M (2010) Opa1 deficiency in a mouse model of dominant optic atrophy leads to retinal ganglion cell dendropathy. Brain 133:2942–2951
54. Carelli V, Morgia CL, Valentino ML, Barboni P, Ross-Cisneros FN, Sadun AA (2009) Retinal ganglion cell neurodegeneration in mitochondrial inherited disorders. Biochim et Biophys Acta 1787:518–528
55. Reske-Nielsen E, Jensen PK, Hein-Sørensen O, Abelskov K (1988) Calcification of the central nervous system in a new hereditary neurological syndrome. Acta Neuropathol 75:590–596
56. Scribanu N, Kennedy C (1976) Familial syndrome with dystonia, neural deafness, and possible intellectual impairment: clinical course and pathological findings. Adv Neurol 14:235–243
57. Conley ME, Howard VC (1993) X-linked agammaglobulinemia (2011). In: www.genetests.org, Pagon RA, Bird TD, Dolan CR et al (eds) Gene reviews. University of Washington, Seattle (WA)
58. Conley ME, Rohrer J, Minegishi Y (2000) X-linked agammaglobulinemia. Clin Rev Allergy Immunol 19:183–204
59. Rendtorff ND, Lodahl M, Boulahbel H, Johansen IR, Pandya A, Welch KO, Norris VW, Arnos KS, Bitner-Glindzicz M, Emery SB, Mets MB, Fagerheim T, Eriksson K, Hansen L, Bruhn H, Möller C, Lindholm S, Ensgård S, Lesperance MM, Tranebjærg L (2011) Missense mutations in WFS1 cause autosomal dominant inherited optic atrophy and hearing loss in eight families. Am J Med Genet 155(6):1298–1313
60. Meyer E, Michaelides M, Tree LJ, Robson AG, Rahman F, Pasha S, Luxon LM, Moore AT, Maher ER (2010) Nonsense mutation in TMEM126A causing autosomal recessive optic atrophy and auditory neuropathy. Mol Vis 16: 650–664
61. Pizzuti A, Fabbrini G, Salehi L, Vacca L, Inghilleri M, Dallapiccola B, Berardelli A (2004) Focal dystonia caused by Mohr-Tranebjaerg syndrome with complete deletion of the DDP1 gene. Neurology 62:1021–1022

62. Ujike H, Tanabe Y, Takehisa Y, Hayabara T, Kuroda S (2001) A family with X-linked dystonia-deafness syndrome with a novel mutation of the DDP gene. Arch Neurol 58:1004–1007
63. Swerdlow RH, Wooten GF (2001) A novel deafness/dystonia peptide gene mutation that causes dystonia in female carriers of Mohr-Tranebjaerg syndrome. Ann Neurol 50:537–540
64. Swerdlow RH, Juel VC, Wooten GF (2004) Dystonia with and without deafness is caused by TIMM8A mutation. In: Fahn S, Hallett M, Mahlon R (eds) Dystonia 4: advances in neurology, vol 94. Lippincott Williams & Wilkins, Philadelphia
65. Plenge RM, Tranebjaerg L, Jensen PK, Schwartz C, Willard HF (1999) Evidence that mutations in the X-linked DDP gene cause incompletely penetrant and variable skewed X inactivation. Am J Hum Genet 64:759–767
66. Orstavik KH, Orstavik RE, Eiklid K, Tranebjaerg L (1996) Inheritance of skewed X chromosome inactivation in a large family with X-linked recessive deafness syndrome. Am J Med Genet 64:31–34
67. Kreisel SH, Binder J, Wohrle JC, Krauss JK, Hofmann S, Bauer MF, Hennerici MG, Bazner H (2004) Dystonia in the Mohr-Tranebjaerg syndrome responds to GABAergic substances. Mov Disord 19:1241–1243
68. Hayes MW, Ouvrier RA, Evans W, Somerville E, Morris JG (1998) X-linked dystonia-deafness syndrome. Mov Disord 13:303–308
69. Ezquerra M, Campdelacreu J, Munoz E, Tolosa E, Marti MJ (2005) A novel intronic mutation in the DDP1 gene in a family with X-linked dystonia-deafness syndrome. Arch Neurol 62:306–308
70. Roesch K, Curran SP, Tranebjaerg L, Koehler CM (2002) Human deafness dystonia syndrome is caused by a defect in assembly of the DDP1/TIMM8a-TIMM13 complex. Hum Mol Genet 11:477–486
71. Roesch K, Hynds PJ, Varga R, Tranebjaerg L, Koehler CM (2004) The calcium-binding aspartate/glutamate carriers, citrin and aralar1, are new substrates for the DDP1/TIMM8a-TIMM13 complex. Hum Mol Genet 13:2101–2111
72. Engl G, Florian S, Tranebjærg L, Mayer TU, Rapaport D (2012) Alterations in expression levels of deafness dystonia protein 1 affect mitochondrial morphology. Hum Mol Genet 21(2):287–299
73. Rohrer J, Minegishi Y, Richter D, Eguiguren J, Conley ME (1999) Unusual mutations in Btk: an insertion, a duplication, an inversion, and four large deletions. Clin Immunol 90(1):28–37
74. Kim HT, Edwards MJ, Tyson J, Quinn NP, Bitner-Glindzicz M, Bhatia KP (2007) Blepharospasm and limb dystonia caused by Mohr-Tranebjaerg syndrome with a novel splice-site mutation in the deafness/dystonia peptide gene. Mov Disord 22:1328–1331
75. Jyonouchi H, Geng L, Toruner GA, Vinekar K, Feng D, Fitzgerald-Bocarsly P (2007) Monozygous twins with a microdeletion syndrome involving BTK, DDP1, and two other genes; evidence of intact dendritic cell development and TLR responses. Eur J Pediatr 167:317–321

Index

A

Accessory subunit, 49, 51, 52, 61, 68, 123, 130, 189
Alpers-Huttenlocher syndrome, 49, 73, 74, 76, 84, 85
Aminoacyl tRNA synthetase, 15, 27, 33, 51, 263, 264, 267, 272, 278, 280
ANT1, 123, 133, 134, 173, 324
Assembly factor, 10, 11, 27, 32, 187–189, 191, 195, 220, 228, 233
Ataxia, 8, 9, 11, 13, 57, 78, 105, 126, 132, 134, 151, 179, 205, 207, 307, 359
Autosomal dominant optic atrophy, 143

B

BOLA3, 303, 305, 310

C

C10ORF2, 131–133
Cardiomyopathy, 10, 12–16, 128, 132, 190, 192, 207, 273, 292, 295, 307, 320–323, 325, 329, 354
Charcot-Marie-Tooth disease, 152, 271
Charcot-Marie-Tooth neuropathy, 264, 356
Chronic progressive external ophthalmoplegia, 57, 142
Clinical manifestations, 10, 73, 104, 125, 126, 210
Complex II, 3, 4, 7, 10, 32, 206, 282, 292
Complex III deficiency, 11, 220, 228–230, 232, 233
Complex IV, 3, 4, 9, 10, 14, 18, 32, 185, 239, 249, 283, 322
CoQ_{10} deficiency, 8, 9, 16, 18, 33
Core subunit, 10, 186, 188, 239, 241
Cytochrome *c* oxidase assembly, 239, 251

D

Defect of the mitochondrial lipid milieu, 17
Defects of intergenomic communication, 5, 15, 17, 18
Defects of mtDNA translation, 14, 16, 18
Deoxyribonucleotide triphosphate, 172
Diagnosis, 12, 14, 16, 28, 39, 50, 75, 80, 97, 105, 110, 190, 320
Dichloroacetate, 195, 196, 302, 313
Direct hits, 4, 5, 8–10
Disorder of intracellular communication, 329
DLAT, 310
DLD, 305, 310, 311
DRP1, 152, 154

E

Electron transfer, 185, 219, 239
Electron transport chain, 50, 80, 97, 118, 163, 205, 277, 308, 311, 312, 325, 327
Elongation factors, 27, 278, 279, 283
Encephalomyopathy, 8, 9, 14, 16, 115, 164, 165, 192, 239, 360
Epilepsy syndrome, 75, 125, 132

G

GDAP1, 154

H

Hepatocerebral mitochondrial disease, 5, 18, 35, 37, 61, 104, 105, 133, 326
Hepatocerebral mitochondrial DNA depletion, 30, 36
HUPRA syndrome, 269

I

Indirect hits, 4, 5, 10, 14, 18

K

Ketogenic diet, 195, 312, 313

L

Lactate, 11, 18, 30, 191, 206, 266, 267, 292, 311, 312, 325, 326
Leigh syndrome, 4, 9, 190, 194, 205, 239, 243, 248, 249, 251, 306, 308, 325, 327
Leigh-like syndrome, *see* Leigh syndrome, 190
Leukoencephalopathy, 7, 11, 16, 18, 103, 128, 209, 213, 266
LIAS, 305, 310
Lipoic acid, 301, 303, 305, 307, 311
Liver failure, 8, 9, 37, 75, 77, 79, 83, 86, 93–95, 98, 106, 109, 127, 227, 292, 293
Liver transplantation, 16, 98, 106, 109

M

Mendelian genetics, 5, 220, 233
MFN1, 153
MFN2, 153–155, 272
Mitochondria, 8, 13, 17, 27, 28, 33, 91, 96, 142, 163, 185, 212, 244, 268, 277, 319
Mitochondrial cardiomyopathy, 270, 273, 327, 329, 330
Mitochondrial complex I, 154, 195, 230
Mitochondrial disease, 3, 17, 65, 74, 75, 103, 125, 206, 239, 272, 277, 281, 282, 350, 352, 354
Mitochondrial disorders, 27, 28, 30, 50, 61, 78, 85, 92, 105, 123, 229, 249, 309, 313, 320, 321, 327, 337
Mitochondrial DNA, 3, 81, 91
Mitochondrial DNA deletion, 150, 358
Mitochondrial DNA depletion syndrome, 113, 165
Mitochondrial DNA polymerase, 81
Mitochondrial DNA replication, 171
Mitochondrial DNA synthesis, 74, 80, 131, 172
Mitochondrial dynamics, 151, 152, 155, 231
Mitochondrial genetics, 5
Mitochondrial myopathy, 8, 114, 115, 164, 291, 293, 326
Mitochondrial respiratory chain, 32, 106, 115, 135, 185, 215, 232, 294, 301, 307, 309
Mitochondrial translation, 14, 15, 242, 267, 268, 271, 278–281, 283, 293
Mitochondrial/Mendelian genetic overlap, *see also* Mitochondrial genetics and Mendelian genetics, 4, 5, 291
MMA, 31, 164
MPV17 gene mutations, 30, 37, 103, 107, 109, 110
MRC, 185, 195, 239, 243, 249
mtDNA copy number, 28, 35, 80, 97, 105, 109, 113, 118, 126, 293

mtDNA deletion, *see also* Multiple mtDNA deletions, 5, 30, 35, 50, 58–60, 99, 128, 133, 151, 154, 173
mtDNA depletion, 4, 5, 13, 15, 27, 31, 35, 37, 65, 75, 80, 95, 114, 115, 119, 120, 123, 128, 166, 173, 179, 293
mtDNA maintenance, 5, 49, 50, 68, 93, 107, 123, 131, 134, 142, 150, 154, 155, 167, 172, 192
Multiple mtDNA deletions, 4, 5, 15, 17, 36, 117, 123, 128, 131, 133, 134, 155, 173, 179, 180
Multisystem disorder, 18, 103
Mutations, 3, 9, 12, 13, 16, 33, 49, 61, 75, 83, 103, 155, 165, 191, 206, 227, 245, 248, 325, 352
Myopathy, 8–10, 16, 17, 36, 60, 75, 94, 116, 130, 147, 207, 268, 294

N

NADH Coenzyme Q reductase, 185
NADH ubiquinone oxidoreductase, 185
Neuropathology, 308, 309, 359
Next generation sequencing, 29, 36, 38, 97, 192, 273, 284
NFU1, 303, 305
Nucleoside modification, 288, 289
Nystagmus, 8, 10, 11, 13, 77, 94, 98, 115, 205

O

OPA1, 147, 149, 151, 153, 155
OPA3, 151
Optic atrophy, 8, 14, 126, 128, 142, 147, 149, 151–154, 205–207, 267, 283, 350, 351, 358, 360
OXPHOS, 3, 185, 190, 192, 219, 229–233, 263, 264, 266–269, 358, 362

P

Paraganglioma, 209, 210, 214, 352
Pathophysiology, 69, 153, 227, 233, 309
PDHA1, 304, 305, 307, 308, 310, 312, 313
PDHB, 310
PDHX, 305, 310
PDP1, 302, 310
PEO, 36, 59, 61, 66, 67, 74, 93, 117, 125–128, 130, 132, 179, 180
Perrault syndrome, 269, 273
Phenotypes, 6, 12, 14, 36, 67, 75, 114, 118, 125, 126, 129, 147, 243, 245, 273, 350
POLG, 74, 85, 129, 130
POLG2, 82, 124, 129, 130
Polymerase gamma, *see also* POLG, 1, 85
Pontocerebellar hypoplasia, 267

Index 369

Prognosis, 98, 106, 110, 192, 197
Progressive external ophthalmoplegia, *see* PEO, 5
Pseudouridine, 289
PUS1, 291, 294, 295
Pyruvate dehydrogenase complex deficiency, 30, 205, 301, 312

R

Respiratory chain, 6, 61, 114, 129, 153, 166, 191, 228, 230–232, 239, 280, 337
Retinal ganglion cell, 144, 147, 150, 151, 153
Ribosomal proteins, 15, 27, 278–281, 283, 337
RNA modification, 33, 287–291, 295

S

SDH, 3, 8, 14, 31, 203–206, 209, 211–214
Sideroblastic anemia, 268, 269, 291, 294, 295
SLC25A, 133
Succinate dehydrogenase, *see* SDH, 3

Succinyl-CoA synthetase, 164, 165, 167
SUCLA2, 30, 31, 103, 164–167, 172, 360
SUCLG1, 31
Survival, 9, 12, 17, 95, 98, 109, 113, 114, 117, 209, 211, 251

T

Therapy for mitochondrial disease, 92, 99, 106, 313, 329, 361
Thiamine, 196, 311, 313
Thiamine pyrophosphate, 301, 302
Thymidine kinase 2, 103, 113
TK2 gene, 113–115, 118, 119
TPK1, 303, 305, 310
Treatment, 77, 79, 85, 99, 106, 115, 156, 167, 195, 196, 231, 312, 313, 328, 329, 351, 357, 361
TRMU, 292, 293, 295
Tumorigenesis, 209, 211–214
Twinkle, 5, 82, 83, 103, 123, 131, 133